Advances in Earth Sciences

Volume - I

ADVANCES IN EARTH SCIENCES

Volume - I

Series Editor : **SATISH C. TRIPATHI**

Guest Editors

S. K. TANDON
A. R. BHATTACHARYA

SSPH

2009

SATISH SERIAL PUBLISHING HOUSE
403, Express Tower, Commercial Complex, Azadpur, Delhi-110033 (India)
Phone : 011-27672852, Fax : 91-11-27672046
E-mail :info@satishserial.com, hkjain1975@yahoo.com
Website : www.satishserial.com

Published on behalf of :

The Society of Earth Scientists

C-207, Indira Nagar, Lucknow-226016 India

E-mail : india.ses@gmail.com

Published by :

SATISH SERIAL PUBLISHING HOUSE

403, Express Tower, Commercial Complex, Azadpur, Delhi-110033 India

Phone : 011-27672852 Fax : 91-11-27672046

E-mail : info@satishserial.com, hkjain1975@yahoo.com

Cover Photo : Balancing Rock : Balanced Granite Rock about 4.5 km from Yadgiri town on Yadgiri-Hyderabad Road, Karnataka, India
Photo Courtesy : Gajanan Hegde

ISBN 81-89304-66-6

Typeset at: Laxmi Art Creations # 9811482328

Printed at: Salasar Imaging Systems, Delhi-35

Preface

In the Earth Sciences, the scientists belonging to entirely different disciplines work for a common objective- understanding the earth system. During the last few decades, Earth Sciences have evolved rapidly and we are now able to provide scientific models to reconstruct and forecast the past and future processes of the planet earth through multidisciplinary research. In order to highlight the contributions of Earth Sciences, particularly for the society, the United Nations declared year 2008 as International Year of Planet Earth. IYPE provided ample opportunity to the scientists to boost scientific temper both for the fundamental researches and for providing its benefits to the society. In response to UN appeal we launched an open access quarterly e-Journal: **Earth Science India**-www.earthscienceindia.info in January, 2008 under the umbrella of The Society of Earth Scientists to fulfill the outreach objective of IYPE. **Advances in Earth Sciences** is a new book series containing selected technical papers of Earth Science India volume-I, 2008. While the e-Journal Earth Science India provides quick publication of scientific researches/data this book is likely to serve the purpose of archival and record. It will also provide an opportunity to scientists of one discipline to know the developments in other disciplines and to explore the possibility of multidisciplinary research.

I am grateful to Prof. S.K. Tandon, University of Delhi and Prof. A.R. Bhattacharya, University of Lucknow, for accepting my request to shoulder the responsibility of editing the first volume and making this endeavour successful. My sincere thanks are due to Dr. Shailesh Nayak, Secretary, Ministry of Earth Sciences and Shri Rasik Ravindra, Director, National Centre for Antarctic and Ocean Research, Goa, for their constant guidance and moral support. Efforts put on by all the members of editorial board of Earth Science India in reviewing the manuscripts, is sincerely acknowledged. Help and support rendered by my colleauges Shri D.D. Bhattacharya and Shri A.C. Pande in every step from initiation of the e-Journal to the publication of this volume, is thankfully acknowledged. My wife Sandhya, deserve praise for allowing me to burn mid-night oil and extending all her support in the completion of this work. The blessings of my parents finally made me realize this dream.

I am thankful to all the scientists who contributed in this volume. My sincere thanks are also to Shri H.K. Jain, Satish Serial Publishing House, Delhi, who offered his services that brought this publication. I hope this book will prove to be useful for the geoscientific community.

Satish C. Tripathi

Introduction

Advances in Earth Sciences : Volume - I, contains research/review papers from almost all the disciplines of earth science - geology, geophysics, oceanography, atmospheric and environmental sciences, thus fulfilling the objective of integrating basic data on earth system at a single platform. Scientists tended to split the earth system into different "spheres". But we have now reached to a point where we must put it back together, to see how it actually works in totality. The modern earth system approach requires a comprehensive integration of existing databases. We hope this book will be a step forward in this direction.

In his paper *O.P. Pandey and P.K. Agarwal* discussed multi-stage crustal evolution of Indian shield both in space and time. The study of south Indian shield reveals segmentation and a secular and progressive lateral zonation of the continental lithosphere around Dharwar nucleus since mid-Archean. They envisaged that the Antongil block of northeast Madagascar seems to correspond to the broken segment of the western Dharwar craton.

Santosh Kumar and Kh. Mohan Singh evaluated Early Ordovician (479±26 Ma) Kyrdem granitoids (KG) and their enclaves in Meghalaya plateau in terms of magnetic susceptibility and whole rock composition in order to understand the redox condition of comingled mafic and felsic melts. They concluded that oxidation state of KG melt was partly elevated under a coeval interacting mafic-felsic magma system. The bulk of oxidized KG melt was reduced locally near the margin of the KG pluton because of country-rock assimilation at shallow emplacement level. *A.R. Bhattacharya* discusses how basement rocks played an important role in the structural architecture and tectonic framework of the Himalayan orogen. The paper critically examines some aspects of the tectonics and structural framework of the Himalaya as related to the basement rocks in the Kumaun - Garhwal sectors. *A.R. Bhattacharya and K.K. Agarwal* discussed their observations on the thrust geometry of the Siwalik rocks of the outer Himalaya and suggested that Siwalik strata may have undergone structural thickening in their internal domain during the overall deformation of these rocks. *Alpa Sridhar* synthesized the status of fluvial palaeohydrological studies in western India and the future prospects. She concluded that palaeoflood history for the past 5000 years in the Narmada, Tapi, Mahi, Luni, Krishna and Godavari rivers appears to be broadly comparable. *K. K. Agarwal and R. Bali* discussed small-scale deformational structures of Almora Crystalline Zone which are significant shear-sense indicators. The shear-sense observed on the northern and southern flanks is found to be - top to north and top to south respectively.

H. S. Saini discusses future impact of climatic changes on Indo-Gangatic plain. He emphasized that a pre-requisite is to develop a high resolution history of climate variations and their impact on landscape and ecology from the geological and historical records of IGP during the past 25 ka. *V. C. Tewari* described oxygen and carbon isotopic composition of Himalayan Speleothems. In this preliminary study negative $\delta^{18}O$ values have been recorded suggesting varying Indian Summer Monsoon strength. *Gandhi et al.* discussed sediment characteristics and heavy mineral distribution in Tamiraparani estuary and off Tuticorin, Tamil Nadu from SEM studies. They concluded that in pre-monsoon, the deposited sediment were transported and shifted due to longshore current action. But, in the post-monsoon period the sediments deposited due to the multi-source like riverine and marine influence.

A.K. Srivastava highlighted the findings of insect wings, insect damaged plant fossils and evidences of fungal and bacterial assault in the flora which revealed the existence of well organized insect-plant relationship and mutualistic, parasitic and saprophytic role of microorganisms in Gondwana flora of India. *D. K. Srivastava and Hukam Singh* recorded for the first time spatangoid genus *Eupatagus* L. Agassiz, 1847, from the sediments of the Khuiala Formation (Early Eocene) exposed near Habur, Jaisalmer district, Rajasthan.

Rajesh Agnihotri and Siby Kurian reviewed some of the recently studied sedimentary records from precipitation dominated eastern Arabian Sea to comprehend past- monsoonal variability in Indian as well as global context. They also focused upon different proxies used so far in the Arabian Sea, their fidelity, limitations and future scope of using novel proxies for a better understanding of past-monsoonal variability especially in the anthropocene epoch. *Koushik Dutta* made in depth analysis of marine ^{14}C reservoir age and Suess effect in the Indian Ocean. The $\Delta^{14}C$ values between 1900 and 1955 from the Indian Ocean and the South China Sea show a decreasing trend, indicating marine ^{14}C Suess effect in the Bay of Bengal. The average magnitude of this Suess effect is -3±2‰ per decade for the Arabian Sea and -2±1‰ per decade for the Bay of Bengal and the South China Sea. He emphasized that reservoir ^{14}C-ages has been reported from many places in the Indian Ocean analyzing known age marine biogenic carbonates, but data on its spatial and temporal variability is still not adequate.

Nandita D. Ganguly studied variation in Erythemal UV Dose (EUVD) for five Indian cities and their expected critical limits using the data obtained from Global Ozone Monitoring Experiment (GOME). She concluded that EUVD variations show a negative correlation with solar activity and are well within the statistically calculated critical limits. *Suresh Tiwari et al.* studied chemical characteristics of water soluble components of fine particulate matter, $PM_{2.5}$, at Delhi. It suggested that particle pollution in Delhi is rather severe, especially due to fine particles and more attention should be paid to the regional sources of secondary pollutants while formulating pollution mitigation programmes at Delhi. *Devendraa Siingh* summarized the present understanding of Earth's atmospheric processes that are affected by Cosmic rays. *S.S. Dugam,* in his paper, analyzed North Atlantic Oscillation Index (NAOI) and Madden Julian Oscillation Index (MJOI) on daily scale in relation to daily Indian summer monsoon rainfall (June-September) for the period 1979-2001.

P.N.Razdan et al. provided an overview on various geothermal fields of India and the status of exploration for future studies. The studies carried out, so far, have clearly pointed out that adequate geothermal potential in respect of direct heat utilization and power generation does exist at several geothermal sites. Reservoir simulation studies have suggested the possibility of generating over 3 MW electric power in Puga (J&K) field. *H. Thomas et al.* described software for classification of fourier shape of the fold taking example from Tamadhaun area, district Almora, Uttarakhand.

S.K. Tandon
A.R. Bhattacharya

Contents

1

Lateral Zonation Around Archean Nucleus of the Dharwar Craton, India: Its Deformation, Segmentation and Subsequent Breakup

O.P. PANDEY[1] and P.K. AGRAWAL[2]

[1] National Geophysical Research Institute, Hyderabad 500 007, India

[2] 5-82/A, Vivekanand Nagar, Street No.8, Habsiguda, Hyderabad 500 007, India

Email: om_pandey@rediffmail.com

Abstract

Regional scale variation of the nature of geological units across the earliest evolved regions of the earth may contain fundamental information about continental evolution. Among the Archean blocks, the Indian shield appears to possess certain unique geotectonic and geodynamical characteristics which provide an opportunity to understand its multi-stage crustal evolution both in space and time. Our study of the south Indian shield reveals segmentation and a secular and progressive lateral zonation of the continental lithosphere around Dharwar nucleus since mid-Archean. However, its cratonic nucleus seems to have ultimately broken due to successive crustal remobilization and foundering of the rheologically and tectonically weak Archean lithosphere. Such weaknesses were caused by (i) episodic plume induced tectonothermal events since 2.7 Ga, and (ii) asthenospheric convective processes associated with a new rifting phase triggered by Marion plume activity along the India's western margin at about 90 Ma. The Antongil block of northeast Madagascar seems to correspond to the broken segment of the western Dharwar craton.

Introduction

The Indian subcontinent with its dynamic history since Late Archean forms one of the most deformed regions among the stable areas of the earth (Pandey and Negi, 1987; Pandey *et al.*, 1996, 2002; Pandey and Agrawal, 1999; Pandey and Agrawal, 2001; Agrawal and Pandey, 2004). It exhibits a variety of geological features formed at different times by different geotectonic processes. It is considered unique in several ways, *viz.* it has been associated with (i) highest rate of mobility (~20 cm /yr.) in the past during 80-53 Ma, (ii) warm and thin lithosphere (~ 105 km on an average), (iii) episodically active rift systems since 1.5 Ga, (iv) several continental breakups, and (v) large basaltic outpourings (McA. Powell, 1979; Negi *et al.*, 1986, 1992; Naqvi and Rogers, 1987; Rogers and Callahan, 1987; Pandey and Agrawal, 1999 etc.). It contains several cratonic blocks (Dharwar, Singhbhum, Aravalli, Bastar, Bundelkhand), characterized by differing style of evolutionary patterns and surrounded by geotectonically weak early to mid-Proterozoic mobile belts and rift systems like Godavari and Mahanadi grabens and Narmada - Son Lineament (NSL) (Fig.1).

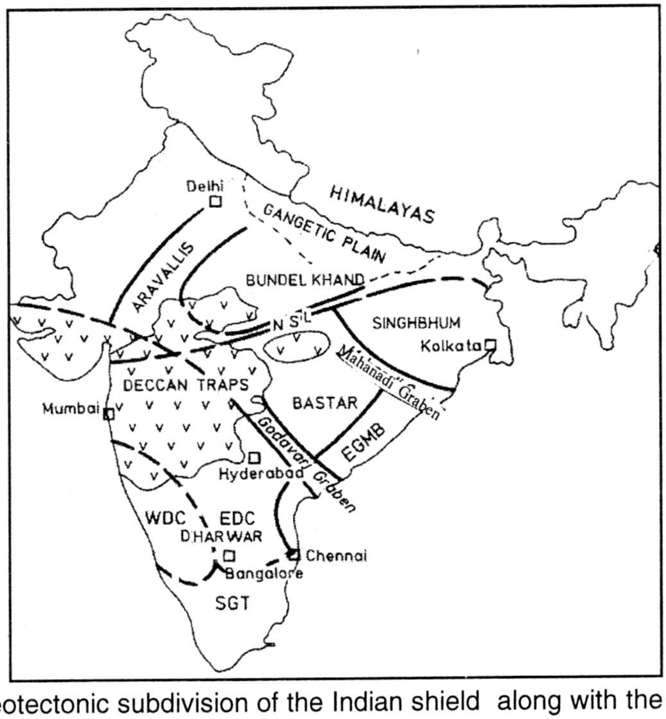

Fig.1 : Geotectonic subdivision of the Indian shield along with the location of Archean-Proterozoic cratonic blocks and major graben structures. NSL: Narmada-Son Lineament; WDC: Western Dharwar Craton; EDC: Eastern Dharwar Craton and SGT: Southern Granulite Terrain; EGMB: Eastern Ghat Mobile Belt.

In the present work, an attempt is made to synthesize different stages of crustal development around the Archean Dharwar nucleus (south India), its segmentation and subsequent breakup which seems to have played an important role in the crustal development of the east Gondwanaland.

Geological Characteristics

The structural, tectonic and geologic patterns among some of the segments of east Gondwanaland are shown in Fig. 2, which reveals a

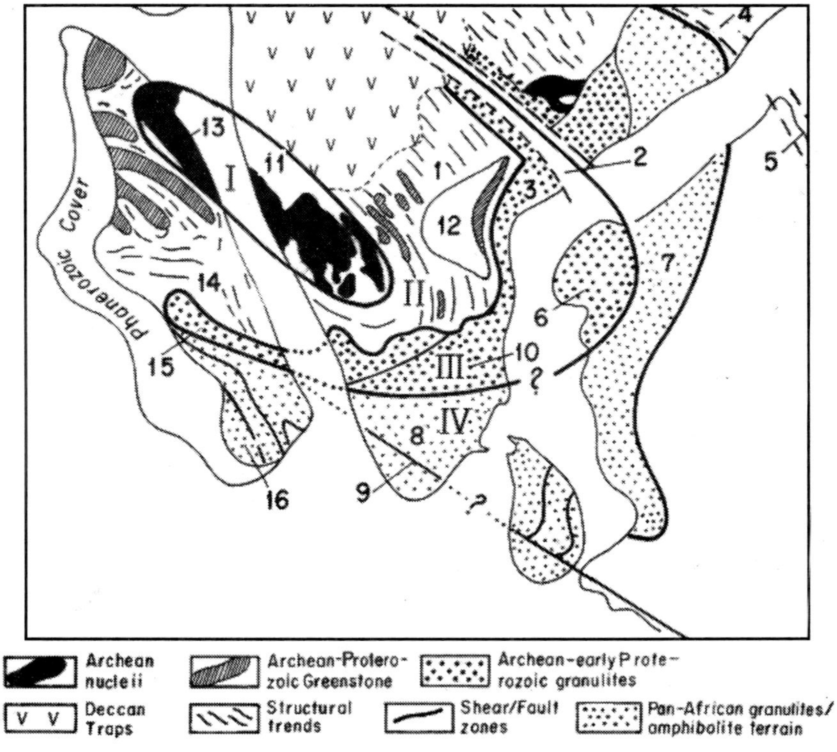

Fig. 2 : Lateral crustal zonation pattern around western Dharwar craton nucleus. I: Cratonic nucleii, II: Late Archean – Early Proterozoic crystalline terrain, III: Late Archean – Early Proterozoic granulite terrain, IV: Pan-African metamorphic terrain, 1: granite-gneiss and metamorphic complexes, 2: Godavari graben, 3: Eastern Ghat Mobile Belt, 4: Mahanadi graben, 5: Lambert rift, 6: Napier complex, 7: Reyner complex, 8: Southern Granulite Terrain (SGT), 9: Achankovil shear zone, 10: Archean – Early Proterozoic granulites, 11: Archean Dharwar nucleii, 12: Cuddapah basin, 13: Antongil cratonic block, 14: granite-gniess-greenstone block of central Madagascar, 15: Ranotsara shear zone, 16: Pan-African granulites of south Madagascar.

systematic and progressive lateral crustal zonation from Zone I to Zone IV around the mid-Archean Dharwar cratonic nucleus. By and large, a broad similarity between the rock types of south India, Madagascar, Sri Lanka and east Antarctica is fairly evident.

Zone I: Cratonic Nucleus

This region corresponds to an elliptical mid to late Archean nucleus situated in the western part of the craton. It is made up of thin primordial basic crust, over which lies vast exposures of probably the oldest and extensively developed greenstone schist belts (3.0–3.4 Ga), granitic and gneissic core, characterised by low temperature metamorphism. This is quite in contrast to those found in the surrounding regions. Most of the ages of gneisses are around 3.0 Ga (Radhakrishna and Naqvi, 1986). The Archean nucleus is covered by Deccan Traps in the north, while in the west, it resembles a passive continental rifted margin. It is now known that until ~90 Ma, India and Madagascar were a single unit in the erstwhile Gondwanaland (Agrawal *et al.*, 1992; Storey *et al.*, 1995; Raval and Veeraswamy, 2003), before they rifted apart to occupy the present positions. We conjecture, that in this rifting process, Dharwar cratonic nucleus was broken, a part of which now forms Antongil cratonic block of the northeast Madagascar (Figs. 2 & 3). This block is also made up of 3.2 Ga migmatitic biotite gneiss (Tucker *et al.*, 1997) to 3.4 Ga old granite (Cohen *et al.*, 1984). Subsequently, it was again intruded by about 2.55 Ga old granitic bodies (Collins, 2000). The entire rock suites of cratonic nucleus of both India and Madagascar have remained literally unaffected by later Proterozoic events (Radhakrishna and Naqvi, 1986; Collins, 2000). As expected, the sub-Moho P_n velocity below the Dharwar cratonic nucleus is found to be very high (8.4km/s) (Reddy *et al.*, 2000). Our calculations indicate presence of a relatively thick (~175 km) and cool lithosphere (Moho temperatures ~ 410° C) below this region compared to adjacent regions in the east and south.

Zone II: Late Archean – Early Proterozoic Crystalline Terrain

This zone is consisting of Late Archean - Early to Mid-Proterozoic cratonic growth. In the Indian subcontinent, it covers the eastern part of the Dharwar craton, which is marked by low pressure regional metamorphism and large scale crustal remobilisation. It experienced extensive plume related tectono-thermal magmatic activity at around 2.7- 2.2 Ga, 1.9-1.8 Ga and 1.2-1.1 Ga (Naqvi and Rogers, 1987; Acharyya, 1997; Anil Kumar *et al.*, 1993; Jayananda *et al.*, 2000;

Radhakrishna and Naqvi, 1986). It contains abundant calc–alkaline to K- rich granitoids (Harish Kumar *et al.*, 2003), gneisses and metamorphic complexes, besides being extensively intruded by mafic dyke swarms (Murthy, 1995). The last major event (~ 1.1 Ga) which may have been related to a deep mantle plume (Mall *et al.*, 2008), resulted in the emplacement of kimberlites at several places (Anil Kumar *et al.*, 1993). In this region, lithosphere is thinner and Moho is warmer compared to that of Dharwar Craton Nucleus. Here, the P_n velocities are much lower at ~ 7.8 km/s (Reddy *et al.*, 2000). On the Madagascar side, this zone is represented by the central region of almost similar age (Late Archean - Proterozoic). Antananarivo and Tsaratanana blocks comprised mainly of greenstones, granites, gneisses and tonalites which were later intruded by Late Proterozoic (~ 500-800 Ma) granite, syenite and gabbro at several places.

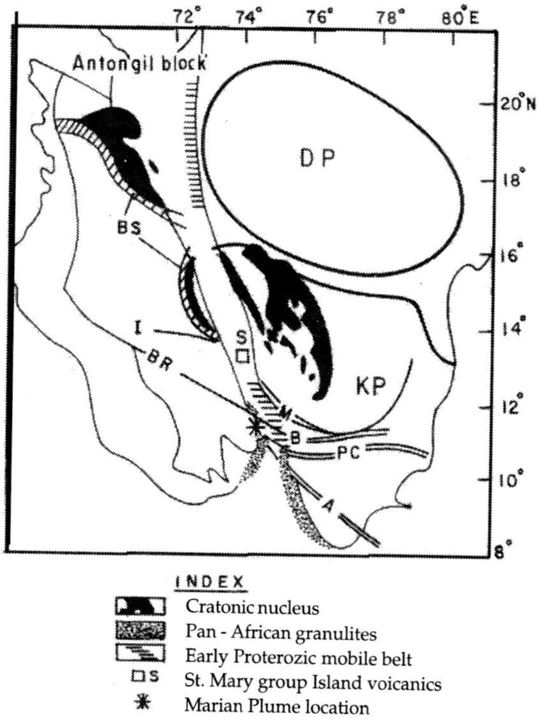

Fig. 3 : Juxtaposition of Madagascar and India before breakup and geotectonic correlation. DP: Deccan Plateau; KP: Karnataka Plateau; M: Moyar shear zone; B: Bhavani shear zone; PC: Palghat-Cauvery shear zone; A: Achankovil shear zone; BS: Betsimisaraka suture zone; I: Itremo lineament; BR: Bongolava-Ranotsara lineament.

Zone III: Late Archean – Early Proterozoic Granulite Terrain

This granulitic terrain is characterised by Late Archean - Early Proterozoic metamorphic thermal events. In Antarctica, it is represented by the Napier complex (Late Archean or even older ; Chetty *et al.,* 2003), while in India, it occurs (i) in the region situated north of the Palghat-Cauvery shear zone (PC in Fig.3) separating Archean granulites in the north from that of Pan-African granulites in the south, (ii) along Eastern Ghat Mobile Belt, and (iii) granulites situated on both the flanks of the Godavari graben (Rajesham *et al.,* 1993; Ramakrishnan *et al.,* 2003). In the granulitic terrain, situated north of PC shear zone, lithospheric thickness, on an average, is ~ 105 km and the mantle lithosphere is much warmer than the adjacent region of the Dharwar craton in the north. Inspite of the inadequacy of the data, there are sufficient indications (Janardhan, 1999; Kroner *et al.,* 1999; Rambeloson, 1999) to suggest that this zone extends into Madagascar and covers an area immediately north of the Ranotsara shear zone (Fig. 2).

Zone IV: Pan-African Metamorphic Terrain

This terrain, criss-crossed by deep-seated faults/shear zones (Figs. 3&4), has undergone extensive high grade regional metamorphism at around 550 ± 30 Ma. It contains granulite - amphibolite facies rocks. In the Indian subcontinent, it is represented by the granulitic terrain occurring south of the Palghat Cauvery (PC) shear zone (Fig. 3). Beneath this terrain, underlying lithosphere is probably thinnest (~ 95 km on an average) and warmest compared to any other part of the Dharwar craton. In Madagascar too, it occupies the southern portion lying south of Ranotsara shear zone (Fig. 2). The entire terrain represents mainly P-T conditions of about 7.5 ± 1.5 kb and 850 ± 150° C respectively, with comparable lithologies on a regional scale. Vast areas of east Antarctica (Reyner complex) and Sri Lanka (Wanni, Highland and Vijayan complexes) are covered by similar grade rocks.

The zonation pattern as described above and shown in Fig. 2, reveals gradual but progressive changes in lithologies around the Archean Dharwar nucleus. Such type of crustal development which started during Middle Archean, continued till the fragmentation of erstwhile Gondwanaland that marked the beginning of a new global plate tectonic era. This new era was extremely dynamic involving extensive plate reorientations and an abrupt increase in spreading rates, besides other important geodynamic events like K-T boundary impact near offshore Mumbai, Deccan and Rajmahal basaltic eruptions and interaction with four mantle plumes. These events completely reshaped the underlying lithosphere (Pandey *et al.*, 1995; Pandey and Agrawal, 1999). This was also the period when the Indian Ocean began to form and the island of Madagascar broke away from greater India.

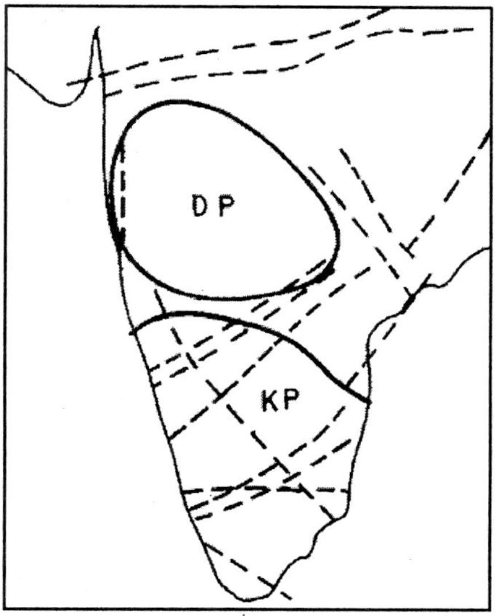

Fig. 4 : Location of major deep-seated faults, taken from the tectonic map prepared by ONGC. DP and KP refer to Deccan and Karnataka plateaus.

Segmentation and Breakup of Dharwar Craton

Cratons are usually considered as regions which have exhibited tectonic stability since Late Archean times. They contain relatively cold and thick keels or roots reaching as deep as 450 km (Polet and Anderson, 1995). Going by this definition, it is difficult to contemplate the breakup of any craton specially the old ones like Dharwar craton. However, a number of studies now suggest that Madagascar was attached to greater India till about 90Ma (Storey *et al.*, 1995; Raval and Veeraswamy, 2003) or it may even have been a splitted segment of Archean Dharwar craton (Agrawal *et al.*, 1992). The probable cause for such a breakup is still debatable. However, what is not debatable is that the underlying Indian lithosphere is not quite as same as other stable regions and certainly it did not remain same as it was during the Early Archean times (Negi *et al.*, 1986; Ramesh *et al.*, 1996; Pandey and Agrawal, 1999 ; Priestley *et al.*, 2006 ; Kumar *et al.*, 2007). It has persistently undergone several cycles of tectono-thermal and geodynamic events during the past *viz.* between 2.7 - 2.2, 1.9-1.8, 1.2-1.1, 0.85 - 0.5 Ga besides several Phanerozoic events (Radhakrishna and Naqvi, 1986; Naqvi and Rogers, 1987; Rogers and Callahan, 1987; Anil Kumar *et al.*, 1993; Acharyya, 1997 ; Jayananda *et al.*, 2000; Harish kumar *et al.*, 2003).

The above events were often related to mantle plumes which time and again reactivated, remobilised and ultimately segmented the entire crustal column. Mid to Neo-Proterozoic reactivations were particularly responsible for the development of uplifted plateaus, rift systems and major shear /fault zones within and around the periphery of Dharwar craton which remained episodically active since then. For example, Dharwar craton is now divisible into three distinct geotectonic regions (Fig. 1): Western Dharwar Craton (WDC), Eastern Dharwar Craton (EDC) and Southern Granulitic Terrain (SGT), all of them exhibiting different geological and geophysical characters. Besides these, it also got segmented into two uplifted plateaus: (i) Deccan Plateau (DP) in the north covering a major part of EDC overlain by Deccan Traps and (ii) Karnataka Plateau (KP) in the south encompassing Archean nucleus (WDC), granulite terrain (SGT) and southern part of EDC (Figs.3-5). Deccan traps covered region of EDC exhibits much higher crustal seismic velocities compared to its adjacent region in the south including SGT. Recent GPS studies indicate a fair possibility of relative motion between DP and KP (Catherine, 2001). These two uplifted blocks also have altogether different geological characters. DP almost solely corresponds to Early Proterozoic mobile belt while southern part of the KP is dominated by Pan-African granulites and numerous mega shear/deep-seated fault zones (Figs. 2&4). Except the cratonic nucleus portion, the entire western margin was occupied either by Early Proterozoic mobile belt terrain (Radhakrishna and Naqvi, 1986) or by Late Archean- Pan-African granulite terrain (Fig. 3), before the break up of the Dharwar craton. It appears that in the past, the Betsimisaraka suture zone (BS) of Madagascar (Collins, 2000) may have extended across the Dharwar craton separating DP and KP (Fig. 3). This is well reflected in the patterns of regional uplift (Fig. 5a) and Bouguer gravity images (Fig. 5b).

It seems the past crustal reactivations resulted into large scale influx of volatiles through weak zones, and greater enrichment of radioactive elements in the already lithophilic element rich Indian crust (Rogers and Callahan, 1987). This made Indian lithosphere rheologically weak and warm besides lowering the viscosity of lithospheric mantle (Negi *et al.*, 1986; Pandey and Agrawal, 1999). It also made the Dharwar craton, in the presence of already existing weak zones (Figs. 3 & 4), vulnerable for breakup in case it was hit by a rising mantle plume.

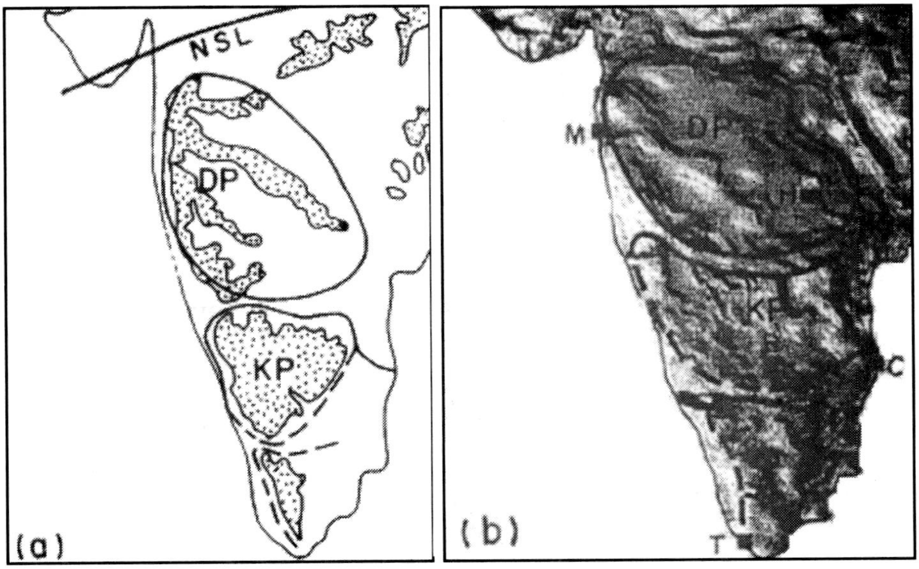

Fig. 5 : (a) Approximate boundaries of Deccan and Karnataka plateaus sitting over the uplifted blocks (shown by dots) of south Indian shield (Raval, 1995). (b) Bouguer gravity anomaly shaded relief images (Sreedhara Murthy, 1999) over DP and KP. C: Chennai; M: Mumbai; T: Trivandrum.

Effect of Marion Plume Interaction

At around 90 Ma, the Marion plume did hit this region (Fig. 6). Location for the plume is said to be quite close to the southeastern margin of Madagascar (Curray and Munasinghe, 1991; Storey *et al.*, 1995; Raval, 1999; Raval and Veeraswamy, 2003), which incidentally falls at the cross section of several mega shear zones and mobile belts (Fig. 3). These shear zones which extended below Moho facilitated magma upwelling up to the surface. Numerous dykes and flows of Late Cretaceous age (like St. Mary Island group of volcanics along the western margin , dated at 86 Ma; Pande *et al.*, 2001; Fig. 3) bear testimony to this magmatic event. Sudden upwelling of magma triggered the ridge, then active between Africa and Madagascar, jump to the east of Madagascar thereby creating a new rifting phase along the India's western margin (Fig. 6). Asthenospheric convective processes associated with the continental breakup are known to give rise to shallow pseudo- plume (Anderson, 2000). Thus, convective processes associated with this new rifting phase together with the past successive plume related thermal events, appear to have degenerated and sheared the once thick cratonic root beneath this craton (Pandey and Agrawal, 1999) which ultimately resulted into the breakup of the Dharwar craton. Progressively sheared lithospheric mantle was then consumed by the asthenospheric mantle. As a consequence, the Dharwar cratonic root has become much thinner compared to 250-450 km found elsewhere in similar terrain (Polet and Anderson, 1995). As stated earlier, our estimated thickness of the lithosphere beneath Western Dharwar Craton (WDC), based on available heat flow data, is only about 175 km while the average Indian shield lithosphere is even thinner at ~ 100 km only (Negi *et al.*, 1986; Pandey and Agrawal, 1999; Kumar *et al.*, 2007). Available geophysical data support these estimates. Although there is very little information available for the lithospheric structure beneath Madagascar, its present lithospheric thickness is expected to be around 100-125 km only (Chapman and Pollack, 1977).

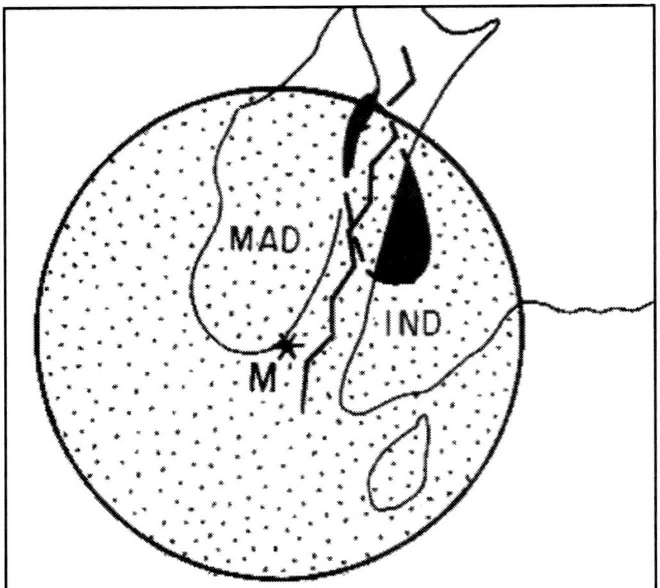

Fig. 6 : Juxtaposition of Madagascar and India before 90 Ma. Star shows the postulated location of Marion Plume (M) before its outburst (Curray and Munasinghe ,1991;Storey *et al.*, 1995; Raval, 1999). Circle with dots indicate the possible area of Marion plume influence.

It is interesting to mention here that based on M - sequence anomalies in the Enderby basin (Antarctica), Royer *et al.* (1992) argued for a large strike slip motion between India and Madagascar. If it was so, then it might have further desegregated the Dharwar craton prior to its breakup by Marion plume.

Geophysical Support and Inferences

The elliptical shaped splitted nucleus of the Dharwar craton (Radhakrishna and Naqvi, 1986) which lies quite close to the western margin (Fig. 7a) is well brought out by the gravity map shown in Fig. 7b. This figure shows modified Bouguer gravity anomaly map of south India contoured at an interval of 50 mgal. Gravity anomalies

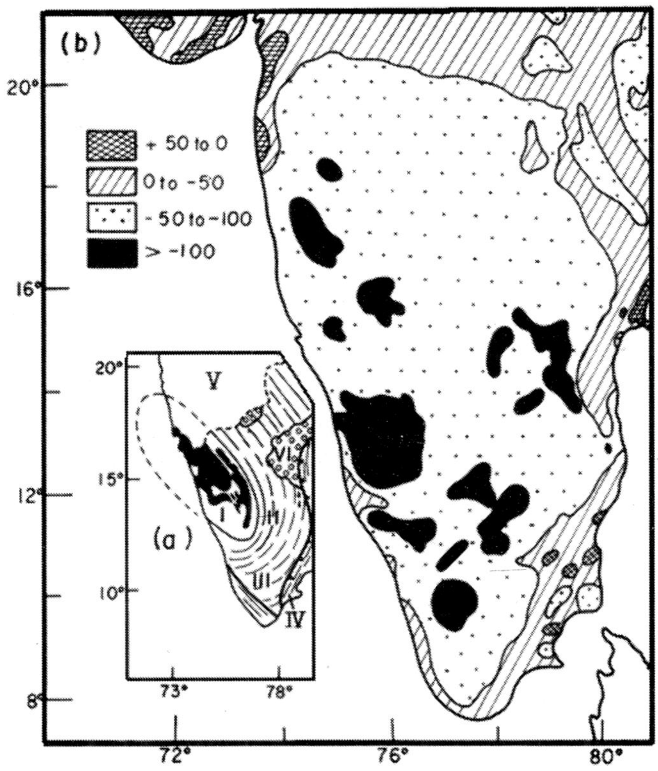

Fig.7 : Generalised geotectonic patterns of Dharwar craton (Fig. a) based on Radhakrishna and Naqvi (1986). I: Western Dharwar craton nucleus, II: Late Archean-Proterozoic crystalline terrain, III: Southern granulites, IV: Eastern Ghat Mobile Belt, V: Deccan traps, VI: Cuddapah basin. Corresponding modified Bouguer gravity anomaly map prepared by NGRI (1975) drawn at an interval of 50 mgal is shown in Fig. b.

show an incomplete gravity closure of unusually low magnitude (50 to 100 mgal) over the Indian side of the Dharwar craton. The high gradient contours around the continental margin represent rifted and possibly uplifted continental blocks. The magnitude and shape of anomalies on Indian side indicates something unusual about the mantle underneath, as such anomalies usually do not occur in other stable regions of the globe. Further, the accurately determined Moho of 33 to 39 km beneath EDC through receiver function (Gupta, 2003) is identical to that of 33 to 39 km found in central Madagascar (Fourno and Roussel, 1994). There is also one to one correspondence between the long wave length MAGSAT anomalies over India and Madagascar (Agrawal *et al.,* 1992) signifying that the two regions have common deep magnetic properties. Regrettably however, compared to Indian shield, not much geophysical information is available over Madagascar.

We have attempted to demonstrate that a multi-stage progressive lateral crustal growth took place encircling the Archaean Dharwar nucleus which lasted till the beginning of Gondwana breakup. Besides, we feel that not all the Archean cratonic nuclei remained stable as hitherto believed. Some of them have definitely been affected by past tectonothermal events. For example, the lithospheric keel beneath the Dharwar craton underwent large scale deformation and foundering due to thermal and geodynamic events which segmented the craton and ultimately led to breakup. Remnants of the broken part could be the Antongil block in northeast Madagascar. Our study refutes the idea of cratonic stability of deep continental blocks which in many cases are no longer associated with the ancient old roots (Polet and Anderson, 1995; Pandey and Agrawal, 1999; Kumar *et al.,* 2007). They are liable to deform and break in the environment of extensional tectonism. Apart from Indian cratons, there are several other cases where Archean cratonic roots have been deformed and destroyed due to Proterozoic tectonothermal/Mesozoic - Tertiary geodynamic events, for example, central African, east African, eastern Sino-Korean, Kaapvaal cratons etc. (Polet and Anderson,1995; Griffin *et al.,*1996, 1998; Prestley *et al.,* 2002). In such areas, the ancient roots have been replaced by hot and thin lithosphere.

Acknowledgements

We are extremely thankful to Dr. U. Raval for his valuable suggestions and involvement in this problem and to Mr. V. Subrahmanyam for his help in the preparation of this manuscript.

Mr. Shyam Vaidya has traced the figures. The permission accorded by the Director, National Geophysical Research Institute, Hyderabad, to publish this work is gratefully acknowledged.

References

Acharyya, S.K. (1997) Evolutionary characters of the Gondwanic Indian crust. Indian Minerals, v.51, pp. 1-24.

Agrawal, P.K. and Pandey, O.P. (2004) Unusual lithospheric structure and evolutionary pattern of the cratonic segments of the south Indian shield. Earth Planets Space, v. 56, pp. 139-150.

Agrawal, P.K., Pandey, O.P. and Negi, J.G. (1992) Madagascar: A continental fragment of the paleo-super Dharwar craton of India. Geology, v. 20, pp. 543-546.

Anderson, D.L. (2000) The thermal state of the upper mantle; no role for mantle plume. Geophys. Res. Lett., v. 27, pp. 3623-3626.

Anil Kumar, Padmakumari, V.M., Dayal, A.M., Murthy, D.S.N. and Gopalan, K. (1993) Rb-Sr ages of Proterozoic kimberlites of India: evidence for contemporaneous emplacement. Precamb. Res., v. 62, pp. 227-237.

Catherine, J.K. (2001) Crustal deformation studies using global positioning system with particular reference to Indian plate. Ph.D. thesis, Osmania Univ., Hyderabad, India, 144 pp.

Chapman, D.S. and Pollack, H.N. (1977) Regional geotherms and lithospheric thickness. Geology, v. 5, pp. 265-268.

Chetty, T.R.K., Vijay, P., Narayana, B.L. and Giridhar, G.V. (2003) Structure of the Nagavali shear zone, Eastern Ghats Mobile Belt, India : Correlation in the east Gondwana reconstruction. Gondwana Res., v. 6, pp. 215-229.

Cohen, L., Snelling, N.J., Delhal, J., Vail, J.R., Bonhomme, M. and Ledent, D. (1984) The geochronology and evolution of Africa. Claredon Press, Oxford, 512 p.

Collins, A.S. (2000) The tectonic evolution of Madagascar: Its place in the east African orogen. Gondwana Res., v. 3, pp. 549-552.

Curray, J.R. and Munasinghe, T. (1991) Origin of the Rajmahal Traps and the 85°E ridge: Preliminary reconstructions of the trace of the Crozet hotspot. Geology, v. 19, pp. 1237-1240.

Fourno, J-P. and Roussel, J. (1994) Imaging of the Moho depth in Madagascar through the inversion of gravity data: geodynamic implications. Terra Nova, v. 6 , pp. 512-519.

Griffin, W.L., Zhang, A., O'Reilly, S.Y. and Ryan, C.G. (1996) Phanerozoic evolution of the lithosphere beneath the Sino-Korean craton. In: *M. Flower,*

S.L. Chung, C.H. Lo and T.Y. Lee (eds), Mantle Dynamics and plate interactions in east Asia. Am. Geophys. Un. Geodyn. Ser., no. 27, pp.107-126.

Griffin, W.L., O'Reilly, S.Y., Ryan, C.G., Gaul, O. and Ionov, D.A. (1998) Secular variation in the composition of our continental lithospheric mantle: geophysical and geodynamic implications. In: *Braun. J. et al. (eds),* Structure and evolution of the Australian continent, Am. Geophys. Un. Geodyn. Ser., no. 26, pp. 1-26.

Gupta, S., Rai, S.S., Prakasam, K.S. and Srinagesh, D. (2003) The nature of the crust in southern India: implications for Precambrian crustal evolution. Geophys. Res. Lett., v. 30, pp. 1419.

Harish Kumar, S.B., Jayananda, M., Kano, T., Shadakshara Swamy, N. and Mahabaleswar, B. (2003) Late Archean juvenile magmatic accretion process in the eastern Dharwar craton: Kuppam-Karimangalam area. Mem. Geol. Soc. Ind., v. 50, pp. 375-408.

Janardhan, A.S. (1999) Southern granulite terrain, south of the Palghat-Cauvery-shear zone: Implications for India-Madagascar connection. Gondwana Res., v. 2 , pp. 463-469.

Jayananda, M., Moyen, J.-F., Martin, H., Peucat, J.-J., Auvrav, B. and Mahabaleswar, B. (2000) Late Archean (2550-2520 Ma) juvenile magmatism in the eastern Dharwar craton: southern India: Constraints from geochronology, Nd-Sr isotopes and whole rock geochemistry. Precamb. Res., v. 99, pp. 225-254.

Kroner, A., Windley, B.F., Jaeckel, P., Collins, A.S., Breuer, T.S., Nemchin, A. and Razakamanana, T. (1999) New Zircon ages for Precambrian granites, gneisses and granulites from central and southern Madagascar: Significance for correlations in east Gondwana. Gondwana Res., v. 2, pp. 351-352.

Kumar, P., Yuan, X., Kumar, M.R., Kind, R., Li, X. and Chadha, R.K. (2007) The rapid drift of Indian tectonic plate. Nature, v. 449, pp. 894-897.

Mall, D.M., Pandey, O.P., Chandrakala, K. and Reddy, P.R. (2008) Imprints of a Proterozoic tectonothermal anomaly below the 1.1 Ga kimberlitic province of southwest Cuddapah basin ,Dharwar craton (southern India) . Geophys. J. Int., v.172, pp. 422-438.

McA. Powell,C. (1979) A speculative tectonic history of Pakistan surroundings : some constraints from Indian ocean. In: *A. Farah and K.A. de jong (eds),* Geodynamics of Pakistan, Geol. Surv. Pakistan. Quetta, pp. 5-24.

Murthy, N.G.K. (1995) Proterozoic mafic dykes in southern peninsular India : A review. Mem. Geol. Soc. Ind., v. 33, pp. 81-98.

Naqvi, S.M. and Rogers, J.J.W. (1987) The Precambrian geology of India. Oxford Univ Press, New York, 223 pp.

Negi, J.G., Pandey, O.P. and Agrawal, P.K. (1986) Super-mobility of hot Indian lithosphere. Tectonophysics, v. 131, pp. 147-156.

Negi, J.G., Agrawal, P.K., Singh, A.P. and Pandey, O.P. (1992) Bombay gravity high and eruption of Deccan flood basalts (India) from a shallow secondary plume. Tectonophysics, v. 206, pp. 341-350.

NGRI (1975) Bouguer gravity anomaly map of India, published by National Geophysical Research Institute, Hyderabad 500 007 (India).

Pande, K., Seth, H.C. and Bhutani, R. (2001) ^{40}Ar -^{39}Ar age of St. Mary's island volcanics, southern India: Record of India-Madagascar break-up on the Indian subcontinent. Earth planet sci. Lett., v.193, pp. 39-46.

Pandey, O.P. and Negi, J.G. (1987) Signals of degeneration of the sub-crustal part of the Indian lithosphere since the break-up of Gondwana land. Phys. Earth Planet. Int., v. 48, pp. 1-4.

Pandey, O.P. and Agrawal, P.K. (1999) Lithospheric mantle deformation beneath the Indian cratons. J. Geology, v. 107, pp. 683-692.

Pandey, O.P. and Agrawal, P.K. (2001) Nature of lithospheric deformation beneath the western continental margin of India. J. Geol. Soc. Ind., v. 57, pp. 497-505.

Pandey, O.P., Agrawal, P.K. and Negi, J.G. (1995) Lithospheric structure beneath Laxmi ridge and late Cretaceous geodynamic events. Geo- Marine Lett., v.15, pp. 85-91.

Pandey, O.P., Agrawal, P.K. and Negi, J.G. (1996) Evidence of low density subcrustal underplating beneath western continental region of India and adjacent Arabean Sea: Geodynamical considerations. J. Geodynamics, v. 21, pp. 365-377.

Pandey, O.P., Agrawal, P.K. and Chetty, T.R.K. (2002) Unusual lithospheric structure beneath the Hyderabad granitic region, eastern Dharwar craton, south India. Phys. Earth Planet. Int., v. 130, pp. 59-69.

Polet, J. and Anderson, D.L. (1995) Depth extent of cratons as inferred from tomographic studies. Geology, v. 23, pp. 205-208.

Priestley, K., Debayle, E., Mckenzie, D. and Pilidou, S. (2006) Upper mantle sructure of eastern Asia from multimode surface waveform tomography. J.Geophys. Res., v. 111, B10304, doi:10.1029/2005JB004082,2006.

Radhakrishna, B.P. and Naqvi, S.M. (1986) Precambrian continental crust of India and its evolution. J.Geology, v. 94 , pp. 145-166.

Rajesham, T., Bhaskar Rao, Y.J. and Murty, K.S. (1993) The Karimnagar granulite terrrain - A new sapphirine bearing granulite provinces, south India. J. Geol. Soc. Ind., v. 41 , pp. 51-59.

Ramakrishnan, M. (2003) Craton-mobile belt relations in southern granulite terrain. Mem. Geol. Soc. Ind., v. 50, pp. 1-24.

Rambeloson. A.R. (1999) Gold in Madagascar. Gondwana Res., v. 2, pp. 423-431.

Ramesh, D.S., Bharthur, R.N., Prakasam, K.S., Srinagesh, D., Rai, S.S. and Gaur, V.K. (1996) Predominance of plate motion- related strain in the south Indian shield. Current Sci., v. 70 , pp. 843-847.

Raval, U. (1995) Seismicity of Koyna region and regional tectonomagmatism of the western margin (India). Pageoph., v. 145, pp. 175-192.

Raval, U. (1999) Some recent geophysical constraints from the central Gondwanaland. PINSA, 65 A, no. 3, pp. 209-260.

Raval,.U. and Veeraswamy, K. (2003) India-Madagascar separation: Breakup along a pre-existing mobile belt and chipping of the craton. Gondwana Res., v. 6, pp. 467-485.

Reddy, P.R., Chandrakala, K. and Sridhar, A.R. (2000) Crustal velocity structure of the Dharwar craton, India. J. Geol. Soc. Ind., v. 55, pp. 381-386.

Rogers, J.J.W. and Callahan, E.J. (1987) Radioactivity, heat flow and rifting of the Indian continental crust. J. Geology, v. 95, pp. 829-836.

Royer, J.Y., Sclater, J.G., Sandwell, D.T., Cande, S.C., Schlich, R., Munschy, M., Dyment, J., Fisher, R.L., Muller, R.D., Coffin, M.F., Patriat, P. and Bergh, H.W. (1992) Indian Ocean plate reconstructions since the late Jurassic, in: *Duncan et al. (eds.),* Synthesis of Results from scientific drilling in the Indian Ocean. Geophys. Monograph no. 70, American Geophys. Union, pp. 471-475.

Sreedhara Murthy, Y. (1999) Images of the gravity field of India and their salient features. J. Geol. Soc. Ind., v. 54, pp. 221-235.

Storey, M., Mahoney, J.J., Saunders, A.D., Duncan, R.A., Kelley, S.P. and Coffin, M.F. (1995) Timing of hot spot-related volcanism and the breakup of Madagascar and India. Science, v. 267, pp. 852-855.

Tucker, R.D., Ashwal, L.D., Handke, M.J. and Hamilton, M.A. (1997) A geochronologic over view of the Precambrian rocks of Madagascar: A record from the middle Archedan to the late Neoproterozoic. In: *Cox, R.M. and Ashwal, L.D. (eds.),* Proc. Unesco-IUGS-IGCP-348/368 Int. Field Workshop on Proterozoic Geology of Madagascar, Antananarivo, Madagascar. 16-30 August, 1997. Gond. Res. Gr. Misc. Pub. 5, Field Sci. Publ., Osaka, pp. 99.

□□□

2

Granite Series Evaluation of Early Ordovician Kyrdem Granitoids and Enclaves, Meghalaya Plateau, Northeast India : Implication on Oxidation Condition of Interacting Mafic-Felsic Magma System

SANTOSH KUMAR[1] and KH. MOHON SINGH[2]

[1]Department of Geology, Kumaun University, Nainital, 263 002, India

[2]Department of Geology, Imphal College, Imphal, 795 001, India

Email: skyadavan@yahoo.com

Abstract

Early Ordovician (479±26 Ma) Kyrdem Granitoids (KG) and their enclaves in Meghalaya plateau have been evaluated in terms of magnetic susceptibility (MS) and whole rock composition in order to understand the redox condition of comingled mafic and felsic melts. KG (MS=15.56-31.55×10^{-3}SI) and microgranular enclaves (ME) (MS=24.47-32.09×10^{-3}SI) are moderately to strongly oxidized, magnetite series granites formed in a late- to post-tectonic calc-alkaline, mafic-felsic magma interacting environment. Oxidizing nature of bulk KG melt was locally reduced to ilmenite series (0.26×10^{-3} SI) granite near the margin of the KG pluton as a result of reaction with pelitic country rocks occurred at shallow emplacement level.

Introduction

Abundance of ferromagnetic and paramagnetic minerals in granites (*sensu lato*) determine the magnetic and weakly magnetic properties of felsic igneous rocks, which can be recorded and evaluated in terms of magnetic susceptibility (MS in 10^{-3}SI unit) and whole rock geochemistry. MS of granites is highly variable and broadly displays bimodal patterns *viz.* magnetite series (MS >3.0×10^{-3}SI) and ilmenite series (MS=3.0×10^{-3}SI) granites (*e.g.* Kanaya and Ishihara, 1973; Ishihara, 1977, 1990; Takagi, 2004), corresponding to oxidized- and reduced-type granites respectively (Takagi and Tsukimura, 1997), which are parallel to metaluminous (I-type) and peraluminous (S-type) granites (Chappell and White, 1974; Takahashi *et al.*, 1980). The dominance of magnetite and ilmenite as an indicator of magnetite series and ilmenite series granites respectively is primarily controlled by oxygen fugacity (fO_2) of magma source, which may be partially or wholly modified by later processes such as assimilation and magma mixing etc. during the ascent and emplacement of felsic magma. Tectonism can also erode out the original magnetic properties of the granites. Consequently, correlation between granite series and granite types need not be similar, and hence MS value as indicator of granite origin should be used with great caution (Gregorová *et al.*, 2003). In this paper, an attempt has been made to record MS value of Kyrdem Granitoids and enclaves, and further to evaluate the granite series combining whole rock geochemical database.

Geology and Field Relation

Assam-Meghalaya plateau, erstwhile known as Shillong plateau, is a Precambrian geotectonic shield of northeast India. Medlicott (1869) reported the occurrence of several felsic plutons, named after prime villages, which are located along the Tyrsad-Barapani lineament of the Meghalaya plateau (Fig. 1a, b). Mazumder (1976, 1986) described the geology of Meghalaya plateau and considered these felsic plutons (Mylliem, Nongpoh, Kyrdem and South Khasi) as late- to post-tectonic, fracture-controlled diapirs resulting from episodic thermal events caused by mantle upwelling. Ghosh *et al.* (1991, 1994, 2005) provided whole rock Rb-Sr isotopic ages for these plutons (Kyrdem 479±26 Ma; Nongpoh 550±15 Ma; Mylliem 607±13 Ma; South Khasi 690±19 Ma), which suggest a protracted thermal event of *ca* 200 Ma (Proterozoic-Early Palaeozoic; *ca* 700-500 Ma) in the Meghalaya, probably related to Pan African-Caledonian orogeny. These felsic plutons appear younging in age from southwest to northeast and contain abundant microgranular enclaves (Kumar, 1998).

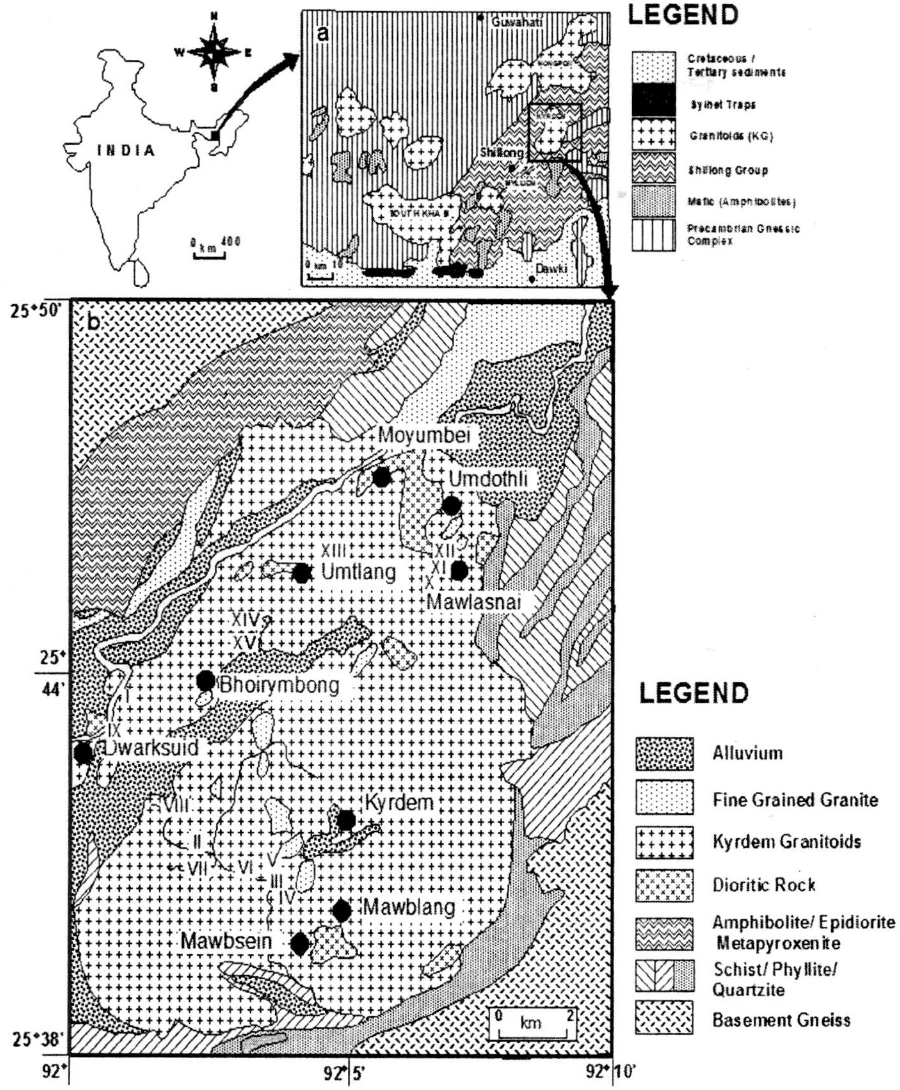

Fig.1: (a) Part of geological map of Khasi Hills of Meghalaya plateau showing locations of various felsic plutons including the Kyrdem Granitoids (KG), drawn after Mazumder (1976). (b) Geological map of Kyrdem and surrounding regions of East Khasi Hills district showing exposures of Early Ordovician felsic magmatism as KG (after Chakraborty, 1989).

Early Ordovician (479±26 Ma) felsic magmatism in the Kyrdem region of Meghalaya plateau, herewith referred to as Kyrdem Granitoids (KG), intrudes the Shillong Group and Precambrian gneissic complex forming an oval-shaped plutonic body with longer axis almost trending N-S (Fig. 1b). Thermal aureole is poorly developed or covered under the alluvium. KG exhibits very coarse grained porphyritic texture with abundant K-feldspar megacrysts (up to 9 cm long) and subordinate amount of amphibole, biotite, plagioclase and quartz. Size of K-feldspar megacrysts increases from margin (Dwarksuid) to the interior (Kyrdem) of the KG pluton. Late felsic pulses as fine grained granite, leucocratic (aplite) and pegmatite veins intrude the KG at several places. Grey and pink varieties of KG can be recognized, but pink colour of KG is the result of post-magmatic fluids, which have not affected the magnetic properties of KG. Modal composition of KG corresponds to quartz monzonite, monzogranite and granodiorite. KG has been geochemically characterized as metaluminous (I-type) to peraluminous (S-type) granitoids (Ghosh *et al.,* 1991). K-Ar ages (521±13 Ma and 457±12 Ma) of biotites from KG are broadly similar to whole-rock Rb-Sr isochron age (479±26 Ma) indicating that the KG pluton has not suffered any post-emplacement tectono-thermal event (Ghosh *et al.,* 1991), and hence redox state of KG should originally correspond to source region and later magmatic processes. Sikdar and Rahman (1998) have studied the structural state of K-feldspar megacrysts from KG, and have concluded that the core of KG pluton has cooled slowly as compared to the margin.

Enclaves hosted in KG can be classified as xenoliths of country rocks (amphibolite and metasedimentary rocks) mostly confined to margin of the KG pluton (Fig 2a, b) and fine to medium grained, mesocratic to melanocratic, mafic-felsic phenocryst-bearing (Fig. 2c) or phenocryst-free microgranular enclaves commonly ubiquitous in porphyritic KG (Fig. 2d), exposed in and around Sohliya, Mawbsein and Mawblang regions. The shape of ME is rounded to elliptical on 2D outcrop, and size varies from a few cm to about one meter across commonly having sharp contacts with felsic host KG. Modal composition of ME belongs to quartz monzonite. ME, as hornblende microdiorite, have been described but have been considered as cogenetic early phase of felsic host KG (Ghosh *et al.,* 1991). Field and petrographic evidences strongly suggest that the ME represent mafic to hybrid coeval magma globules, which were mingled and undercooled into a partly crystalline, relatively cooler felsic host KG, as similarly noted elsewhere (Kumar *et al.,* 2004, 2005a; Kumar and Rino, 2006). Magnetic susceptibility parameter of KG and ME has been preferably investigated to evaluate redox series of comingled felsic melt and ME magma globules.

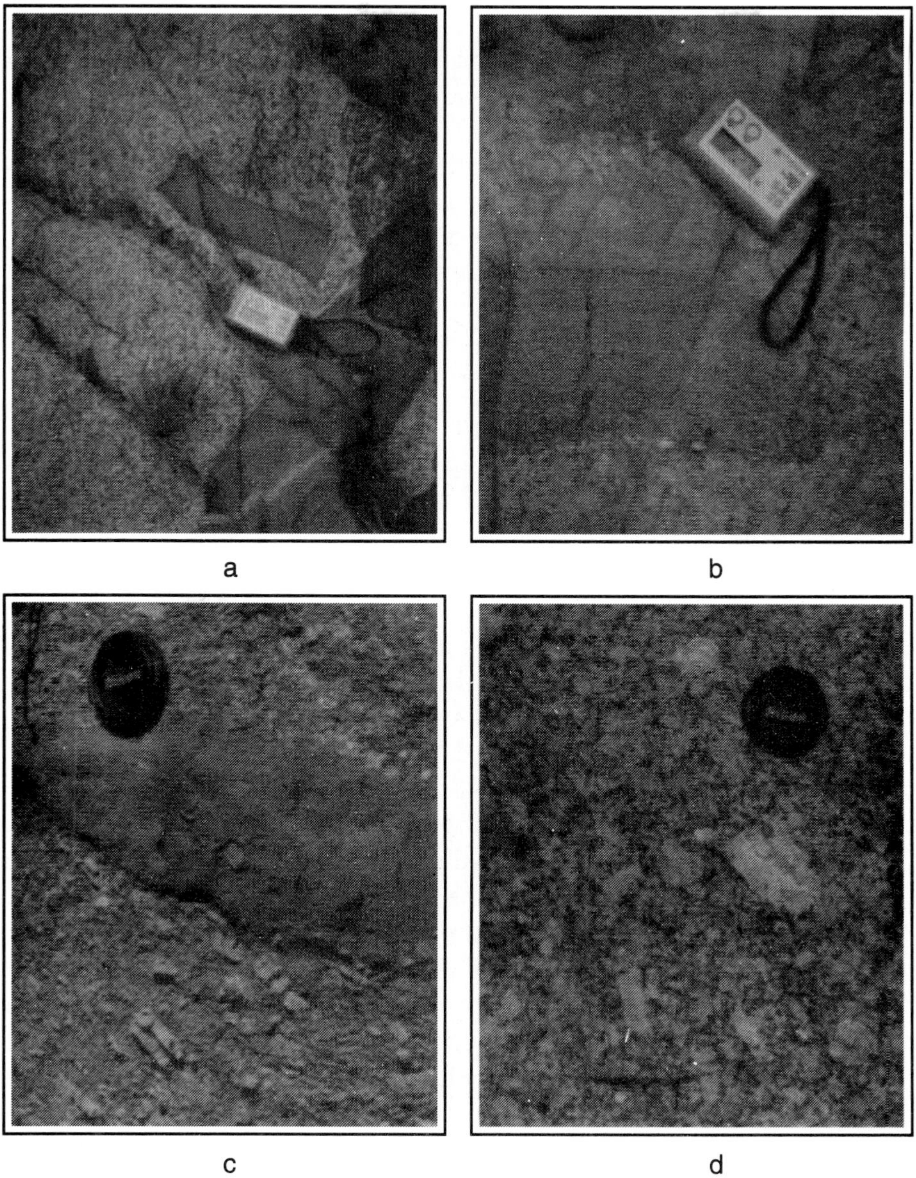

Fig. 2: (a) Amphibolite xenoliths of various shapes and size in porphyritic (K-feldspar megacrysts) Kyrdem Granitoids (KG). Note the felsic net-veining of xenolith. Length of MS meter equals 9.7 cm. (b) Gneissic xenolith in KG. Scale is the same as in (a). (c) Elongated K-feldspar megacrysts bearing microgranular enclave (ME) in KG. Alignment of K-feldspar megacrysts in ME correlate with that of host KG. Length of lens' cap equals 5.5 cm. (d) Randomly oriented K-feldspar megacrysts of various shapes and sizes. Note a small, platy, pelitic xenolith in KG. Scale is the same as in (c).

Analytical Method

MS measurements were taken using SM-20 model of MS meter on smooth rock surfaces of fresh outcrops across the Kyrdem pluton, and the obtained MS values were further corrected according to rock surface unevenness using the recommended factors: 1mm-1.07, 2mm-1.15, 3mm-1.23, 4mm-1.32, 5mm-1.41, 6mm-1.51, 7mm-1.61, 8mm-1.72, 9mm-1.84, 10mm-1.96. The measuring range of MS meter is between 0.000 and 999×10^{-3}SI, and sensitivity of instrument is 1×10^{-6} SI units.

Results and Interpretation

Magnetic Susceptibility Variation

The corrected MS values of Kyrdem Granitoids and enclaves measured in and around various localities are given in table-1. At western contact (Dwarksuid) of the Kyrdem pluton, xenoliths of amphibolite (Khasi greenstone), gneiss, schist and phyllite (Members of Tyrsad Formation belonging to Shillong Group) are found hosted in KG. These xenoliths vary in shape and size from a few cm to several meters across. Amphibolite xenoliths mostly occur as isolated large block in KG, which are cross-cut by several leucocratic felsic veins. Amphibolite xenoliths measure MS values in the range of 28.98-68.20×10^{-3}SI with an average of 43.56×10^{-3} SI (N=9), which relate to the presence of high amount of amphibole and magnetite in xenolith. On the other hand, average MS values for gneissic (0.28-0.50×10^{-3} SI) and pelitic (0.33×10^{-3} SI) xenoliths have been measured very low. Alongwith these xenoliths, KG have also shown very low average MS values ranging between 0.22 and 0.31×10^{-3} SI, which are related to reduced type, ilmenite series granites. The presence of large-size amphibolite xenoliths in KG suggests shallow-level emplacement of KG pluton and probably the stopping of xenoliths from wall and/or roof of the magma chamber. The temperature of KG melt was not high enough to assimilate the amphibolite (metabasalt), whereas gneissic and pelitic xenoliths (as deeper-derived lithology and/or country rocks) have been partially assimilated, and consequently KG melt near the margin of pluton has reduced to ilmenite series granite as similarly observed in the case of South Khasi felsic pluton (Kumar *et al.*, 2005b). However, average MS values of KG away from the contacts in various localities (spot I to XV, table-1) have increased by many folds, mostly varying between 15.56 and 31.55×10^{-3} SI, which are moderately to highly oxidized, magnetite series granite. It is interesting to note that KG with less abundant K-feldspar, in the Mawlasnai locality (spot X), measures relatively lower average MS values (15.78×10^{-3} SI, N=4) as compared to KG with abundant K-feldspar megacrysts ($21.46-31.55\times10^{-3}$ SI). This is because K-feldspar megacrysts poikilitically enclose granular magnetite and ferromagnesian minerals in a zoned manner suggesting its magmatic origin and participation in mafic-felsic magma mixing and ME mingling events (*e.g.* Kumar *et al.*, 2005a, Vernon and Paterson, 2008). Aplite (at Mawblang, spot III) and pegmatite (at Mawbsein, spot IV) veins measure average MS values of 4.53×10^{-3}SI (N=14) and 5.97×10^{-3} SI (N=5) respectively both corresponding to low oxidized, magnetite series granites. KG containing variable amount of mafic minerals on a single outcrop (at Mawbsein, spot IV) have shown variable MS values ranging from 18.35 to 47.92×10^{-3}SI with an average MS of 25.81×10^{-3} SI (N=14). The maximum recorded MS value (47.92×10^{-3} SI) of KG contains highest amount of mafic minerals. It is therefore likely that modal abundance of magnetite, biotite, and amphibole in KG control the MS variations.

Microgranular enclaves (ME) of various shapes and sizes are more frequent in northeastern part of the KG pluton, in and around the Sohliya region. Broadly two types of ME (>20 cm across) with variable MS values of 25.23×10^{-3} SI (N=6) and 32.09×10^{-3} SI (N=8) can be recognized, which are related to their mesocratic and melanocratic nature respectively. A relatively small size ME (~10cm across) occuring in Mawbsein area (spot IV) measures MS in the range of $14.65-34.45\times10^{-3}$ SI with an average MS value of 24.47×10^{-3} SI (N=10). Majority of MS values for KG and ME represent moderately to strongly oxidized, magnetite series granites. High MS values of ME at least suggest the origin of ME unrelated to xenoliths (amphibolite, gneiss etc.) hosted in KG. However, solely based on MS values a magma-mixed or autolith (cognate) origin for ME in KG cannot be established. The hybridization (active mixing region) commonly prevails in open magma system mostly common in a calc-alkaline, I-type felsic pluton, and it is probable that mafic and felsic magma interaction, as evident by the presence of ME, might have also played an important role in raising the oxidizing level from moderately to strongly oxidized, magnetite-series of bulk KG pluton.

Biotite in Granite

Magnetite series granites may contain biotite (Z colour: greenish brown) rich in Mg because iron is consumed to form early magnetite rich in Fe^{3+} whereas Z colour in biotite of ilmenite series granites is reddish-brown because of its Fe^{2+} rich and Fe^{3+} poor compositions (Ishihara, 1998). In Kyrdem pluton, mostly KG and ME contain abundant greenish-brown biotites, and are found associated with magnetite, which have high MgO/FeO ratio, and are therefore derived originally from oxidizing source region. In Dwarksuid region near the margin of KG pluton the ilmenite series granite contains high amount of reddish brown biotite (low MgO/FeO) dominating over greenish brown variety, which have been precipitated from a reducing KG magma as a result of country rock assimilation by originally oxidized KG melt.

Whole-Rock Fe_2O_3/ FeO Ratio

Whole-rock Fe_2O_3/FeO ratio of 0.5 at 70% SiO_2 of granitoids can be empirically used to distinguish between magnetite series (Fe_2O_3/FeO>0.5) and ilmenite series (Fe_2O_3/FeO<0.5) granites because oxidation state of iron is a primary expression of redox state of magmas commonly demarcated at fayalite-magnetite-quartz buffer (Ishihara, 1977; Ishihara *et al.*, 2000; Ishihara *et al.*, 2002a, b; Hart *et al.*, 2004). Ghosh *et al.* (1991, Table-2) have presented major oxide data including Fe_2O_3 and FeO contents of sixteen KG samples. Fe_2O_3/FeO ratio of KG samples varies from 0.09 to 5.25 in the silica range of 63.01-74.15 wt%, with an average Fe_2O_3/FeO of 1.71 and average SiO_2 of 68.53 wt%. Six KG samples have Fe_2O_3/FeO ratios below 0.50 (0.09 to 0.44) and ten other KG samples have Fe_2O_3/FeO ratio more than 0.5 in the range of 0.65 to 5.25, representing to ilmenite- and magnetite-series granites respectively. This is further corroborated in terms of Fe_2O_3/FeO vs FeO* (Fig. 3) and (FeO+MgO)-(Fe_2O_3+TiO_2)-MgO ternary components (Fig.4), which equivocally suggest moderately to very strongly oxidized nature of KG. It is noteworthy that a few KG samples (*e.g.* 8, 3, 10/ km, Fig. 2 of Ghosh *et al.*, 1991) are from the contact with country rocks (schist, phyllite and quartzite). It is, therefore, likely that originally oxidized, magnetite series nature of KG melt has locally reduced to ilmenite series granite because of assimilation of the metasediments near western (Dwarksuid) and northern (Raitong-Mawhati) marginal parts of the KG pluton, which are consistent with the observed low MS values and nature of biotite in KG.

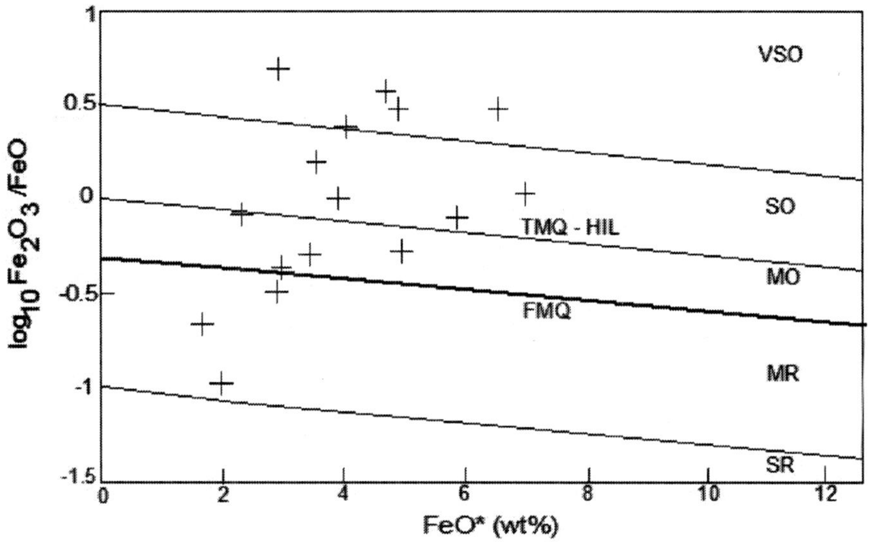

Fig. 3 : Redox classification scheme plotted for Kyrdem Granitoids (+). FMQ and TMQ-HIL buffer lines and fields of VSO: very strongly oxidized, SO: strongly oxidized, MO: moderately oxidized, MR: moderately reduced, SR: strongly reduced, are taken from Hart *et al.* (2004).

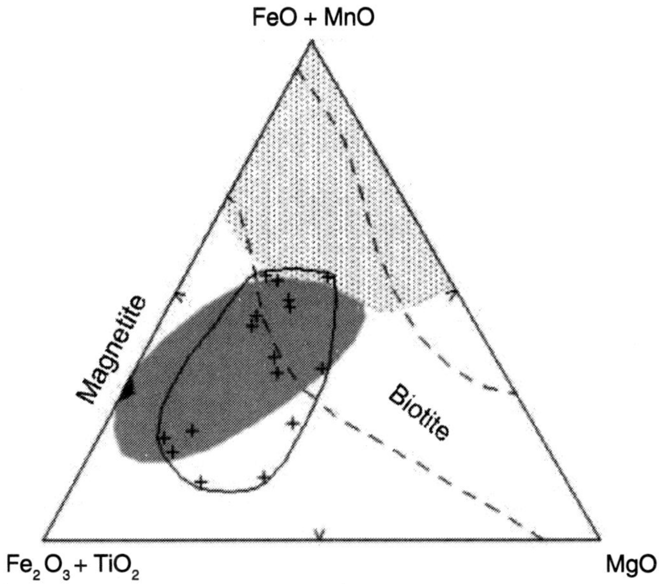

Fig. 4 : Magnetite series (grey shaded region) and ilmenite series (dotted region) granites of Japan as shown by Shunso Ishihara (pers. comm. 2005), are compared with KG (+) encircled by solid line. Magnetite and region of biotite and amphibole (as shown by broken lines) compositions are taken from Ishihara (1971).

Alumina Saturation Index

Alumina saturation index (*i.e.* A/CNK=Molar Al_2O_3/CaO+Na_2O + K_2O) of granitoids has been commonly used to distinguish between metaluminous (I-type, A/CNK<1.05) and peraluminous (S-type, A/CNK>1.05) granitoids (Chappell and White, 1974). About 60% of I-type granitoids of Lachlan Fold Belt, Australia belong to ilmenite series granite, however there should be a parallelism between ilmenite-/magnetite-series and S-/I-type granitoids (Takahasi *et al.,* 1980). In Japan nearly all the ilmenite series granites are metaluminous in nature, and hence magnetite-/ilmenite-series classification does not exactly correspond to the S-/I-type classification scheme (Takahasi *et al.,* 1980; Ishihara, 1998). In Kyrdem pluton, nine KG samples (Table-2, Ghosh *et al.,* 1991) correspond to I-type (metaluminous) granitoids (A/CNK=0.77-1.02), and their Fe_2O_3/FeO ratios mostly comply with magnetite-series granites (Fe_2O_3/FeO >0.5) except for one KG sample. Seven other KG samples are peraluminous (S-type) granitoids (A/CNK=1.08-1.15), which mostly correspond to ilmenite-series granites (Fe_2O_3/FeO <0.5) except two KG samples with substantially high Fe_2O_3/FeO ratios (5.25 and 2.54). For the KG samples, which do not exactly comply with classification parameters in terms of A/CNK, Fe_2O_3/FeO, MS values etc., it is likely that some factors such as assimilation of carbonaceous pelitic materials at local shallow emplacement level, other than protolith involvement, might have played significant role in the evolution of reduced, ilmenite series KG melt during its shallow emplacement level.

Conclusions

Kyrdem Granitoids (KG) and micrgranular enclaves (ME) are moderately to strongly oxidized, magnetite series granites, which are mostly inherited from their source regions. However, hybrid and mingled nature of ME indicate that oxidation state of KG melt was partly elevated under a coeval interacting mafic-felsic magma system. The bulk of oxidized KG melt was reduced locally near the margin of the KG pluton because of country-rock assimilation at shallow emplacement level.

Acknowledgements

DST-New Delhi grant (ESS/23/VES/046/98) supported this work. Dr. V. Rino, Dr. Brajesh Singh, Mr. N. Surdas Singh and Ms. Manjari Pathak are thanked for extending help in the preparation of the manuscript. Dr. Shunso Ishihara is thanked for providing valuable literature on granite series.

Table 1: Magnetic susceptibility (MS) measurements of Kyrdem Granitoids (KG) and associated rocks

Locality (in and around)	Characteristics of rock types	Corrected MS values ($\times 10^{-3}$ SI)	Average MS value ($\times 10^{-3}$ SI)
I. Dwarksuid N 25°41'36.7" E 91°59'51.6"	? KG containing K-feldspar megacrysts	**0.24**, 0.25, 0.26, 0.26, **0.27**	0.26 (N=5)
	? Small pelitic xenoliths hosted in KG	**0.31**, **0.35**	0.33 (N=2)
	? KG containing K-feldspar megacrysts	**0.21**, 0.22, 0.22, 0.22, 0.23, **0.23**	0.22 (N=6)
	? Small gneissic xenoliths in KG	**0.40**, 0.47, 0.53, 0.54, **0.57**	0.50 (N=5)
	? KG containing K-feldspar megacrysts	**0.24**, 0. 32, 0. 31, 0.36, 0.32, 0.27, 0.26 , 0.28, 0.35, 0..31, **0.38**	0.31 (N=11)
	? Large amphibolite xenolith in KG	36,.23, **28.98**, 52.44, 30.94, 30.71, **68.20**, 45.43, 52.21, 46.81	43.56 (N=9)
	? Rectangular shaped gneissic xenoliths in KG	0.26, **0.25**, 0.30 , 0.27, **0.31**	0.28 (N=5)
II. From Dwarksuid to Kyrdem N 25°41'20.7" E 92°02'32.3"	? KG containing K-feldspar megacrysts	**24.82**, 28.64, 24.95, 27.72, **29.04**, 27.59	27.13 (N=6)
III. 2 km NW of Mawblang N 25°40'40" E 92°04'19.3"	? Aplite vein cross-cutting the KG	4.47, 3.97, **3.65**, 4.38, 5.15, **5.37**, 4.53, 5.18, 4.98, 4.69, 3.92, 4.67, 3.88, 4.58	4.53 (N=14)
IV. Mawbsein N 25°40'27.8" E 92°04'23.7"	? KG containing some mafic aggregates	20.59, 33.53, 23.10, 26.93, 18.88 . **18.35**, 24.02, 22..31, 24.82, 23.76, 23.76, 27.46, 25..87, **47..92**	25.81 (N=14)
	? Microgranular enclave in KG (10cm size across)	**34.45**, 26.93, 33.40, 25.21, 33.13, 25.08, 18.48, 14.92, 18.48,**14.65**,	24.47(N=10)
	? Pegmatitic vein cross cutting the KG containing less mafic aggregates	5.12, **4.67**, 6.51 , 6.37, **7.15**	5.96 (N=5)
	? KG containing K-feldspar megacrysts	24.29, 31.81, 25.48, **15.71**, 23.76, 19.01, 25.74. 23.10, **41.05**, 17.16, 19.01	24.20 (N=11)
V. Kyrdem-Mawpun N 25°41'13.3" E 92°04'17.8"	? KG containing K-feldspar megacrysts	21.65, 22.04, 16.37, 33.00, 28.51, 27.06, 29.96, 19.40, **15.84**, 30.10, 25.74, **34.72**, 32.60, 26.40, 19.40	25.52 (N=15)
VI. Mawpun N 25°40'55.7" E 92°03'35.6"	? KG containing K-feldspar megacrysts	27.19, 20.06, 33.92, **35.38**, 24.82, 23.10, **18.88**, 22.18, 21.78, 22.84, 20.20	24.58 (N=11)
	? Microgranular enclave in KG	**45.41**, 45.14, 45.14, 45.28, 19.27, 18.48, **15.84**, 22.18	32.09 (N=8)
	? KG containing K-feldspar megacrysts	19.27, 18.48, **15.84**, **22.18**	18.94 (N=4)
VII. 4.5 km NW of Mawblang N 25°41'12" E 92°02'19"	? KG containing K-feldspar megacrysts	15.44, 17.82, 14.78, 14.78, 21.91, 29.83, 18.48, 26.80, **23.23**, 30.49, 24.82, 19.27, **13.33**, 25.87, 13.99	20.72 (N=15)
VIII. 5 km NW of Mawbsein N 25°42'33.6" E 92°00'36.8"	? KG containing K-feldspar megacrysts	**12.22**, 17.95, 20.33, 19.14, 20.86, 19.01, 13.46, **25.48**, 13.20, 16.24, 18.35, 14.65, 16.50, 12.84, 17.95	17.21 (N=15)

Locality (in and around)	Characteristics of rock types	Corrected MS values (×10-3 SI)	Average MS value (×10-3 SI)
IX. Bhorymbong - Dwarksuid N 25°42'37" E 91°59'56.5"	• KG containing less mafic aggregates	**0.89**, 0.55, 0.62, 0.17, 0.27, 0.20, 0.19, 0.22, 0.15, 0.25, 0.18, 0.28, 0.29, 0.29, 0.21, **0.14**	0.31 (N=16)
X. Mawlasnai N 25°44'46.5" E 92°06'49.5"	•KG containing K-feldspar megacrysts	24.16, **27.32**, **20.33**, 24.55, 23.76, 25.74, 25.74, 23.50	24.39 (N=8)
	•Small microgranular enclave in KG	27.19, **21.38**, 23.36, 27.19, **27.98**, 24.29	25.23 (N=6)
	•KG containing K-feldspar megacrysts	23.89, **35.24**, 32.47, 20.59, **18.35**, 21.12	25.28 (N=6)
	•KG containing K-feldspar megacrysts	**37.75**, 26.93, 33.66, 34.32, **25.08**	31.55 (N=5)
	•KG containing K-feldspar megacrysts	15.97, **15.84**, 24.82, 19.01, **27.46**, 19.80, 27.32	21.46 (N=7)
	•KG containing K-feldspar megacrysts	17.69, **15.05**, 17.95, 19.67, 27.32, 35.90, 19.54, 19.40, 20.59, 29.44, 29.04, 30.76, **43.03**, 25.21, 14.65, 22.31, 20.46	24.00 (N=17)
	•KG containing less K-feldspar megacrysts	13.99, **13.17**, **18..08**, 17.95	15.78 (N=4)
XI. Sohlia - Mawlasnai N 25°44'45.2" E 92°06'42.6"	•KG containing K-feldspar megacrysts	24.98, 26.00, **28.91**, 20.99, 26.40, **19.67**	24.39 (N=6)
XII. Sohlia N 25°44'56.4" E 92°06'2.2"	•KG containing K-feldspar megacrysts	13.20, **10.47**, 12.50, 18.61, 16.10, **21.12**, 17.56, 16.63, 13.86	15.56 (N=9)
XIII. Larbang N 25°45'15" E 92°05'3.5"	•KG containing K-feldspar megacrysts	22.18, 26.14, 21.52, 21.12, 22.44, 21.52, 25.74, 17.42, 18.22, 16.50, 20.86, 17.95, 18.48, 20.59, 19.40, 20.33, 33.92, **34.72**, 22.97, **16.37**	21.93 (N=20)
XIV. Mawbri N 25°43'58.6" E 92°03'22.4"	•KG containing K-feldspar megacrysts	24.29, 15.97, 24.95, 19.93, **29.57**, 19.40, 20.19, 26.80, 17.56, **15.44**, 24.68, 26.14	22.16 (N=12)
XV. Umchaproh N 25°43'6.2" E 92°02'5.9"	•KG containing K-feldspar megacrysts	**10.86**, 18.08, 20.33, **24.42**, 22.31, 14.52, 21.25, 18.74, 21.91, 18.22, 12.50, 21.12, 19.14, 10.88, 13.59, 19.01	17.95 (N=16)

The MS values given as bold represent range of min-max MS values. N is number of MS measurements

References

Chappell, B.W. and White, A.J.R. (1974) Two contrasting granite types. Pac. Geol., v. 8, pp. 173-174.

Chakraborty, S. (1989) Report on systematic geological mapping of Kyrdem pluton in East Khasi Hills district. Unpublished GSI report (Geological map published in Ghosh *et al.*, 1991).

Ghosh, S., Bhalla, J.K., Paul, D.K., Sarkar, A., Bishui, P.K. and Gupta, S.N. (1991) Geochronology and geochemistry of granite plutons from East Khasi Hills, Meghalaya. J. Geol. Soc. Ind., v. 37, pp. 331-342.

Ghosh, S., Chakraborty, S., Paul, D.K., Bhalla, J.K., Bishui, P.K. and Gupta, S.N. (1994) New Rb-Sr isotopic ages and geochemistry of granitoids from Meghalaya and their significance in Middle to Late Proterozoic crustal evolution. Ind. Minerals, v. 48, pp. 33-44.

Ghosh, S., Fallick, A.E., Paul, D.K. and Potts, P.J. (2005) Geochemistry and origin of Neoproterozoic granitoids of Meghalaya, Northeast India: Implications for linkage with amalgamation of Gondwana Supercontinent. Gond. Res., v. 8, pp. 421-432.

Gregorová, G., Hrouda, F. and Kohût, M. (2003) Magnetic susceptibility and geochemistry of Variscan West Carpathians granites: implications for tectonic setting. Phys. Chem. Earth, v. 28, pp 729-734.

Hart, C.J.R., Goldforb, R.J., Lewis, L.L. and Mair, J.L. (2004) The northern Cordillerian Mid-Cretaceous plutonic province: ilmenite/magnetite-series granitoids and intrusion-related mineralization. Resour. Geol., v. 54, pp. 253-280.

Ishihara, S. (1971) Modal and chemical composition of the granitic rocks related to the major molybdenum and tungsten deposits in the Inner zone of southwest Japan. J. Geol. Soc. Japan, v. 77, pp 441-452.

Ishihara, S. (1977) The magnetite-series and ilmenite-series granitic rocks. Mining Geol., v. 27, pp. 293-305.

Ishihara, S. (1990) The inner Zone batholith vs. the Outer Zone batholith of Japan: Evaluation from their magnetic susceptibility. In: *Shimizu, M. and Gastil, G. (eds.)*, Recent advances in concepts concerning Zoned Plutons in Japan and Southern and Baja California. Univ. Tokyo, Nature and Culture, No. 2, pp. 21-34.

Ishihara, S. (1998) Granitoid series and mineralization in the Circum-Pacific Phanerozoic Granitic Belts. Resour. Geol., v. 48, pp. 219-224.

Ishihara, S., Anhaeusser, C.R. and Robb, R.L. (2002a) Granitoid-series evaluation of the Archean Johannesburg Dome granitoids, South Africa. Bull. Geol. Surv. Japan, v. 53, pp. 1-9.

Ishihara, S., Hashimoto, M. and Machida, M. (2000) Magnetite/ilmenite-series classification and magnetic susceptibility of the Mesozoic-Cenozoic batholiths in Peru. Resour. Geol., v. 50, pp. 123-129.

Ishihara, S., Robb, L.J., Anhaeusser, C.R. and Imai, A. (2002b) Granitoid series in terms of magnetic susceptibility: a case study from the Barberton region, South Africa. Gond. Res., v. 5, pp. 581-589.

Kanaya, H. and Ishihara, S. (1973) Regional variation of magnetic susceptibility of the granitic rocks in Japan. J. Japan Assoc. Mineral. Petrol. Econ. Geol., v. 68, pp 219-224. (In Japanese with English abstract)

Kumar, Santosh (1998) Granitoids and their enclaves from east Khasi Hills of Meghalaya: Petrogenetic and Geochemical reappraisal. Workshop on Geodynamics and natural Resources of Northeast India, Dibrugarh, Assam. Abstract volume, pp. 17-18.

Kumar, Santosh and Rino, V. (2006) Mineralogy and geochemistry of microgranular enclaves in Palaeoproterozoic Malanjkhand granitoids, central India: evidence of magma mixing, mingling and chemical equilibrium. Contrib. Mineral. Petrol., v 152, pp 591-609.

Kumar, Santosh, Rino, V. and Pal, A.B. (2004) Field evidence of magma mixing from microgranular enclaves hosted in Palaeoproterozoic Malanjkhand granitoids, Central India. Gond. Res., v. 7, pp. 539-548.

Kumar, Santosh, Pieru, T., Rino, V. and Lyngdoh, B.C. (2005a) Microgranular enclaves in Neoproterozoic granitoids of south Khasi Hills, Meghalaya Plateau, northeast India: field evidence of interacting coeval mafic and felsic magmas. J. Geol. Soc. India, v. 65, pp. 629-633.

Kumar, Santosh, Pieru, T. and Rino, V. (2005b) Evaluation of granitoid-series and magmatic oxidation of Neoproterozoic South Khasi granitoids and their microgranular enclaves, Meghalaya: constraints from magnetic susceptibility and biotite composition. J. Appl. Geochem., v. 7, No. 2, pp. 175-194.

Medlicott, H.B. (1869) Geological sketch of the Shillong Plateau in N.E. Bengal, Mem. Geol. Surv. Ind., v. 7, pt. 1, pp. 151-207.

Mazumder, S.K. (1976) A summary of the Precambrian geology of the Khasi Hills, Meghalaya. Geol. Surv. India, Misc. Publ. No. 23, pp. 311-324.

Mazumder, S.K. (1986) The Precambrian framework of part of the Khasi Hills, Meghalaya. Rec. Geol. Surv. Ind., v. 117 (2), pp. 1-59.

Sikdar, D. and Rahman, S. (1998) Triclinicity of the potash feldspar and the petrochemistry of the Kyrdem granite, Khasi hills, Meghalaya. Proc. Reg. Sem. Dev. Geol. Res. NE India, Gauhati Univ., Guwahati, pp. 45-55.

Takagi, T. (2004) Origin of magnetite- and ilmenite-series granitic rocks in Japan arc. Amer. J. Sci., v. 304, pp 169-202.

Takagi, T. and Tsukimura, K. (1997) Genesis of oxidized- and reduced-type granites. Econ. Geol., v. 92, pp. 81-86.

Takahashi, M., Aramaki, S. and Ishihara, S. (1980) Magnetite-series/Ilmenite-series vs. I-type/S-type granitoids. Mining Geol. Spec. Issue, No. 8, pp. 13-28.

Vernon, R.H. and Paterson, S.R. (2008) How late are K-feladspar megacrysts in granites? Lithos, v. 104, pp. 327-336.

□□□

3

Basement Rocks of the Kumaun - Garhwal Himalaya : Implications for Himalayan Tectonics

A. R. BHATTACHARYA
Centre of Advanced Study in Geology,
University of Lucknow, Lucknow, (U.P.), India
Email: arb65k@rediffmail.com

Abstract

The Himalayan orogenic belt is characterized by the presence of basement rocks. These rocks got involved in the evolution of the mountain belt and, as such, have played a great role in the structural architecture and tectonic framework of the orogen. These rocks have evolved from depth due to concentration of compressive stresses generated by plate collision and rose vertically up because of buoyancy up to the surface and were then pushed southwards over the less- or un-metamorphosed, younger rocks of the sedimentary belt of the Lesser Himalaya. In order to make room for such a vast chunk of the basement, the sedimentary pile had to undergo large scale deformation sympathetically. As such, the present tectonic setting, structural architecture and deformation pattern of the upper crustal rocks of the Himalaya owe much to the rise of the basement rocks during evolutionary stages of the Himalaya. The paper examines some aspects of the tectonics and structural framework of the Himalaya as related to the basement rocks in the Kumaun - Garhwal sectors.

Introduction

Basement rocks constitute an integral part of crustal rocks. These constitute the foundation on which thick pile of sedimentary rocks are deposited. The basement rocks mostly constitute an assemblage of crystalline and metamorphic rocks that behave in accordance with the overall tectonic behaviour of the crust, *i.e.* whether the crust is undergoing vertical or epeirogenic movements or compressional or orogenic movements. Also, there are situations when the crust undergoes extension or buckling to give rise to intracratonic basins that in turn give rise to a thick pile of sedimentary rocks deposited unconformably over the basement rocks. Thus, both stratigraphy and structure of a region are, by and large, controlled by the presence or absence of the basement rocks. Because of all these factors, coupled with some others also, the regional stratigraphic setting, structural design and tectonic framework of the cratons of the Peninsular India, Proterozoic basins of Central India and that of the Himalayan region are different from each other. In this paper, the role of basement rocks in controlling the structure and tectonics of the Himalaya is discussed. The work is based on author's studies in the Kumaun-Garhwal sectors of the Himalaya.

The Himalaya has been geologically divided (Fig. 1) into four lithotectonic subdivisions (Gansser, 1964). From south to north these are: (1) *Outer Himalaya,* that mostly includes the molassic Siwalik Supergroup of Mio-Pliocene ages and is demarcated by two tectonic planes, the Himalayan Frontal Thrust (HFT) to the south and the Main Boundary Thrust (MBT) to the north. (2) *Lesser Himalaya,* that exposes a thick pile of highly folded Proterozoic sedimentary strata together with a few outcrops of older crystalline rocks; this subdivision is bounded by the MBT to the south and the Main Central Thrust (MCT) to the north. (3) *Greater* or *Higher Himalaya,* that exposes a massive, north-dipping pile of metamorphic rocks - the Central Crystalline Zone - and is demarcated by the MCT to its south and the Dar-Martoli Fault or Tethys Fault or the South Tibetan Detachment (STD) to the north. (4) *Tethys Himalaya* that includes with a thick pile of sedimentary rocks of Cambrian to Lower Eocene ages.

Heim and Gansser (1939) and Gansser (1964) presented an overall geological framework of the Himalaya. Le Fort (1975) presented some specific structural and petrological aspects of the Himalaya. Molnar *et al.* (1977) interpreted the structure and tectonics of the Himalaya in the light of some relevant geophysical observations. Valdiya (1981) described the geology of the Kumaun region. Bouchez and Pecher (1981) studied fabrics of the Main Central Thrust pile and its quartz-rich tectonites in Central Nepal. Burg *et al.* (1984) discussed the deformation pattern of leucogranites of the Crystalline Main Central Thrust Sheet in southern Tibet. Burg and Chen (1984) described the tectonics and structural zonation of southern Tibet. Colchen *et al.* (1986) discussed some aspects of collision tectonics in the Indus Suture Zone, Ladakh. Gruijic *et al.* (1996) studied quartz microfabrics of Bhutan Himalaya and discuused the nature of ductile extrusion of the Higher Himalayan Crystallines. Grasemann *et al.* (1999) presented a kinematic flow analysis from the Main Central Thrust of NW Himalaya in the light of implications for a decelerating strain path and extrusion of orogenic wedges. Beaumont et al. (2001) explained Himalayan tectonics as an extrusion of low-viscosity crustal channel coupled to focus surface denudation. Vannay and Grasemann (2001) discussed Himalayan inverted metamorphism and syn-convergence extension as a consequence of a general shear extrusion.

The present work is based on the author's study in the Kumaun and Garhwal sectors of the Himalaya (Fig. 2). These two sectors represent all the four major lithotectonic subdivisions of the Himalaya. As such, interpretations made from these sectors should have regional implications for the Himalaya.

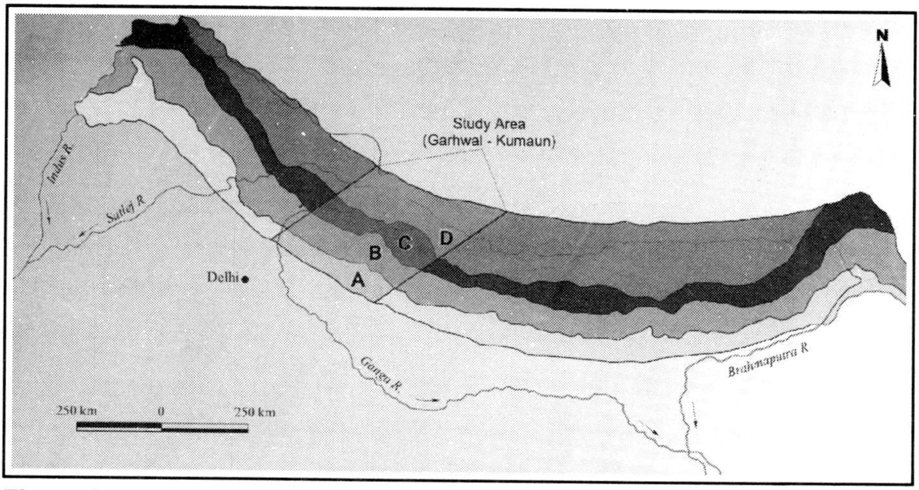

Fig. 1: Geological sketch map of the Himalaya. A - Outer Himalaya, B - Lesser Himalaya, C - Greater Himalaya, D - Tethys Himalaya. The Main Central Thrust (MCT) is defined by the contact between the Lesser and the Higher Himalayas (After Gansser, 1964).

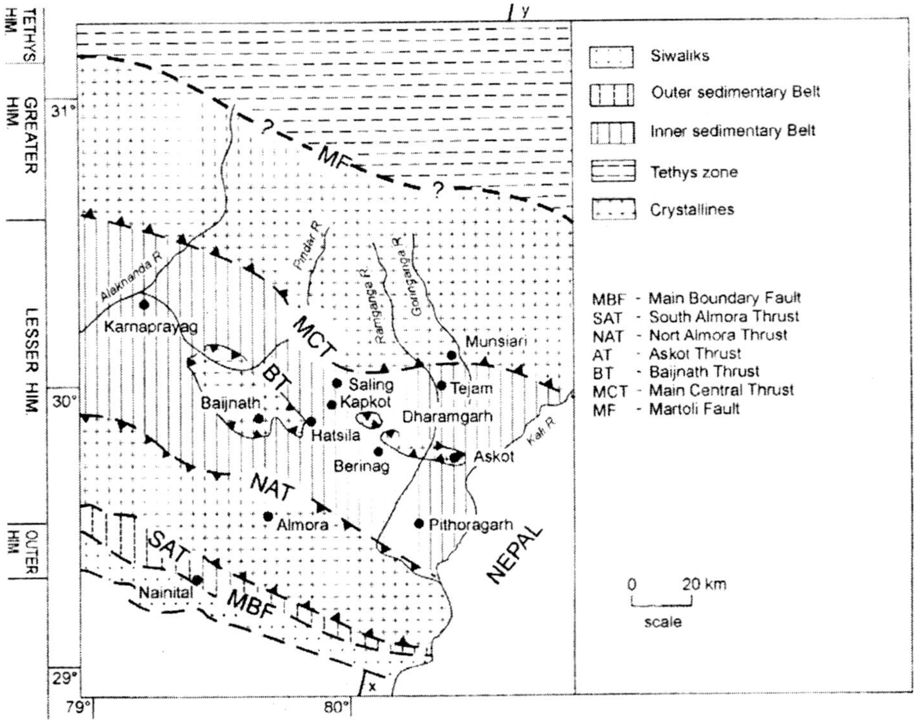

Fig. 2: Geological sketch map of Kumaun-Garhwal Himalaya (After Bhattacharya, 1999).

Basement Rocks of the Himalaya

In the Himalayan domain, the basement rocks appear to have played a great role in controlling and governing the overall stratigraphy, structure and tectonics. The vast stretch of Greater Himalaya exposes a massive pile of crystalline-metamorphic rocks that are designated as the Central Crystalline Zone (CCZ). It includes a remobilized basement of about 30 km thick medium to high grade metamorphic sequence. The dominant rock types include augen gneiss, granite gneiss, garnetiferous biotite gneiss, biotite-muscovite gneiss, calcsilicates and quartzites. Dikes of aplites and pegmatites are common in the rocks of higher (northern) horizons of this zone. The major rock types are gneisses, quartzites, variety of schists, calcsilicates, and leucogranite dikes especially at higher (northern) horizons.

Barring the vicinity of the MCT, the observed strain levels of the folds of the vast stretch of the Central Crystalline Zone (CCZ) of the Greater Himalaya, has been found to remain rather moderate.

Flattening strain in folds range from about 15 to about 95 percent (Bhattacharya and Siawal, 1985; Bhattacharya, 1999), and the strain data show a rather random pattern of distribution. Finite strain (shown by minor folds) shows a progressive increase towards the MCT such that the trace of the MCT shows the highest concentration of ductile strain. Also, the vicinity of the MCT shows the development a variety of shear-sense markers and the stretching lineation becomes progressively stronger towards the MCT. The deformational features around the MCT show marked differences with those developed in areas away from it, thus, suggesting marked differences in the deformational processes operative in the MCT zone and those away from it. The vicinity of the MCT has been regarded here as the center of the shear zone (see Bhattacharya and Weber, 2004).

Barring a zone of 10 to 12 km width from the MCT northwards, the structural features do not show any specific orientation or regularity of their occurrence. The rocks show moderate strain levels.

Structural Architecture of Basement Rocks

The Central Crystalline Zone of the Greater Himalaya can, in a generalized way, be described to be the representative of basement rocks in the Himalayan domain. The structural/tectonic disposition of this zone in the framework of the overall structure of the Kumaun-Garhwal Himalaya has been shown in a generalized cross- section (Fig. 3). This zone exposes a massive pile of medium to high grade metamorphic rocks together with younger intrusives of aplite and pegmatite.

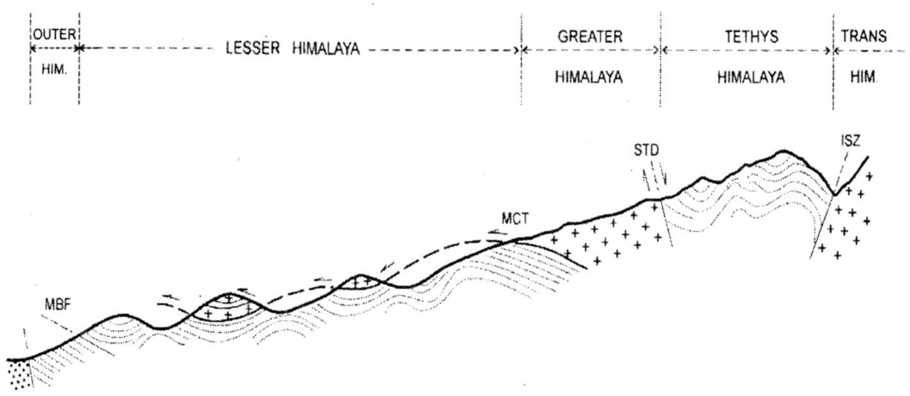

Fig. 3 : A generalized geological cross section of the Kumaun Himalaya (After Bhattacharya, 1999).

Typically, the foliation of these massive pile of metamorphic rocks show moderate dips northwards. Apparently, no large scale or kilometric scale folds are noticed, at least in the Kumaun-Garhwal sectors. As such, the overall structure of the Central Crystalline Zone can be described to be a homocline (Fig. 4). This general lithological sequence is noticed practically in all the sections of Uttarakhand (Kumaun-Garhwal) Himalaya.

The base of the crystalline pile includes mylonites and sheared rocks that have been included here to constitute the MCT shear zone. Physically, over these rocks lie a pile of medium to high grade rocks with gneisses and schists.

In contrast to the overall simple homoclinal structure for the CCZ, the rocks show complicated structures in their internal domain, *i.e.* on meso- to micro- scales. The folding is, especially, complicated. The rocks of the higher horizons have been found to show at least three episodes of folding.

Ductile Shear Zones

The southern part (10-12 km) of the Central Crystalline Zone is characterized by the development of a large scale shear zone formed due to large scale movement /translation of a part of this zone along the Main Central Thrust Zone. The rocks of the northern parts of the CCZ apparently do not show effects of shear deformation. Moving southwards, the rocks gradually show effects of shear deformation, as revealed by the development of a variety of structures/micro-structures and textures in the rocks, sigmoidal foliation, shear folds etc. All these structures gradually show an asymmetry with respect to the earlier or domainal foliation. This asymmetry has developed as a result of shearing, and this asymmetry becomes progressively stronger southwards, *i.e.* progressively with increase of ductile shear strain. The shear zone, thus developed at the southern fringe of the CCZ, is about 12-15 km wide and is mainly recognized by:

(i) Progressive development of a variety of minor structures within the shear zone.

(ii) Progressive development of fabric and grain orientations.

(iii) Spatial variation of strain within the shear zone.

(iv) Structures/micro-structures and shear-sense markers show progressive increase of asymmetry suggesting a southward sense of movement.

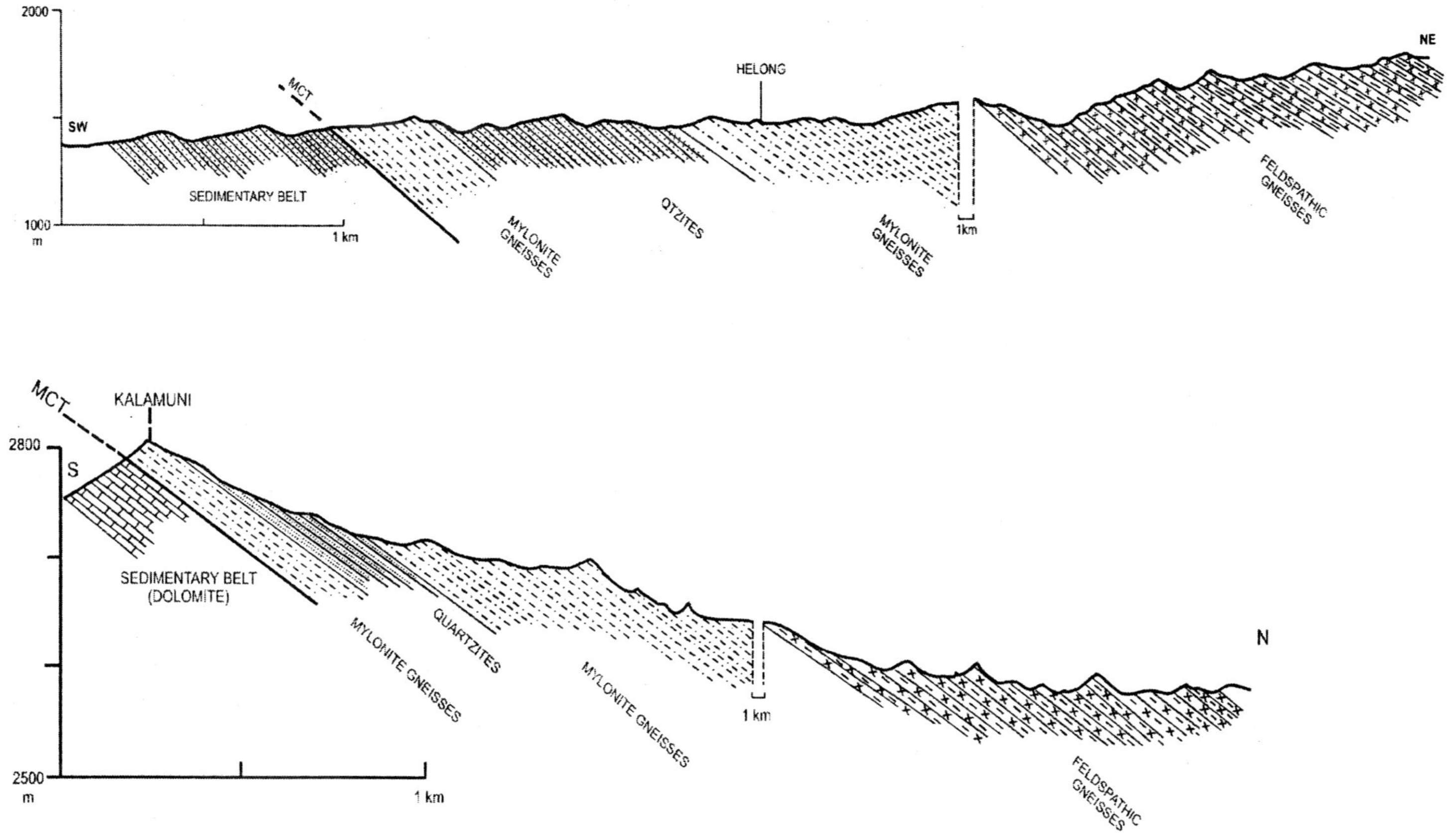

Fig. 4: Structure of the Central Crystalline Zone in the Garhwal Himalaya in the Heong transect (Joshimath area) and in the Kumaun Himalaya in the Kalamuni transect (Munsiari area).

(v) Rocks of the higher (northern) horizons gradually become mylonitized towards the MCT and show sequential changes in a variety of microstructures:

(a) Quartz shows progressive increase of dynamic recrystallization and decrease of grain size (down to 0.01 mm)

(b) Feldspar shows progressive recrystallization to muscovite

(c) Quartz c- and <a> axis fabrics gradually become prominent and evolve from a random fabric outside the shear zone through coaxial to non-coaxial strain path close to the MCT.

Thus, three major deformation zones can be distinguished in the Central Crystalline Zone:

(a) A northern, less deformed sector, with fewer deformation features

(b) A central sector showing incipiently to weakly developed deformation features that progressively intensifies towards south

(c) The southern sector (MCT zone) showing, from north to south, a sequential development of shear- related deformation features

Tectonic Framework

It has been indicated by the author (Bhattacharya,1999) that as a result of collision of the Indian plate against the Asian plate in the Eocene time, very high ductile strain was concentrated along a narrow, linear zone at depth where the rock mass was highly ductile. As such, with continued collision, this mass could rise vertically up due to its buoyancy. In response to the persistent collisional stresses, this mass was brought to the surface and then pushed southwards along the MCT (Fig. 5). This long transport is evidenced by the occurrence of a thick zone of mylonites at the base of the CCZ as well all along the crystalline units physically occurring at the top of the sedimentary sequence of the Lesser Himalaya at several places (see Bhattacharya and Agarwal, 1985). These crystalline masses have been given different names in the Lesser Himalaya. In the Kumaun-Garhwal Himalaya, these include the Almora, Baijnath, Askot, Chhiplakot and Dharamgarh crystalline units. These mylonites represent products of very high ductile deformation under deep seated conditions.

According to Heim and Gansser (1939) and Gansser (1964), the above crystalline units of Lesser Himalaya show an inverted sequence, *i.e.* inverted metamorphism, in which the grade of metamorphism increases from base to top of the sequence. They interpreted the present disposition of the crystalline units of the Lesser Himalaya as due to a gigantic recumbent anticline of the crystalline axis of the Himalaya (*i.e.* the crystalline rocks of the Greater Himalaya) whose lower inverted limb is represented by the present-day, synformally-disposed crystalline rock units while the upper, normal limb might have been eroded away. Later workers, including the present one, are, on the other hand, of the opinion that there is no specific order of distribution of the metamorphic minerals and rocks within the sequence and hence there is no inversion of the sequence. Further, the occurrence of a thick pile of mylonites and highly sheared rocks at the base of all the crystalline units of Lesser Himalaya, and so also at the base of the Central Crystalline Zone, clearly indicate long transport of these rock masses. As such, the crystalline masses of the Lesser Himalaya can be described to constitute *thrust sheets* that have been transported from the north, *i.e.* from the Greater Himalaya (see also Bhattacharya, 1987, 1990, 1999; Bhattacharya and Siawal, 1985; Bhattacharya and Weber, 2004).

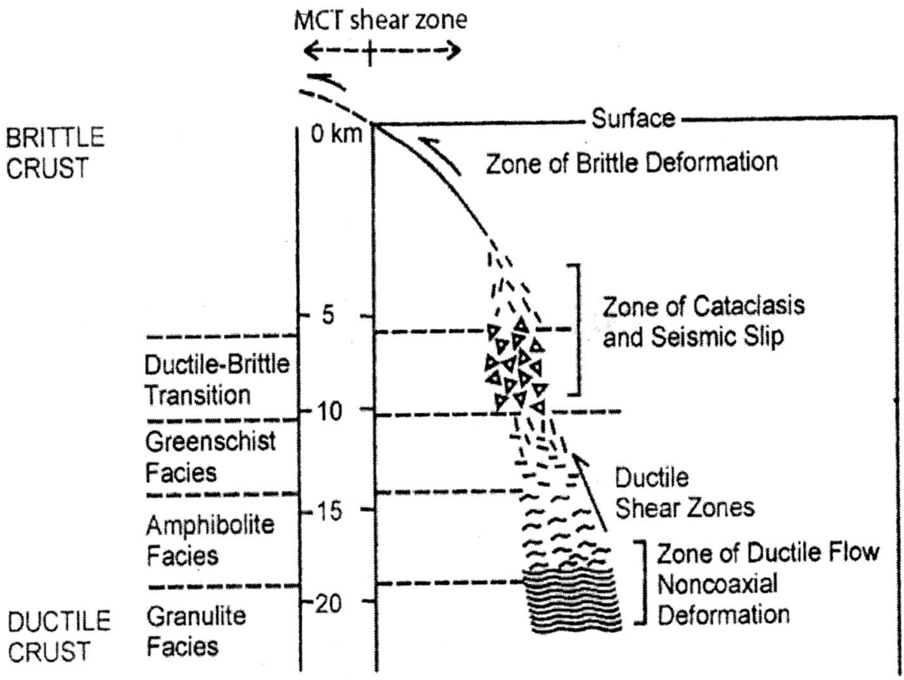

Fig. 5: An evolutionary model of the Main Central Thrust showing how it brings a part of the basement rocks up to the surface. (After Bhattacharya, 1990).

Discussions

(1) Why does the Central Crystalline Zone show a simpler (homoclinal) structure?

As we have indicated above, the CCZ shows a monotonously north-dipping sequence of crystalline-metamorphic rocks, thus displaying a homoclinal structure on a regional basis. It seems quite unusual that such a major subdivision of the Himalaya shows a relatively simpler structural architecture especially keeping in view of the complicated post-collisional structural-tectonic history of the Himalayan region. Unfortunately, there appears to be a lack of literature on this aspect. The author suggests one possible explanation for this, as outlined below.

It has been mentioned earlier that as a result of India-Asia collision, a part of the basement rocks was lifted up from the depth due to its buoyancy. The rock mass moved along the MCT that brought it up on to the surface and then pushed it southwards over the un-metamorphosed, younger sedimentary belt of the Lesser Himalaya. It is suggested that most of the collisional stresses that were of compressive nature were utilized in moving the mass up from the depth. This caused shearing process to be operative in the moving rock mass and this, in turn, triggered a series of deformational processes in the internal domain of the crystallime mass. This is also evidenced by the fact that while the overall structure of the CCZ looks homoclinal, the rocks show a variety of shear-generated structures in their internal domain. The intensity and complexity of shear-generated structures show a progressive increase towards the trace of the MCT, *i.e.* in the direction of higher ductile strains. All this possibly explains why the CCZ shows a rather simpler (homoclinal) structure while the occurrence of large scale folding and the related structures are common features of all the other three lithotectonic subdivisions of the Himalaya.

(2) Why is the contact of the Central Crystalline Zone with the Tethys zone a fault?

An important geotectonic element of the Himalaya is an E-W striking, N-dipping normal fault at the base of the Tethys zone and this fault separates the Tethys sequence with the crystalline zone of the Greater Himalaya. According to Burg *et al.* (1984) and Burg and Chen (1984), this fault and a few other similar type of faults in the

Higher Himalaya and southern Tibet were formed during the post-collisional convergence of India and Tibet. These faults extend for at least 600 km along strike. Royden and Burchfiel (1987) interpret them as possible late (?) Miocene extensional features with perhaps several tens of kilometers downward northerly displacement and they are of the opinion that N-S extension produced in this way is probably confined to upper crustal levels only and are developed during convergence in many orogenic belts.

In the Himalayan scheme of events, the culmination of deformation and metamorphism is marked by the formation and subsequent development of the MCT that lifted a part of the basement on to the surface and then it was pushed southward over the less- to un-metamorphosed, younger sedimentary belt of the Lesser Himalaya (see Le Fort, 1975; Burg *et al.*, 1984; Grujic *et al.*, 1996). This was also a period (mainly Oligocene to Middle Miocene) during which the Himalaya was lifted much higher. In the Himalayan region, this uplift could be considered as an up-buckling of the upper crust that was possible only because the crust has some 'flexibility' to large scale compressive stresses that, in turn, allowed a part of the collisional stresses could be accommodated within the dome-shaped Himalayan crust through deformation. However, after the optimum limit of this flexibility was attained, the outer part of this dome-shaped structure developed tensional stresses leading to the formation of a crack or a fault because of its brittleness. Since the contact of two different lithological assemblages, in general, constitutes a regional anisotropy, it was the contact of the Central Crystalline Zone and the sedimentary rocks of the Tethys zone that became more vulnerable or susceptible to brittle deformation by the formation of a fault. As the process of uplift, supported by mass transport along the MCT from below continued, the crystalline part continued to rise up (hanging wall) while, in order to counterbalance, the Tethyan pile continued to go down along the fault (down-faulting, foot wall). Thus, the throw of the fault - the Martoli Fault or the South Tibetan Detachment - continued to increase progressively yielding, as of now, a throw of some 7 to 10 km.

The formation of a major normal fault in a regime of compressive stresses, as described above, is a unique example of operation of extensional processes during an overall compressional regime (see also Royden and Burchfiel, 1987).

(3) Why are brittle faults rare in the Central Crystalline Zone ?

The rocks of the CCZ, as has been mentioned earlier, show a relatively simpler (homoclinal) structure. Faults are practically absent in the rocks of this zone, while all the other lithotectonic subdivisions of the Himalaya are characterized by the occurrence of numerous faults on various scales. This draws attention also due to the fact that during the evolution of the MCT, the rocks of the MCT zone have been subjected to ductile deformation at depth during earlier stages and later on when the rock masses came up nearer, or on to, the deformation regime was dominantly brittle in which faults could have formed. The author would like to suggest the following causes.

The rocks of the CCZ are characterized by strong development of foliation mainly due to the dominance of micaceous and other platy minerals. The MCT zone had evolved through very strong compressive stresses. Because of the strong foliation, these compressive stresses have easily been transmitted to the rock masses along the foliation planes. Under the circumstances, movement or transport of rock masses are rather more facilitated along the foliation planes than across them. Also, the micaceous minerals have some tolerable ductility that could prevent development of cracks across their overall alignment, *i.e.* foliation. This suggests that mainly because of the strong foliation due to abundance of micaceous minerals, the cohesive forces are stronger than the frictional forces. As such, shear stresses could not develop cracks easily and instead, the component shear stress had allowed slip or movement along, or within, the layers of the micaceous minerals even at the later stages of evolution of the MCT zone when the rock masses were on, or very near to, the surface. This possibly explains why faults are rare in the rocks of CCZ.

Acknowledgements

The work has been carried out under facilities of the Centre of Advanced Study in Geology, University of Lucknow, Lucknow. The concepts and interpretations as incorporated in the paper are culmination of several years of work. Discussions with Dr. K.K. Agarwal at various stages over the past several years are thankfully acknowledged.

References

Beaumont, C., Jamieson, R.A., Nguyen, M.H. and Lee, B. (2001) Himalayan tectonics explained by extrusion of low-viscosity crustal channel coupled to focus surface denudation. Nature, v. 414, pp. 738-742.

Bhattacharya, A.R. (1987) A 'Ductile Thrust' in the Himalaya. Tectonophysics, v. 135, pp. 37-45.

Bhattacharya, A.R. (1990) Crustal processes associated with the Main Central Thrust of the Himalaya. Zeitschr. Geowissen. Berlin, v. 18(4), pp. 293-299.

Bhattacharya, A.R. (1999) Deformational regimes across the Kumaun Himalaya: A study in strain patterns. Gondwana Research Mem. (Japan), v. 6, pp. 81-90.

Bhattacharya, A.R. and Agarwal, K.K. (1985) Mylonites from the Kumaun Lesser Himalaya. Neues Jahrb. Mineralogie Abh., v. 152 (1), pp. 65-77.

Bhattacharya, A.R. and Siawal, A. (1985) A phenomenon of unusual flattening in natural folds associated with a Himalayan thrust. Geol. Mijnbouw, v. 64, pp. 159-165.

Bhattacharya, A.R. and Weber, K. (2004) Fabric development during shear deformation in the Main Central Thrust Zone, NW-Himalaya, India. Tectonophysics, v. 387, pp. 23-47.

Bouchez, J.L. and Pecher, A. (1981) The Himalayan Main Central Thrust pile and its quartz-rich tectinites in Central Nepal. Tectonophysics, v. 78, pp. 23-50.

Burg, J.P., Brunel, M., Gapais, D., Chen, J.M. and Liu, G.H. (1984) Deformation of Leucogranites of the Crystalline Main Central Thrust Sheet in southern Tibet (China). J. Struct. Geol., v. 6, pp. 535-42.

Burg, J.P. and Chen, J. M. (1984) Tectonics and structural zonation of southern Tibet, China. Nature, v. 311, pp. 219-23.

Colchen, M., Mascle, G. and Van Haver, T. (1986). Some aspects of collision tectonics in the Indus Suture Zone, Ladakh. In: *M.P. Coward and Ries, A.C. (eds.).* Collision Tectonics, Geological Society Spec. Publ. No. 19, pp. 173-184.

Gansser, A. (1964) Geology of the Himalayas. Interscience. 289 p.

Grasemann, B., Fritz, H. and Vannay, J.-C. (1999) Quantitative kinematic flow analysis from the Main Central Thrust Zone (NW-Himalaya): implications for a decelerating strain path and extrusion of orogenic wedges. J. Struct. Geol. v. 21, pp. 837-853.

Grujic, D., Casey, M., Davidson, C., Hollister, L.S., Kundig, R., Pavlis, T. and Schmid, S. (1996) Ductile extrusion of the Higher Himalayan Crystalline in Bhutan: evidence from the quartz microfabrics. Tectonophysics, v. 260, pp. 21-43.

Heim, A. and Gansser, A. (1939) Central Himalaya: geological observations of the Swiss expedition 1936. Mem. Soc. Helv. Sc. Nat., v. 73, 246 p.

Le Fort, P. (1975). Himalayas: the collided range. Present knowledge of the continental arc. Amer. Jour. Sci., v. 275A, pp. 1-44.

Molnar, P., Chen, W.P., Fitch, T.J., Tapponier, P., Warsi, W.E.K. and Wu, F.T. (1977) Structure and tectonics of the Himalaya: a brief summary of relevant geophysical observations. Coll. Intern. C.N.R.S., no. 268: Eclogie et Geologie de l'Himalaya, Sevres, 1976. Sc. de la Terre, pp. 269-285.

Royden, L.H. and Burchfiel, B.C. (1987) Thin-skinned N-S extension within the convergent Himalayan region : gravitational collapse of a Miocene topographic front. Geol. Soc. London, Spl. Pub., v. 2B, pp. 611-619.

Valdiya, K.S. (1981) Geology of Kumaun Himalaya. Wadia Inst. of Himalayan Geol., Dehradun, 291 p.

Vannay, J.-C. and Grasemann, B. (2001) Himalayan inverted metamorphism and syn-convergence extension as a consequence of a general shear extrusion. Geol. Mag. v. 138, pp. 253-276.

□□□

4

Some Observations on the Thrust Geometry of the Siwalik Rocks of the Outer Himalaya, India

A. R. BHATTACHARYA and K. K. AGARWAL
Centre of Advanced Study in Geology,
University of Lucknow, Lucknow, (U.P.), India
Email: arb65k@rediffmail.com; kamalagarwal73@gmail.com

Abstract

The rocks of the Siwalik Supergroup of the Outer Himalaya have been found to show a complicated system of thrust geometry in their internal domain. The structures, which the lithic layers commonly show, include the following: duplex, imbricate fan system, thrust propagation structures, fault propagation folds, pop-up structures, antiformal stack, snakehead anticline, snakehead duplex, overstep thrust system and the related types. Occurrence of these structures suggests, amongst others, the active role, and/or reactivation, of the anisotropy planes of the rocks during deformation, and that the deformation was progressive in nature. Development of most of these structures may have involved addition of mass to the moving thrust complexes. Presence of the above structures possibly indicates that the Siwalik strata, in general, may have undergone structural thickening in their internal domain during the overall deformation of these rocks.

Introduction

The Himalayan mountain belt is an excellent example of collision of two continental plates during the Eocene times. The present-day tectonic setup and the structural architecture of this mountain chain is a direct reflection of the deformational processes that operated during and after the collision. The emerging Himalayan mountain belt and the thrust sheet loading caused flexural subsidence of the Indian lithosphere creating a foreland basin in front of it, in which the Siwalik sediments were deposited. Thus, the Siwalik sequence constitutes the outermost part of the Himalayan mountain chain, and extends almost all along the strike length, *i.e.* from Potwar Plateau in the west to Arunachal Pradesh in the east. Lithologically, the Siwalik unit is dominantly constituted of sandstone, siltstone, mudstone, and locally developed conglomerate horizons. This lithological assemblage represents typical foreland deposits. The Outer Lesser Himalaya in the vicinity of the Siwalik zone is dominantly constituted of a thick calcargillaceous sequence (Krol Group) of Proterozoic age.

The present study is based on detailed investigation of a typical Siwalik section as exposed in the Mohand area located south of Dehradun (Fig. 1). The geology of the area has been worked out in detail by Raiverman *et al.,* (1994). Burbank *et al.,* (1996) presented a regional tectonic framework of the foreland basin of the Himalaya. Srivastava and John (1999) studied deformation pattern in the Himalayan Frontal Fault zone of the adjoining Mohand-Khara area. Mishra and Mukhopadhyay (2002) worked out balanced structural cross-section of the Mohand Anticline and they modeled this structure as multi-bend fault-bend fold related to the Main Frontal Thrust. Bhattacharya (2005) proposed an "Expanding Mountain Front Hypothesis" on the basis of the complicated thrust geometry of the Siwalik rocks in their internal domain. He suggested that in order to maintain the critical taper of the Siwalik wedge and to maintain the mountain topography in an equilibrium state, the Himalayan mountain front must adjust the additional topography (caused to structural thickening), by spreading or expanding the mountain front towards the foreland, *i.e.* southwards, as such, the Himalayan mountain front could be considered to be in a state of expanding southwards.

The present work incorporates a systematic study of the internal structures of the Siwalik rocks as mainly carried out along the Mohand-Dehradun traverse that closely follows the strike of the Siwalik lithounits.

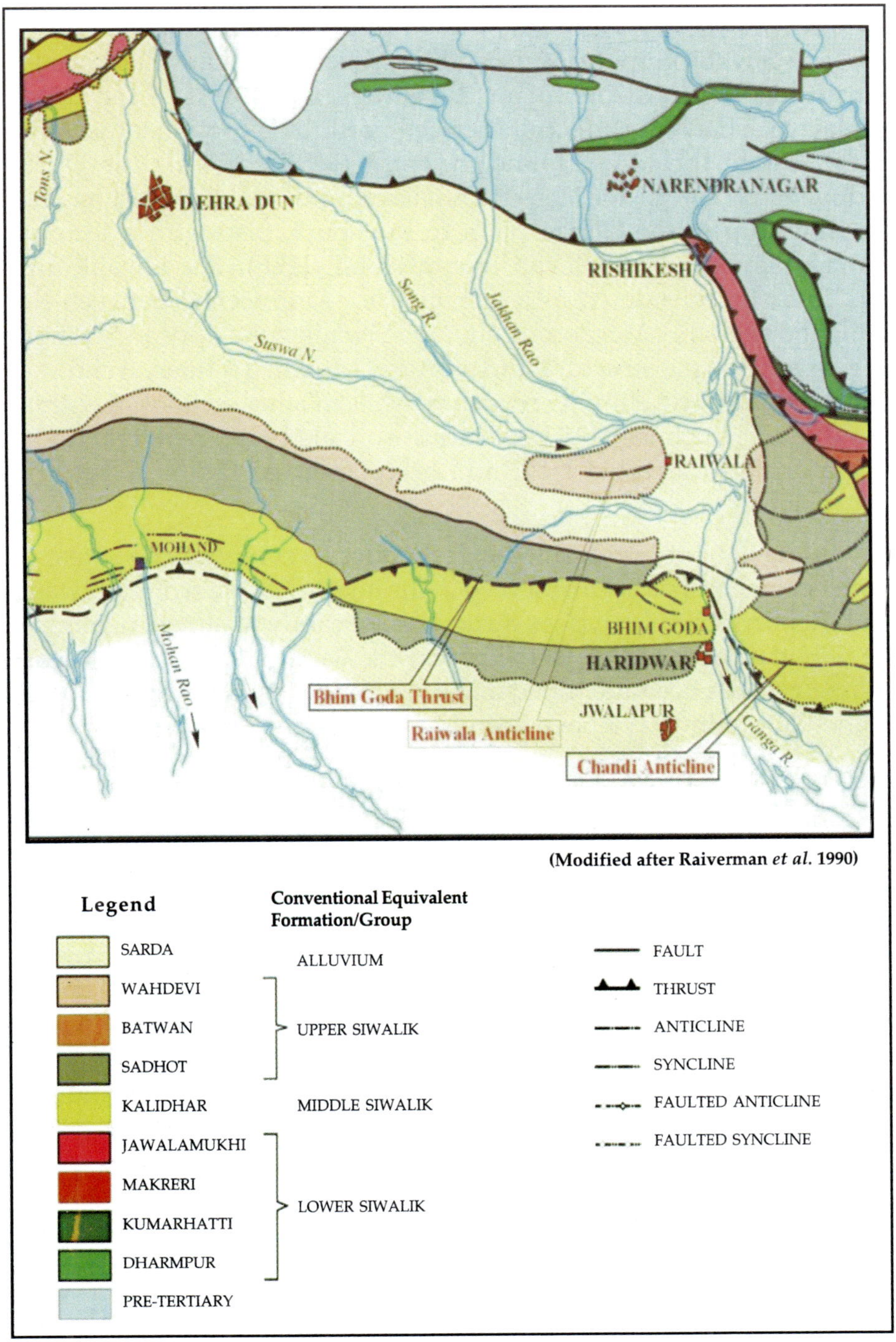

Fig. 1: Geological map of the Mohand area. The stratigraphic units of the area have been given in the legend (Modified after Raiverman *et al.*, 1990)

Structure of the Siwalik Zone

The Siwalik Supergroup (Middle Miocene to Late Pleistocene) constitutes the most dominant lithostratigraphic unit of the Outer Himalaya. This unit is delimited to the north by a prominent tectonic plane, called the Main Boundary Fault (MBF), against the Outer Sedimentary Belt of the Lesser Himalaya, and is juxtaposed against the alluvium of the Ganga plain to the south, also with a tectonic contact named as Himalayan Frontal Fault (HFF). The Siwalik unit can, thus, be considered as a tectonic slice sandwiched between the MBF and the HFF. Megascopically, the Siwalik zone appears to show a relatively simple structural architecture. In some sections, it shows a rather simple north-dipping structure, while in other sections, the strata show open synclines and anticlines - so called Jura-type folds. The strata are also affected by a number of normal faults of varying dimensions.

In the light of our detailed investigation, a complicated system of thrust geometry has been observed in the internal domain of the Siwalik unit. Report of similar type of structures is scantly to almost negligible (see Bhattacharya, 2005). In the paper, the various types of structural geometry have been identified and their implications for the structure of the Outer Himalaya have been highlighted.

Internal Structures and Thrust Geometry

Duplex

Duplex seems to be the most common type of structure of the Siwalik rocks in the internal domain. In its most general form, the structure shows two parallel thrust faults (the roof and floor thrusts), and the intervening area shows a few parallel to sub- parallel faults. The latter commonly occur in imbricate fashion and are known as horses (Boyer and Elliott, 1982; Mitra and Boyer, 1986). The individual subsidiary faults (horses) commonly occur as planar units with straight sides (Fig. 2 A, B). The beds above and below the duplex, are more or less undisturbed.

The duplex structures in the area have significant bearing on the structure of the Siwalik rocks in the internal domain. According to Boyer and Elliott (1982), duplexes represent a mechanism for slip transfer from one glide horizon at depth to another at shallower levels. This slip transfer and creation of new horses cause structural thickening, duplex growth, and addition of mass to the moving thrust complex. It is therefore, quite possible that the Siwalik strata showing duplex structures have undergone structural thickening in their internal domain.

Imbricate Fan System

Occasionally, a few faults join up together so as to give a branching array. Thus, a lower common thrust - sole thrust - can be identified. From the sole thrust, a few thrust slices spread upward such that they are asymmetrically shaped downward, thus giving rise to an imbricate fan system. Dip of individual thrust is in the same direction and increases from the sole thrust upward, and thus the structure looks like roof tiles in which the individual horses take up a more or less triangular shape. At some later stage, the thrusts are tangential to both the roof and sole thrusts (Fig. 2C). All these suggest progressive deformation and development of foreland directed thrusts (cf. Boyer and Elliott, 1982).

With the development of the above geometries, the imbricate structures appear to have caused thickening of the lithic units in which they are developed. The adjacent members of the lithic unit, where imbricate fan structure is not developed, however, do not show thickening. Development of the imbricate fan systems, as described here can thus be considered to constitute a mechanism of thickening and shortening of the lithologic sequence of the Siwalik Supergroup.

Fault-Bend Folds

At places, an antiform is formed due to riding of a thrust slice above a ramp. Occasionally, this gives rise to the development of a flat-topped anticline on the hanging wall, and an antiform–synform structure is also developed above a ramp. Formation of such folds - also called fault-bend-folds - has been commonly observed in relatively thin layers (upto few cm) of the Siwalik rocks (Fig. 2 D). With progressive deformation, the early-formed structure (antiform, in this case) continued to deform to give rise to more complicated structures. The structure in its present form has been described here as a fault propagation fold.

Antiformal Stack

The antiformal stack (Fig. 2E) seems to be a common structure developed in the Siwalik rocks. It contains an anticline, which possibly has developed by curving up or up arching of the sole thrust. This appears to be a case when an early-formed duplex has undergone relatively larger displacement.

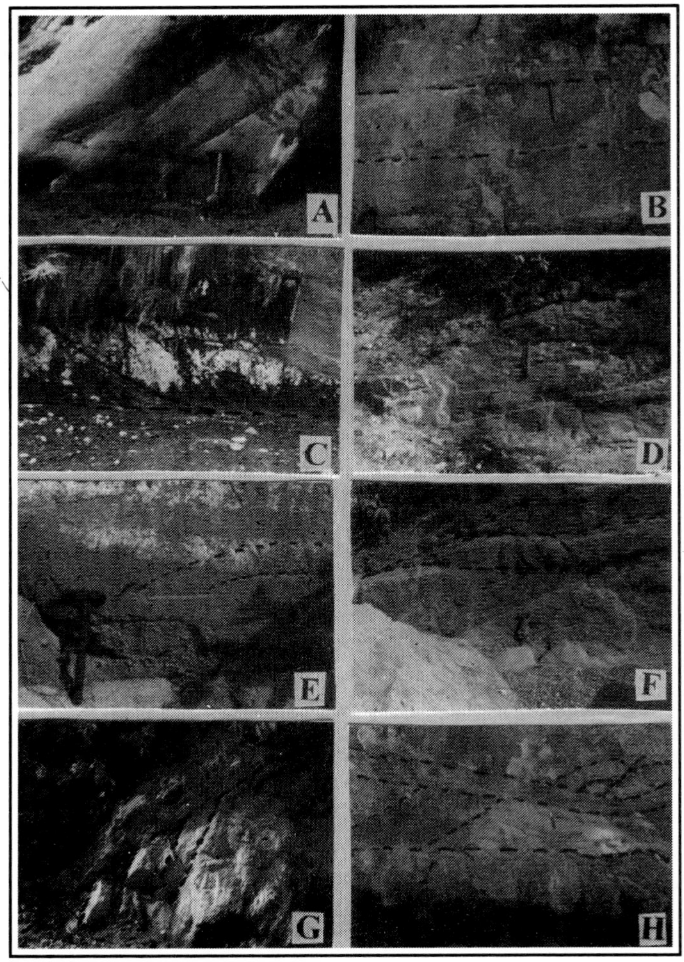

Fig. 2 : Field photographs of the structures/thrust geometries observed in the Siwalik rocks of the Mohand-Dehradun section. (A) A thick sandstone layers showing *duplex* structure. (B) *Duplex* structure in sandstone layers showing the horses within the lithic layers. (C) *Imbricate fan system* in which the thrusts within a sandstone layer taper tangentially at the sole thrust below. (D) A *Fault bend fold* is developed (central right side of the photograph) when a sub-horizontal fault is obstructed during its movement. (E) *Antiformal stack* in the moderately bedded sandstone layers. (F) *Snakehead anticline* in the moderately dipping sandstone beds. (G) *Overstep thrust system* in the thickly bedded sandstone layers. A number of near-vertical thrusts are abruptly truncated against a moderately dipping (to the left) thrust. (H) *Composite structure* developed due to progressive deformation and rotation of an initial lithic layers; the structure is characterized by the presence of more than one type of structure, *e.g.* a partly developed pop-up structure (upper right) enclosed within an open synform (central right), *duplex* with *horses*, etc.

Presence of the antiformal stacks in the Siwalik rocks possibly indicates that the formation of the thrusts progressed in time downward and toward the foreland. As a result, most duplexes appear to have been folded by displacement on later faults/thrusts resulting in culminations in the earlier thrusts (cf. Twiss and Moores, 1992, p. 105).

Snakehead Anticline

The snakehead anticline (Fig. 2F) is also a common type of structure in the Siwalik rocks. The structure is in the form of an anticline that rests over a ramp. Apparently, the anticline has not moved very far past the ramp. The limbs of the anticline show the development of a set of contraction faults oriented perpendicular to the layering. These faults may have developed during the formation of the snakehead anticline. The structure in the present form also looks like a domed unit that encloses another unit with a few curved duplex. Thus, the snakehead anticline shows an anticline resting over a ramp. But the presence of internal duplex system below the domed unit (or anticline) and above an apparently undisturbed bed seems to be quite typical, as such, this structure can also be designated here as a snakehead duplex.

Overstep Thrust System

Occasionally, a few layers of low-dipping beds get abruptly truncated by another unit, which contains few parallel and steep faults dipping in the opposite directions (Fig. 2G). As a result, new thrusts appear to have developed in the hanging wall and the individual horses are disposed steeply. The structure, thus developed, has been designated here as a overstep thrust system.

Composite Structures

In addition to the above, a few other types of structures are also developed in the Siwalik rocks. These structures appear to occur in a combination of a few structures. Since the authors have not seen any report of such structures in the literature, no specific nomenclature has been assigned to these structures in the present work. Instead, these have simply been grouped here as composite structures. A partly developed pop-up structure is seen enclosed within an open synform (Fig. 2H). The whole structure appears to have developed in a few stages. Initially, a pop-up structure may have developed as described above. At a later stage, the strata may have undergone gentle synformal folding. Some lower layers show duplex structures with horses within the bedding layers. Formation of such a (composite) structure clearly indicates considerable progressive deformation that may have caused quite a good deal of shortening in the associate lithologic units.

Discussion

Presence of a variety of structures and thrust systems in the internal domain of the Siwalik strata, as described in the paper, suggest a complicated structural history of these rocks. The geometry of the individual structures indicates progressive deformation of the early-formed structures. It is quite possible that initiation of deformation may have taken place by the reactivation of the anisotropies existing in the rock layers, mainly bedding planes. This may have given rise to the formation of bedding-plane thrusts. The latter, with progressive deformation, may have formed a series/variety of structures depending upon the local stress field and rheological factors, thus giving rise to ramps, horses, etc. The progressive deformation also induced progressive reorientation of structures, thus leading to local reversal of directions of the early-formed minor thrusts giving rise to backthrusts, reversed horses, etc. At some advanced stages of deformation, reversal of the original thrusts, horses, etc. may have also taken place mainly due to space problem and to accommodate the additional stresses in the compressional regime. All this ultimately gave rise to a complicated system of structures and thrust geometries to the Siwalik rocks.

Most of the structures, thus appear to have caused local thickening of the original lithologic layers by progressive and persistent internal deformation. If this is visualized on a regional scale, the entire Siwalik succession appears to have undergone thickening of its original lithologic layers to accommodate the additional and persisting stresses. Thus, the present-day Siwalik sequence may be described to occur in a structurally thickened state.

Acknowledgements

We express our sincere thanks to Prof. A. K. Jauhri, Head, Centre of Advanced Study in Geology, Lucknow University, for providing working facilities. Dr. Axel Vollbrecht of Göttingen University, (Germany) is thanked for his suggestions and comments. This study is a part of the project awarded to KKA by DST (ESS/16/242/2005/ subansiri (04).

References

Bhattacharya, A.R. (1983) On the nature of the Main Boundary Fault in Kumaun Garhwal Himalaya with special reference to measurement of flattening in folds. Neues Jahrb. Geol. Palaont. Mh., v.4, pp. 193-204.

Bhattacharya, A.R. (2005) An "Expanding Mountain Front" Hypothesis for the tectonic evolution of the Siwalik zone, Outer Himalaya. Palaeont. Soc. India, spl. publ. No. 2, pp. 73-80.

Boyer, S.E. and Elliott, D. (1982) Thrust Systems. Amer. Assoc. Petrol. Geol., v.66, pp. 1196-1230.

Burbank, D.W., Beck, R.A. and Mulder, T. (1996) The Himalayan foreland basin. In: *A. Yin and M. Harrison (eds.),* The tectonic Evolution of Asia, Cambridge University Press, pp.149-188.

Dubey, A.K., Misra, R. and Bhakuni, S.S. (2001) Erratic shortening from balanced cross-sections of the western Himalayan foreland basin: causes and implications for basin evolution. J. Asian. Earth. Sci., v. 19, pp. 765-775.

Kumar, R. and Nanda, A.C. (1989) Sedimentology of the middle Siwalik Subgroup of Mohand area, Dehra Dun valley, India. J. Geol. Soc. India, v.34, pp. 597-6616.

Misra, P. and Mukhopadhyay, D. (2002) Balanced structural models of Mohand and Santurgarh Ramp Anticlines, Himalayan Foreland Fold-Thrust Belt, Dehradun Re-entrant, Uttaranchal. J. Geol. Soc. India, v. 60, pp. 649-661.

Mitra, G. and Boyer, S. E. (1986) Energy balance and deformation mechanisms of duplexes. J. Struct. Geol., v.8, pp. 291-304.

Raiverman, V., Srivastava, A. K. and Prasad, D. N. (1993) On the foothill thrust of the Northwestern Himalaya. J. Him. Geol., v. 4, pp. 237-256.

Raiverman, V., Srivastava, A. K. and Prasad, D. N. (1994) Structural style in Northwestern Himalayan foothills. Him. Geol., v.15, pp. 263-280.

Raiverman, V., Ganju, J.L., Ram, J. and Misra, V.N. (1990). Geological Map of Himalayan Foothills between Ravi and Yamuna rivers. Oil and Natural Gas Corporation, Dehra Dun.

Srivastava, D. C. and John, G. (1999) Deformation in the Himalayan Frontal Fault Zone: evidence from small-scale structures in Mohand-Khara area, NW Himalaya. Mem. Gondwana Res. Gr., v.6, pp. 273-284.

Thakur, V.C., Sati, D., Rautela, P., Bist, K.S. and Chaudhury, B.K. (1995) The recent rise of the river bed near Mohand, Uttar Pradesh. Curr. Sci., v.68, pp 588-590.

Twiss, R. J. and Moores, E. M. (1992) Structural Geology. W. H. Freeman & Co., 532 p.

Valdiya, K.S. (1992) The Main Boundary Thrust Zone of the Himalaya, India. Annales Tectonicae, Spl. Issue, 6, pp. 54-84.

Yeats, R.S. and Lillie, R.J. (1991) Contemporary tectonics of the Himalayan frontal fault system: folds, blind thrusts and the 1905 Kangra Earthquake. J. Struct. Geol., 13, pp. 215-225.

Yeats, R. S., Nakata, T., Farah, A., Fort, M., Mirza, M. A., Pandey, M. R. and Stein, R. S. (1992) The Himalayan Frontal Fault system. Annales Tectonicae, Spl. Issue, 6, pp. 85-98.

□□□

5

Fluvial Palaeohydrological Studies in Western India : A Synthesis

ALPA SRIDHAR
Department of Geology
The M.S.University of Baroda, Vadodara 390 002, India
Email: alpasridhar@rediffmail.com

Abstract

The river basins of India are important repository of the late Quaternary hydrological and climatic changes. The long term (millennial scale) and short term (centennial scale) hydrological changes in these river basins can be reconstructed through the records of planform alterations and sediment deposits that occur during the rare extreme flood events. The status of fluvial palaeohydrological studies in western India and the future prospects are discussed in this synthesis. Almost continuous palaeoflood records spanning two thousand years is available from the bedrock rivers on western India, whereas older events dating back to ~ 5 ka have also been documented from the alluvial reaches. Though bedrock gorges are considered to be ideal sites for the preservation of palaeoflood deposits and related discharge estimations, the alluvial reaches of the rivers in western India owing to their unique geomorphic setup are equally promising. Regime-based palaeodischarge estimations for the average flows in these alluvial reaches would be significant in providing a regional history of the hydrological changes that have occurred in the river basins of western India.

Introduction

The fluvial hydrologic regime is sensitive to climate change and the system responds by altering its morphology and sedimentation pattern. Palaeohydrological studies are thus significant in deciphering climatic changes. The fluvial deposits preserved in the overbank sites are largely the records of flood events and detailed studies of such deposits provide information on the flood frequency, character and flood magnitude. The channel planform alterations on the other hand, indicate the changing average flow regime in the fluvial system. In India, where the hydrological records are extremely short, the value of historical and palaeohydrological data is substantial. Palaeoflood investigations in some of the Indian rivers, namely, Narmada (Ely *et al.*, 1996; Kale *et al.*, 1997, 2003), Tapi (Kale, 1999) , Godavari, Krishna and Pennar (Kale *et al.*, 1996), Luni (Kale *et al.*, 2000) and Mahi (Alpa, 2007a,b) have been carried out and the slackwater records, augmented by gauge and historical records indicate significant changes in the frequency and magnitude of large floods during the last two thousand years. All rivers studied barring Luni and Mahi are known for their spectacular monsoon discharges and occur in the sub-humid to humid climatic zones and most of these palaeoflood sites have been located in the bed rock gorges. The status of fluvial palaeohydrological studies in western India and the future prospects is discussed here.

Palaeohydrological Studies: the Concept

Palaeohydrological interpretations are concerned with the reconstruction of former river flows, be they average flows of low to moderate magnitude and frequency, or infrequent high magnitude events. The latter are commonly reconstructed in the bedrock settings using slackwater deposits and other palaeostage indicators (Baker, 1987), whereas the discharge estimations of average flows is based on the palaeochannel dimensions and sedimentological characteristics in the alluvial reaches (Patton, 1988; Reinfelds, 1995). The most commonly used techniques in palaeohydrological investigations include palaeostage estimates and regime-based palaeoflood estimates (Baker, 2000). The palaeostages can be estimated by documenting flood-induced erosion or deposition upto the water level reached by a flow. The slackwater deposits and palaeostage indicators occurring along stable- boundary fluvial reaches are the best preserved records of palaeoflows. Unconsolidated sands and silts that accumulate relatively rapidly from suspension during major floods, particularly where flow boundaries result in markedly reduced local flow velocities are described as slackwater deposits (Baker, 1987). For palaeoflood studies, a slackwater sedimentation site should be optimum for both the accumulation and preservation of the relatively fine-grained sediments carried high in flood flows at maximum stage. A few ideal sites of deposition and subsequent preservation of slackwater deposits have been described (Kochel and Baker, 1982; Baker, 1987; Benito *et al.*, 2003); amongst these the most common are the tributary mouths, channel-margin alcoves, caves and rock shelter deposits. The local hydraulic conditions at these sites result into a drop in the flow velocity and deposit flood sediments. Therefore, most palaleoflood studies are concentrated along the bedrock reaches of the rivers. Stratigraphical, sedimentological and chronological studies of the slackwater deposits provide reasonably accurate information about palaeofloods -their timing and magnitude. The palaeoflood deposits can be dated using radiocarbon or optically simulated luminescence dating. The palaeoflood magnitudes are determined by comparing the palaeostage indicator levels with water surface profiles generated by step-backwater computer program (O'Connor and Webb, 1988). The regime-based palaeoflood estimation procedures use empirically derived relationships to relate high-probability flows to palaeochannel dimensions, sediment types and other geomorphic expressions. The relations apply in particular to alluvial channels which adjust their planform to high flood discharges. The Indian river systems, thus, promise ample scope for the alluvial as well as bedrock palaeohydrological investigations; however, so far most of the studies remained focused on the palaeoflood estimations.

Palaeoflood Records in Western India

The western Indian region discussed here includes the southern Thar Desert, the Alluvial Plains of Gujarat and a part of Deccan Peninsula. The rivers Luni, Sabarmati, Mahi, Narmada and Tapi flowing through the southern Thar Desert and the Gujarat Alluvial Plains originate in the Aravalli and Vindhyan ranges and flow westward into the Arabian sea, whereas the rivers Godavari and Krishna arising in the Deccan Peninsula flow eastward into the Bay of Bengal (Fig.1). Palaeoflood records are available from the bedrock reaches of the Krishna, Godavari, Tapi, Narmada, Mahi and Luni rivers and from alluvial reaches of Mahi and Narmada (Fig.1).

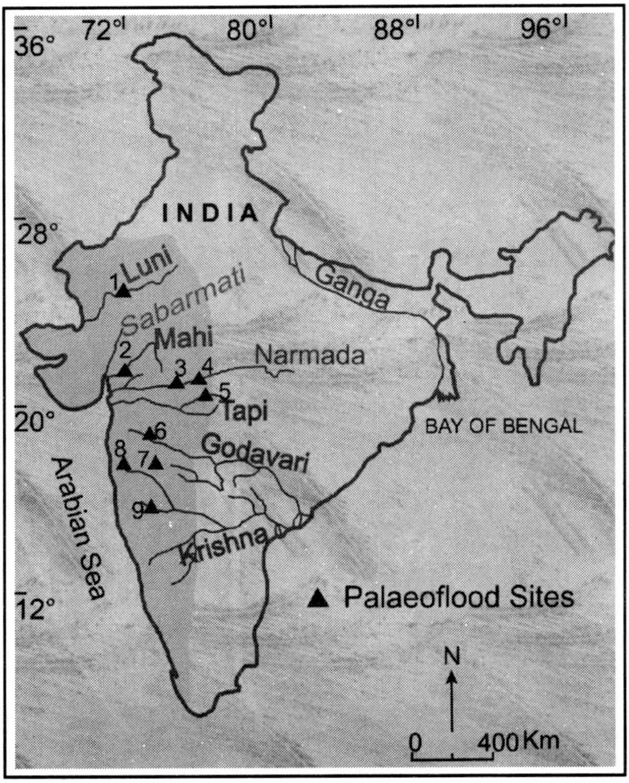

Fig. 1 : Palaeoflood sites in western India 1. Bhuka on Luni River, 2. Dodka on Mahi River, 3. Barjar on Choral River, 4. Sakarghat on Narmada River, 5. Guttigarh on Tapi River, 6. Daimabad on Godavari River, 7. Annapur on Krishna River, 8. Bibi on Krishna River, 9. Manjri on Krishna River.

Narmada is the most studied river basin for its palaeoflood record owing to its intense flood regime. Almost a 2000 year record of large floods has been assembled in the bedrock gorges of Narmada river basin (Kale *et al.*, 2003). The oldest record of catastrophic floods in the Narmada several millennia before the Last Glacial Maximum is the coarse gravel layer 7m above the bed of the palaeochannel at Bhedaghat (Rajaguru *et al.*, 1995). A record of the mid Holocene floods is documented at Barjar in the Choral River, a tributary of Narmada. About seven flood units have been identified and the lowest unit has been dated to 5170±135 years BP indicating that half-a-dozen large floods have occurred on the Choral River during the last 5 ka (Kale *et al.*, 1993). Continuous sequences of late Holocene floods have also been documented along the main river at Punasa (Kale *et al.*, 1994, Ely *et al.*, 1996) and Sakarghat (Kale *et al.*, 1997). The slackwater deposit sequence at Punasa is dominated by multiple flood units of sand and clay and has preserved an evidence of at least 14 floods that have occurred during the last 1700 years. The sequence at Sakarghat (Fig. 2) comprises about 26 flood units that have been grouped into four distinct series and spanning over 2000 years. Kale *et al.,* 2003 have inferred from the total palaeoflood record of the Narmada River that a number of extraordinary floods occurred during 400-1000 AD and post-1950 AD and that floods of lower magnitude and frequency generally characterized the period 1400-1950 AD.

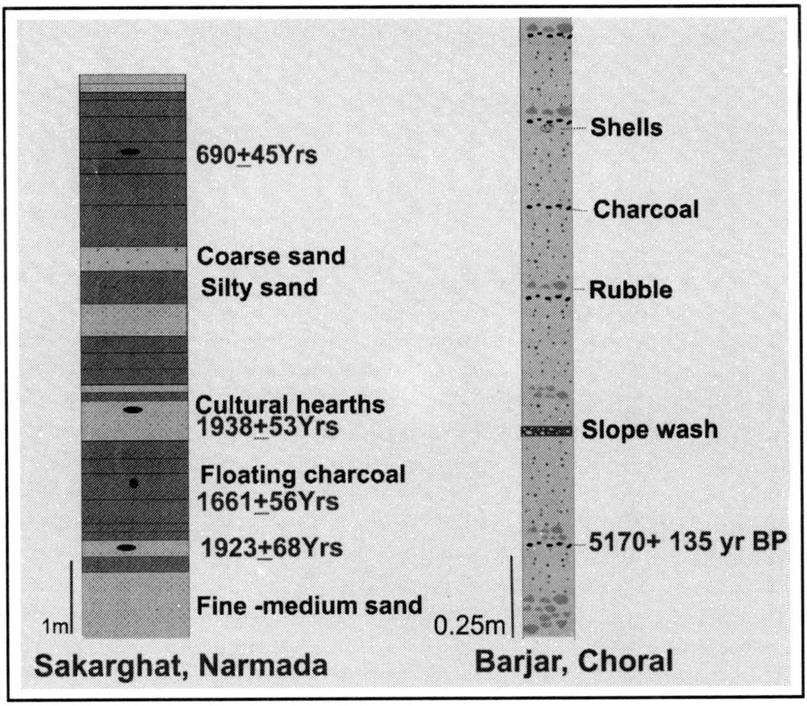

Fig. 2 : Litholog of slackwater deposits at Sakarghat on Narmada River (Kale *et al.*, 1997) and Barjar on Choral River, a tributary of Narmada (Kale *et al.*, 2000).

The Tapi River, also known to have a history of extreme floods, has preserved palaeoflood records in its bedrock gorges. The older flood deposits have been located at Guttigarh (Kale *et al.,* 1994) comprising a 4m thick sequence of silty clay sediments. A radiocarbon date of the shells occurring about 2m below the surface is about 8510±100 years BP in age. Late Holocene flood events have also been documented at Guttigarh, Teska and Khapa. At Guttigarh the upper units have yielded post 1950 radiocarbon dates (Fig.3) and the bottom most units a maximum age of 240± 50 years BP. Based on the hydraulic modeling Kale *et al.* (1994) have suggested that the highest flood deposits at this site were associated with floods close to $4000 m^3 s^{-1}$. The 2m thick sequence at Teska contains evidence of 6-9 floods and the oldest event has been dated back to 225±40 years BP. At Khapa, the 2.5 m thick section reveals the presence of 13 flood units emplaced by floods later than 380±45 years BP (Kale *et al.*, 1994).

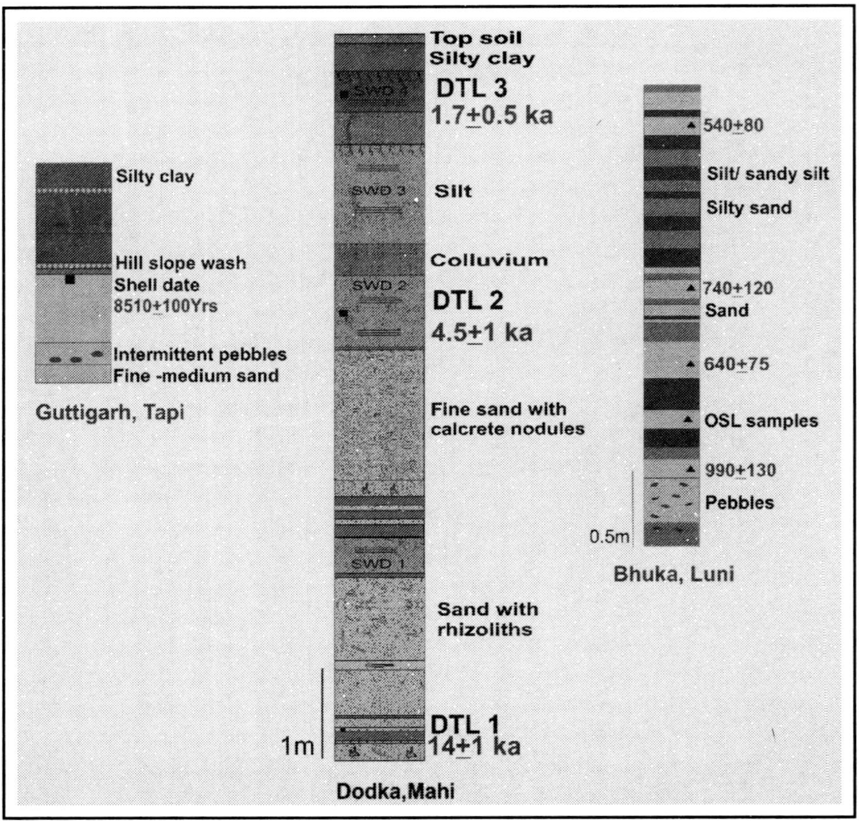

Fig. 3 : Lithologs of slackwater deposits at Guttigarh on Tapi River (Kale *et al.*, 1994), at Dodka in the alluvial reach of Mahi River (Alpa, 2007a) and at Bhuka in the Sindari gorge on Luni River (Kale *et al.*, 2000) monsoon. The other two units, however, represent extreme high magnitude floods in a period of low average precipitation.

Slackwater deposits in the alluvial reaches of the Mahi River occur in the ravines incised during the early Holocene on an alluvial surface comprising sediments of Late Pleistocene age (Alpa, 2007a). The slackwater deposits occur at elevations up to 20 m from the present river level and extend to about 500 m inland. The carbonate rich sediments forming the ravine cliffs have provided bank stability and the dissections in the ravines have helped in the accumulation of slackwater deposits due to backflooding of the floodwater from the main channel. Recent gullies have incised the sediments and exposed deposits related to major flood events. The best exposures of slackwater deposits have been observed at Dodka (Fig.3). The sediment succession of the slackwater deposits is dominated by bedsets and laminasets of silt and sand separated by colluvial sediments. Four events of flood deposition occurred during the mid to late Holocene. Two units of slackwater deposits, SWD 2 and 4 have been dated by IRSL at 4.6 ± 1 ka and 1.7 ± 0.5 ka (Alpa, 2007a). The stratigraphy of these deposits indicates that the first two slackwater units (SWD1 and SWD2) have resulted due to flooding in a regime of intense monsoon. The other two units, however, represent extreme high magnitude floods in a period of low average precipitation.

Regime based palaeoflood estimation has also been attempted in the alluvial reach of the arid Mahi River basin to quantify contemporary and palaeo-discharges and changes in the hydrologic regime through the mid-late Holocene (Alpa, 2007b). The occurrence of terraces and pointbars high above active river levels (Fig. 4) and change in the width/depth ratio can be regarded as geomorphic responses to changes in discharge. Discharge estimates are made based on the channel dimensions and established empirical relations for the mid-late Holocene, historic (the channel that deposited extensive pointbars above the mid-late Holocene channel was ~55 000 m^3s^{-1} and that of the historic channel was ~9 500 m^3s^{-1}, some ~25 times and ~5 times greater than that of the present river (2000 m^3s^{-1}), respectively.

The ephemeral Luni River flows through the Thar Desert of India and has been catastrophically flooded in the recent past. Evidences of such extreme events in the past have been recorded in the slackwater deposits preserved in the back-flooded tributary at Bhuka (Kale *et al.*, 2000). The study suggests that the Luni River has experienced at least 17 extreme floods during the past millennium. The hydraulic modeling carried out for the top most flood units indicate the related peak discharges to be close to 12 000 m^3s^{-1} (Kale *et al.*, 2000).

Palaeoflood records spanning almost 2000 years have been documented from the Krishna and Godavari rivers by Kale *et al.* (1996). The slackwater deposit sequences in the upper Krishna basin have preserved eighteenth century pottery. Similarly, radiocarbon dates from Annapur (770±80 years BP) and Bibi (180±90 years BP) represent high magnitude floods in the upper Krishna basin. At Manjri, human skeletons have been discovered below 8m flood deposits of about 4000 years in age. High magnitude floods in the Chalcolithic period ~ 3000 years BP have been reported at Daimabad in the upper reaches of the Godavari River (Rajaguru, 1986).

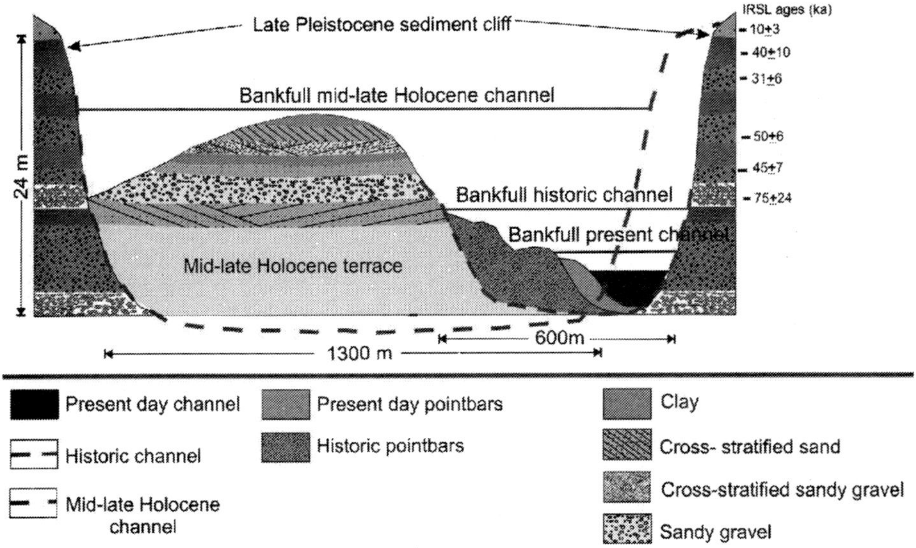

Fig. 4 : Section across the Mahi River exhibiting the channel dimensions and bankfull stages during the mid-late Holocene, historic time and present day (Alpa, 2007b).

Palaeodischarge Estimations

Estimation of discharges associated with extreme events is based on the elevations of the slackwater deposits or other stage indicators and step-backwater modeling is performed to generate related water surface profiles. Palaeodischarge calculations have been carried out at a few select sites in the Narmada, Tapi, Luni and Mahi rivers using HEC-2 computer program. The highest slackwater deposits at Sakarghat in the Narmada basin post-1900 are associated with discharges between 55,000 and 60,000 m^3s^{-1} and at Barjar in Choral river, a tributary of Narmada are found to be associated with discharges >4 500 m^3s^{-1} (Kale *et al.,* 1992, 1994). Similar estimation carried out at Guttigarh on Tapi river suggests that the highest flood deposits here probably post 1950 can be related to floods close to 4,000 m^3s^{-1}. The estimation of peak discharges using similar techniques has been attempted by Kale *et al.* (2000) in the Luni River and the results indicate peak discharges closer to 12,000 m^3s^{-1}. The peak discharge associated with the highest slackwater deposits dated to 1.7 ka at Dodka in the Mahi basin has been estimated to be around 55,000 m^3s^{-1} (Alpa, 2007a). Estimations on average flow discharges in the alluvial Mahi River have also been attempted by Alpa (2007b) and these are based on the channel dimensions and established empirical relations. Accordingly bankfull discharges for the Mahi river channel during three distinct time periods *viz.* mid-late Holocene, historic and present have been calculated to ~ 55,000 m^3s^{-1}, 9,500 m^3s^{-1} and 2,000 m^3s^{-1} respectively.

Discussion

All the major Indian rivers are rain fed and monsoon related storms are the major cause of high magnitude floods in the modern period. The most well documented and continuous palaeoflood records related to these extreme events span about 2000 years. The detailed analysis of these records on the humid to sub-humid Narmada, Tapi, Krishna and Godavari show clustering of events in certain time periods. According to Ely *et al.* (1996), a major break in the flood frequency occurred between ca. 400 to 0 ^{14}C years BP. There is also evidence of clustering of low frequency, extreme floods between 400 and 1000 AD and post 1950. The palaeoflood record spanning 800 years in the Luni River also shows similar clustering between 1000 and 500 years of high floods and was followed by a period of reduced floods (Kale *et al.*, 2000) corresponding to the Medieval Warming and Little Ice Age epochs respectively.

The slackwater deposits of the post-1950 flood events are found to rest over the older units in these river basins suggesting a higher magnitude of modern floods. Kale *et al.* (2000) also suggest that the post 1950 floods were the largest during the historical period while the early Holocene floods were the largest at least in the Holocene on the Narmada River. The palaeoflood records, as located so far, on the semi-arid Mahi River however, do not confirm to this fact. Whereas the older flood deposits are located in the ravines further away from the main channel, the historic and modern floods appear to have been deposited in the proximity to the main channel. However, the study is still incipient and needs further confirmation of ages for the recent flood sediments. The alluvial reach of the Mahi River has in fact preserved a spectacular record of mid-late Holocene flood events dating back to about 5000 yrs as compared to those in the bedrock gorges. Similar record dating back to ~ 8000 yr BP in Tapi and 5170 yr BP in Choral River has been reported.

The palaeoflood records available in the western India and discussed here suggest distinct linkages between individual river basins. The palaeoflood history for the past 5000 years in the Narmada, Tapi, Mahi, Luni, Krishna and Godavari rivers appears to be broadly comparable. A review by Kale (2008) also suggests that changes in the flood regime conditions of most of the Indian rivers are linked to fluctuations in the monsoon intensity and the associated phenomena on different timescales. Qualitative data on the hydrological changes in the various river basins is available in terms of the palaeoflood records however, quantification of discharge variance is lacking. A few quantitative estimations of peak discharges though available do not even help in subtle inferences as they are not very time specific. The available palaeodischarge estimates have been used to develop a regional flood envelop curves for the Indian Peninsula (Kale, 2008). The curves suggest that the largest palaeofloods were lower than the modern (post 1950) extreme floods.

Palaeoflood hydrology and palaeodischarge quantification provides information on the extreme flood events for the gauged and ungauged rivers and extends the records of largest floods.

Palaeohydrological studies, therefore, have lot of relevance for the river systems of India where the available gauge records are very short. Also, most of the lower order streams are ungauged and have lot of bearing on the hydrologic regime of the river. Palaeohydrological studies in the Indian rivers over the last two decades have shown immense scope for further palaeoflood analysis. Although the potential for palaeoflood studies in alluvial rivers is limited, it has been possible to reconstruct the palaeoflood history of the Mahi River due to its peculiar geomorphic setting (Alpa, 2007a). Almost all the major rivers of the Gujarat Alluvial Plains offer similar boundary conditions and hence, can be extensively explored for palaeoflood reconstructions. It would also be interesting to carry out discharge estimates for the average and bankfull flows of these alluvial rivers during distinct time periods using the unique geomorphic indicators. Such estimates along with the ones of extreme events from the bedrock reaches of Peninsular rivers would provide a regional picture of the hydrological variations that have occurred in western India and their possible links to climate change.

Acknowledgement

The author is grateful to Prof. L.S.Chamyal for his invaluable guidance and for the constructive suggestions for improving this manuscript. The financial assistance provided by DST (Project No. DST/WOS-A/ ES-35/2003) is gratefully acknowledged.

References

Alpa, Sridhar (2007a) A mid-late Holocene flood record from the alluvial reach of the Mahi River, Western India. Catena, v.70, pp. 330-339.

Alpa, Sridhar (2007b) Mid-late Holocene hydrological changes in the Mahi River, arid western India. Geomorphology, v.88, pp. 285-297.

Baker, V.R. (1987) Palaeoflood hydrology and extraordinary flood events. J. of Hydrology, v.96, pp.79-99.

Baker, V.R. (2000) Palaeoflood hydrology and the estimation of extreme floods. In: *Wohl, E.E. (editor)* Inland Flood Hazards. Cambridge University Press, pp. 359- 377.

Benito, G., Yolanda, S.M. and Alfonso, S. (2003) Sedimentology of high-stage flood deposits of the Tagus River, Central Spain. Sedimentary Geology, v.157, pp. 107-132.

Ely, L.L., Enzel, Y., Baker, V.R., Kale, V.S. and Mishra, S. (1996) Changes in the magnitude and frequency of late Holocene monsoon floods on the Narmada River, central India. Geological Society of America Bulletin, v.108, pp. 1134-1148.

Kale, V.S., Mishra, S., Baker, V.R., Rajaguru, S.N., Enzel, Y. and Ely, L.L. (1993) Prechistoric flood deposits on the Choral River, Central Narmada Basin, India. Current Science, v. 65, pp. 877-878.

Kale, V.S. (1999) Long- period fluctuations in monsoon floods in the Deccan Peninsular, India. J. Geol. Soc. India, v. 53, pp. 5-15.

Kale, V. S. (2008) Palaeoflood hydrology in the Indian context. J. Geol. Soc. India, v.71, pp. 56-66.

Kale, V. S., Mishra, S., Enzel.Y., Ely, L., Baker, V. R. and Rajaguru, S. N. (1992) Geomorphic investigations of floods in Central Narmada Basin, India. First Nat. Symp. On Environmental Hydraulics, C.W.P.R.S., Pune, pp.211-220.

Kale, V.S., Ely, L.L., Enzel, Y. and Baker, V.R. (1994) Geomorphic and hydrologic aspects of monsoon floods on the Narmada and Tapi rivers in central India. Geomorphology, v.10, pp.157-168.

Kale, V.S., Ely, L.L., Enzel, Y. and Baker, V.R. (1996) Palaeo and historical flood hydrology, Indian peninsula. Geological Society of India Special Publication, v.115, pp.155-163.

Kale, V.S., Mishra, S. and Baker, V.R. (1997) A 2000- Year palaeofloods record from Sakarghat on Narmada, central India. J. Geol. Soc. India, v. 50, pp.283-288.

Kale, V.S., Singhvi, A.K., Mishra, P.K. and Banerjee, D. (2000) Sedimentary records and luminescence chronology of Late Holocene palaeofloods in the Luni River, Thar Desert, northwest India. Catena, v. 40, pp.337-358.

Kale, V.S., Mishra, S. and Baker, V.R. (2003) Sedimentary records of palaeofloods in the bedrock gorges of the Tapi and Narmada rivers, central India. Current Science, v. 84, pp.1072-1079.

Kochel, R.C. and Baker, V.R. (1982) Palacoflood hydrology. Science, v. 215, pp. 353-361.

Kochel, R.C. and Baker, V.R. (1988) Palaeoflood analysis using slackwater deposits. In: *Baker, V.R., Kochel, R.C., Patton, P. C. (eds.),* Flood geomorphology, Wiley, New York, pp.357-376.

O'Connor, J.E. and Webb, R. H. (1988) Hydralic modeling for palaeoflood analysis. In: *Baker, V.R., Kochel, R.C., Patton, P.C. (eds.)* Flood Geomorphology, New York: Wiley, 383-402.

Patton, P. C. (1988) Geomorphic response of streams to floods in the glaciated terrain of southern New England. In: *Baker, V.R., Kochel, R.C., Patton, P. C., (eds.),* Flood geomorphology, Wiley, New York, pp.261-277.

Rajaguru, S. N. (1986) Geoarchaeology of Daimabad. In: *Sali, S.A. (editor)* Daimabad 1976-79. Archaeological Survey of India. New Delhi. pp. 580-587.

Rajaguru, S., Gupta, A., Kale, V. S., Ganjoo, R., Ely, L., Enzel, Y. and Baker, V.R. (1995) Channel form and processes of the flood-dominated Narmada river, India. Earth Surface Processes and Landforms, v. 20, pp. 407-421.

Reinfelds, I. (1995) Evidence for high magnitude floods along Waimakariri River, New Zealand. J. of Hydrology, v. 34, pp. 95-110.

6

Small-Scale Deformational Structures as Significant Shear-Sense Indicators: An Example from Almora Crystalline Zone, Kumaun Lesser Himalaya

K. K. AGARWAL and R. BALI
Centre of Advanced Study in Geology
University of Lucknow, Lucknow-226 007, India
Email: kamalagarwal73@gmail.com

Abstract

Small-scale deformational structures *e.g.* folds, asymmetric boudins, thin veins in an en-echelon pattern, small-scale duplex, shear zones and shear bands are the most common structural elements used as reliable shear-sense indicators in crystalline rocks. Most of these structures are observed in the Almora Crystalline Zone (ACZ) in Kumaun Lesser Himalaya, which is disposed in the form of a huge asymmetric synform. The shear-sense observed on the northern and southern flanks is found to be - top to north and top to south respectively.

Introduction

The study of small-scale structural elements in the recent years has drawn much attention especially to be used as reliable shear-sense indicators (Simpson and Schmid, 1983; Weijermars and Rondeel, 1984; Brunel, 1986; Ildefonse and Caron, 1987; Mawer, 1987; Choukroune *et al.*, 1987; Agarwal, 1994; Passchier and Trouw, 1996; Bali and

Agarwal, 1999) in deformed and tectonically transported rocks as thrust sheets.

The Almora Crystalline Zone (ACZ) occurs as a thrust-sheet over the Precambrian-Lower Palaeozoic sedimentary sequences of Lesser Himalaya (Heim and Gansser, 1939; Gansser, 1964; Valdiya; 1980). The two tectonic contacts it has with the surrounding sedimentaries are the North Almora Thrust (NAT) and the South Almora Thrust (SAT), which dip to the south and to the north respectively (Fig. 1). The characteristic asymmetric nature of the ACZ and the two thrusts have been discussed by many earlier workers (Misra and Sharma, 1972; Mehdi *et al.*, 1973; Srikantia, 1988; Agarwal, 1994).

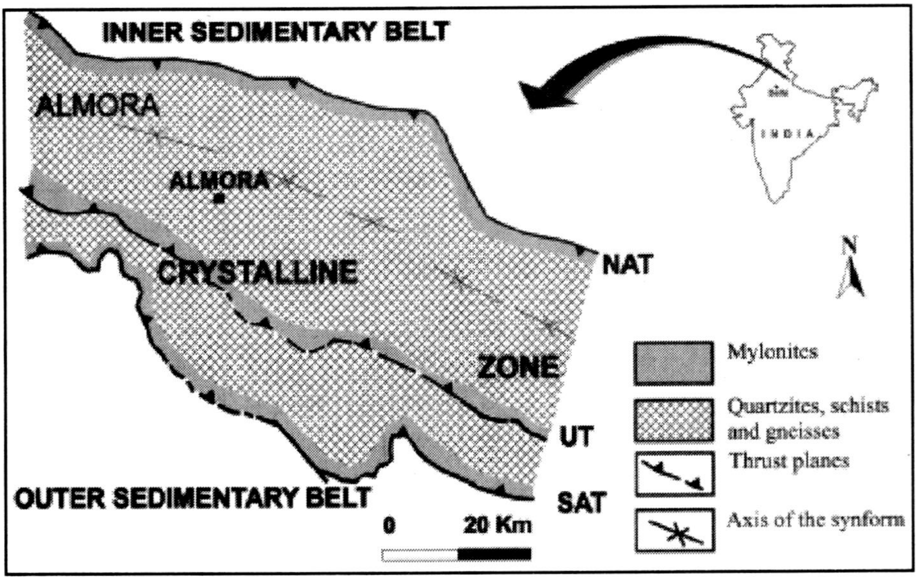

Fig.1: Simplified sketch map of the Almora Crystalline Zone (ACZ), Kumaun Lesser Himalaya, showing the location of the ACZ in the map of India (inset) (based on Valdiya, 1980).

The present paper is an attempt to understand the significance of the small-scale deformational features as reliable shear-sense indicators based on the field observations along a complete north-south traverse across the ACZ.

Deformational Structures as Shear-Sense Indicator

Several small-scale structures are found to be very good indicators of shear-sense in the orogenic belts, as these not only provide the information about the direction of tectonic transport but also help to understand the later tectonic adjustments.

In the ACZ, the dominant lithologies are slate, phyllite, various types of schist, gneiss and mylonite (Bhattacharya and Agarwal, 1985). The structural features that have been used as shear sense indicators in the ACZ are: folds, en-echelon veins, boudins (also pinch and swell structures), shear band foliation, stretching lineation, small shear zones, small duplex and ramp and flat structures. The sense of shear is deciphered by determining first the direction along which the observation is to made, and the plane along which the shear-sense indicator is observed; this can be referred to as the sense-of-shear plane ("SOS" plane, Davis and Reynolds, 1996). A brief description of the shear-sense indicators observed in the ACZ is as follows:

Small-Scale Folds

Asymmetric folds are observed at many places, which vary in size (ranging from few meters to a few cm in size). The fold vergence is used as shear-sense indicator (*cf.* Mawer, 1987) as these tend to have rotated towards the dominant lineation present in the rocks. The folds observed on the southern flank essentially verging to the SW to SSW. These folds are normally seen in thin quartzites interbedded with schists and are mainly isoclinal close to tight folds. The folds found on the northern flank commonly show similar geometry but verging mainly to the NE direction.

Boudins

These structures are commonly formed due to layer parallel extension or layer perpendicular shortening in the rocks with considerable difference in the viscosity across the layering. In the present area boudins and pinch and swell structures are found developed in the sequences with interbedded character of relatively soft (schists) and hard (quartzites) rocks. The orientations of the boudins at several locations (Fig. 2, 3D) reveal a top to south sense of displacement on the southern flank of the Almora synform. These have developed parallel to S_1, and the S_1 leans over the sense of shear (Davis and Reynolds, 1996).

Shear Zones

Small-scale shear zones which at places oriented both synthetic and antithetic, especially in the mylonites (Bali and Agarwal, 1999), are used as shear-sense indicators (Weijermars and Rondeel, 1984; Mawer, 1987). Shear zones in the present area indicate a top to south movement on the southern limb of the ACZ and on the northern limb give a top to the north movement (Fig. 3A).

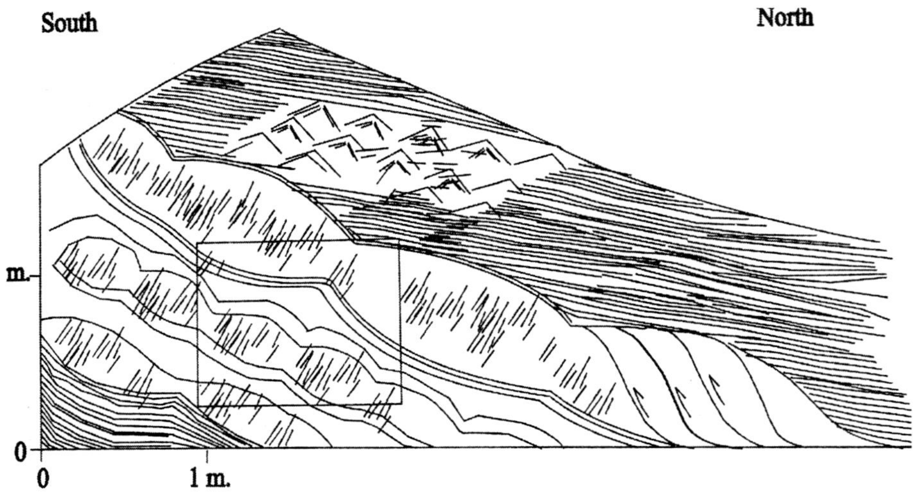

Fig. 2 : Diagrammatic sketch of the structural features as exposed on the southern limb of the ACZ (19 km south of Almora town, on the main highway). Note the development of boudins; the foliation is dipping to the north (to the right).

Stretching Lineations

Stretching lineations in deformed rocks are most often marked by the development of a new mineral (mostly sericite and chlorite in the present case) on the foliation (XY) plane with gentle plunge down the dip. Though these have not been suitably used to determine the shear sense but are good indicators of the tectonic transport direction in the orogenic areas (Brunel, 1986; Ildefonse and Caron, 1987; Daly, 1988, Jain and Anand, 1988; Agarwal, 1994; Passchier and Trouw, 1996).

In the ACZ the stretching lineations are observed throughout the section and also studied on the foot-wall-rocks (the surrounding sedimentaries) in the north as well as in the south. Along the two tectonic contacts *viz.* NAT and SAT, the stretching lineations are oriented almost perpendicular to the traces of these thrusts, and plunge down the dip. The attitude of the stretching lineations in the southern side varies between NE-ENE to NW-NNW, whereas in the northern side it is SW (*cf.* Agarwal, 1994).

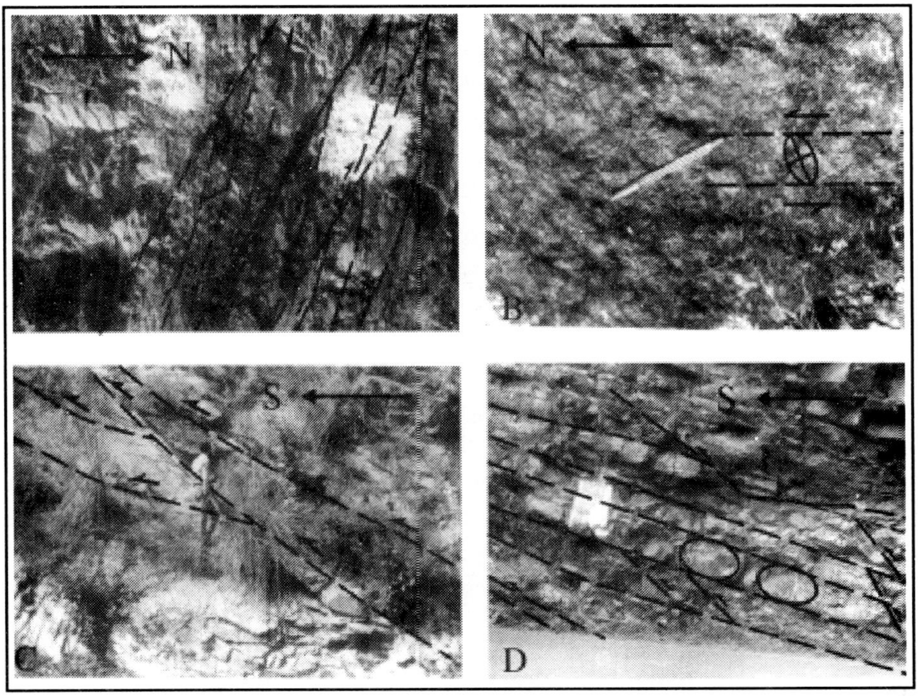

Fig. 3 : Field photographs of few structural features. (A) Small-scale shear zones on the northern limb (near the NAT, on the Almora-Kausani highway), note the shear movement is top to the north (right side of the photograph); (B) Fine tension veins arranged in en echelon pattern, the shear movement is top to the N (to the left of the photograph); (C) Ramp–flat structure within the small duplex on the southern limb, the shear-sense is top to south (to the left); (D) Boudins developed on the southern limb of the ACZ (location as in fig.2).

En Echelon Veins

These veins result from the filling-up of the tension gashes or fractures (Davis and Reynolds, 1996), and are very reliably used as shear-sense indicators. These veins are observed at many places, and give top to north sense of shear consistently for the northern flank (Fig. 3 B).

Duplex, Ramp and Flat Structures

These help a great deal in interpreting the bulk shear displacement in the fold-thrust belts. Duplex results from the progressive cutting of the foliations by later shear planes. Most often these planes show the development of slickenside striations and fibers. Duplex of varying dimensions (ranging from a few dcm. to a few meters) have been observed throughout the ACZ. The floor thrusts are exposed almost everywhere (Fig. 3 C), while the roof thrusts are rarely seen. The individual horses are not much thick (~ a meter), and are bounded between the s planes of the main basal thrust.

Discussion and Conclusions

A variety of small-scale deformational structures has been observed in the ACZ and is used as shear-sense indicators. The ACZ occurs as a large thrust sheet with its roots along the Main Central Thrust (MCT, Misra and Sharma, 1972; Mehdi *et al.*, 1973; Valdiya, 1980; Srikantia, 1988; Agarwal, 1994), and the major tectonic transport direction remains as N-S. The shear-sense as revealed by the field analysis of the structural features indicate that there is a distinct top to the south shear movement and top to the north shear movement on the southern and the northern flanks respectively of the Almora synform. This is due to the late stage tectonic adjustment of this crystalline thrust sheet after it was emplaced over the sedimentaries. There has been a long pending debate over the synformal character of these crystalline thrust sheets occurring in the Lesser Himalaya. The top to south and top to north shear sense observed on the two limbs supports the back thrusting model proposed earlier by Agarwal (1994).

Acknowledgements

The authors are grateful to Prof. A. K. Jauhri, Head, Centre of Advanced Study in Geology, Lucknow University, Lucknow, for providing working facilities, and to Prof. A. R. Bhattacharya for fruitful discussions. Thanks are also due to the anonymous referee for suggesting corrections/modifications in the earlier version of thems.

References

Agarwal, K. K. (1994) Tectonic evolution of the Almora Crystalline Zone, Kumaun Lesser Himalaya: A reinterpretation. J. Geological Society of India, v. 43 (1), pp.5-14.

Bali, R. and Agarwal, K. K. (1999) Microstructures of mylonites in the Almora Crystalline Zone, Kumaun Lesser Himalaya. Gondwana Research Group Memoir, v. 6, pp. 111-116.

Bali, R. and Bhattacharya, A. R. (1988) Geological and structural studies of the rocks of the Dwarahat-Chaukhutia area, Kumaun Lesser Himalaya with special reference to the North Almora Fault. Geoscience Journal, v. ix, pp. 215-230.

Bhattacharya, A. R. and Agarwal, K. K. (1985) Mylonites from the Kumaun Lesser Himalaya. Neues Jahrbuch fur Mineralogie Abhandlungen, v. 152 (1), pp. 65-77.

Brunel, M. (1986) Ductile thrusting in the Himalayas: Shear sense criteria and stretching lineations. Tectonics, v. 5, pp. 247-265.

Choukroune, P., Gapais, D. and Merle, O. (1987) Shear criteria and structural symmetry. J. Structural Geology, v. 9, pp. 525-530.

Coward, M. P., Windley, B. F., Broughton, R., Luff, I., Petterson, M. U., Pudsey, C., Rex, D. and Khan, M. A. (1986) Collision Tectonics in NW Himalaya. In: *Coward, M. P. and Ries, A. (eds.)*, Collision Tectonics. Geological Society of London, Special Publication, no. 19, pp. 203-219.

Daly, M. C. (1988) Crustal shear zones in Central Asia: A kinematic approach to Proterozoic tectonics. Episodes, v. 11, pp. S-11.

Davis, G. H. and Reynolds, S. J. (1996) Structural Geology of rocks and regions. 2nd edition, John Wiley & Sons, Inc. New York, 776p.

Gansser, A. (1964). Geology of the Himalayas. Interscience, London, 289p.

Heim, A. and Gansser, A. (1939) Central Himalayas: Geological observations of the Swiss Expedition 1936. Mem. Soc. Helv. Sci. Nat., v. 73, pp. 1-245.

Ildefonse, B. and Caron, J. M. (1987) The significance of stretching lineations in terms of progressive deformation and finite strain. Geodinamica acta, v. 1 (3), pp. 161-170.

Jain, A. K. and Anand, A. (1988) Deformation and strain pattern of intra-continental collision ductile shear zone - an example from the Higher Himalayas. J. Structural. Geol., v. 7, pp. 717-737.

Mawer, C. K. (1987) Shear criteria in the Grenville Province, Ontario, Canada. J. Structural. Geol., v. 9, pp. 531-539.

Mehdi, S. H., Kumar, G. and Prakash, G. (1973) Tectonic evolution of eastern Kumaun Himalaya: A new approach. Him. Geol., v. 2, pp. 481-501.

Merh, S. S. and Vashi, N. M. (1976) The problem of the South Almora Thrust. Him. Geol., v. 6, pp. 508-516.

Misra, R. C. and Sharma, R. P. (1972). Structure of Almora Crystalline, Lesser Kumaun Himalaya: An interpretation. Him. Geol., v. 2, pp. 330-341.

Passchier, C. W. and Trouw, R. A. J. (1996). Microtectonics. Springer-Verlag, Heidelberg. 289p.

Simpson, C. and Schmid, S. M. (1983) An evaluation of criteria to deduce the sense of movement in sheared rocks. Bulletin Geological Society of America, v. 94, pp. 1281-1288.

Srikantia, S. V. (1988) Himalayan Thrusts and structural belts. J. Geological. Society of India, v. 31, pp.210-229.

Valdiya, K. S. (1980) Geology of the Kumaun Lesser Himalaya. Wadia Institute of Himalayan Geology, Dehradun, 291p.

Weijermars, R. and Rondeel, H. E. (1984) Shear band foliation as an indicator of sense of shear: Field observations in central Spain. Geology, v. 12, pp. 603-606.

□□□

7

Climate Change and its Future Impact on the Indo-Gangetic Plain (IGP)

H. S. SAINI
Geological Survey of India, Faridabad, India
Email: hssainigsi@gmail.com

Abstract

The Indo-Gangetic plain (IGP) is an environmentally sensitive, socially significant and economically strategic domain of India where landscape, hydrology and fertility are threatened by climate warming and anthropogenic pressure. Irregular availability of water is going to be the biggest hazard in future. In case of increased water supply, the piedmont zone and river lowlands are threatened by erosion and sedimentation while in case of decreased water supply, the upland surface is endangered by salinization, desertification and drying-up of aquifers. Decline in food production will be the major problem. In order to make assessments of the environmental changes, concerted efforts should be initiated to understand the geological past and model the future. A pre-requisite is to develop a high resolution history of climate variations and their impact on landscape and ecology from the geological and historical records of IGP during the past 25 ka. Such data can help in the evaluation of forecasting scenarios, and thereby assist in developing mitigation plans regarding the environment. Artificially induced recycling of the water can be one of the possible ways to maintain a minimum availability of water.

Introduction

A major concern stands emphasized further after the release of the assessment reports of the Intergovernmental Panel on Climate Change (IPCC) in February and May, 2007. It is difficult to ignore the scientific conclusions drawn in the reports for the globe in general and India in particular. Today, an ordinary person is aware of the extreme events of unseasonal rains, monsoonal drought and rapid temperature fluctuations. He interprets them as the advance indications of an impending change in climate. Climate is not changing for the first time. It has changed several times in the geological past. The most recent of such changes were during the Quaternary period, which began with a major cooling of the earth's environment about 1.8 m.y. ago. Since then, there have been nine phases of cooling when glacial masses on the earth advanced significantly. The warm intervals between successive glaciations too had pulses of cooling. We live in the interglacial period *i.e.* Holocene. Since its inception, around 11,000 years ago, it has witnessed six climatic changes (Mayewski et *al.*, 2004). Among them, the last spell of cooling known as *little Ice Age,* between 15th and 19th Century, and the last spell of warming *Medieval Warm Epoch* between10th and 13/14th centaury are historically known (Wigley *et al.,* 1981; Lamb, 1984).

Human societies have suffered enormously due to climatic changes but ultimately most were able to adjust by changing their life style and migrating to the hospitable areas. Few centuries back, the impact of climate was not so severe because the population was less, it lived in small groups and economic stakes were low. Their migration to the fertile areas was easier because large areas on the earth were yet to be habited. Today, the societies have grown tremendously and are anchored in areas that are difficult to abandon because of the high economic stakes. Most of the habitable land is occupied. So if a population is forced to migrate under the pressure of global warming, there is not much place left for resettlement.

In the past the earth's climate changed due to natural reasons. Milankovitch effect, plate tectonics and rise of mountain chains are considered as the major driving forces (Ruddiman and Kutzbach, 1989; Ruddiman, 2001). But the present cause of global warming is, in part, man made due to the increased concentration of carbon dioxide in the atmosphere (IPCC, 2007). The fear of environmental 'losses' has stimulated the world to strategize the reduction of carbon emission in the atmosphere. Our past actions have already perturbed the earth climate system, and the impact of which is inevitable in future. The projected global warming has the potential to change the course of history, if not tackled in right earnest.

Today, the scientific community faces two primary tasks: a) to stabilize and reduce the carbon dioxide build-up in the atmosphere and b) to develop the hazard specific mitigation measures so that the earth's surface can sustain the societies. For the first task, affordable technologies have to be evolved and dependence on fossil fuel to be minimized in energy and transport sectors. This topic, however, is beyond the scope of this article. The second task requires the acquisition of specific knowledge about the types of hazard likely to develop, delineation of areas to be affected by them and working out suitable mitigation strategies for each one of them.

The Indian Scene

The Indian landmass consists of three physiographic domains: the Himalayas, Indo-Gangetic Plain (IGP), and Peninsular Shield (Fig.1). The Himalaya is the highest domain of rugged topography with deep valleys and high hill ranges. It is a tectonically active terrain shaped by the dissection of crystalline and sedimentary ranges by glacial, glacio-fluvial and fluvial agencies. It houses several hundred glaciers that feed and maintain the perennial flow of the Indus, Ganga and Brahmputra systems and is the main contributor of detrital loads to rivers landscaping the IGP, deltas and sub-marine floors. Erosion of its landforms plays an important role in the carbon contribution to atmosphere and has a direct bearing on the global climate change (Raymo and Ruddiman, 1992).

The Indo-Gangetic Plain is a 400-800 km wide, low relief, east-west zone between the Himalaya in the north and the Peninsula in the south (Fig.1). It is a sinking basin that came in to being about 50 Ma ago due to epiorogenic movements of Himalaya and was subsequently filled up by the sediments deposited by northerly and southerly drainage under the influence of climate changes, mainly from the Middle Miocene (Rowley, 1996).

The present surface configuration of IGP was fashioned about 100 ka BP (Gibling *et al.,* 2005). Its western part - the Punjab and Haryana Plains, constitute the Indus regime which has a southwesterly master slope and is drained by a perennial Satluj system in the extreme west and by short rain- fed, southerly -southwesterly flowing, ephemeral streams like Ghagghar and Markanda in the east. A subordinate system of northerly flowing seasonal streams namely Krishnawati and Sahibi originating from the northern part of Aravali ranges drain this region from the south although they carry only a small sediment load from the adjoining alluvial-aeolian plain. This is an arid to semi-arid, water-starved region with annual precipitation of 300-600mm and an evaporation of 1200 mm. In spite of the dry climate, a network of canals sustains the intensive cultivation of the entire surface. Natural forest is absent. Groundwater is fresh and shallow in the vicinity of canals but saline and deep near the margin of the Thar Desert. The surface is densely populated and cultivated.

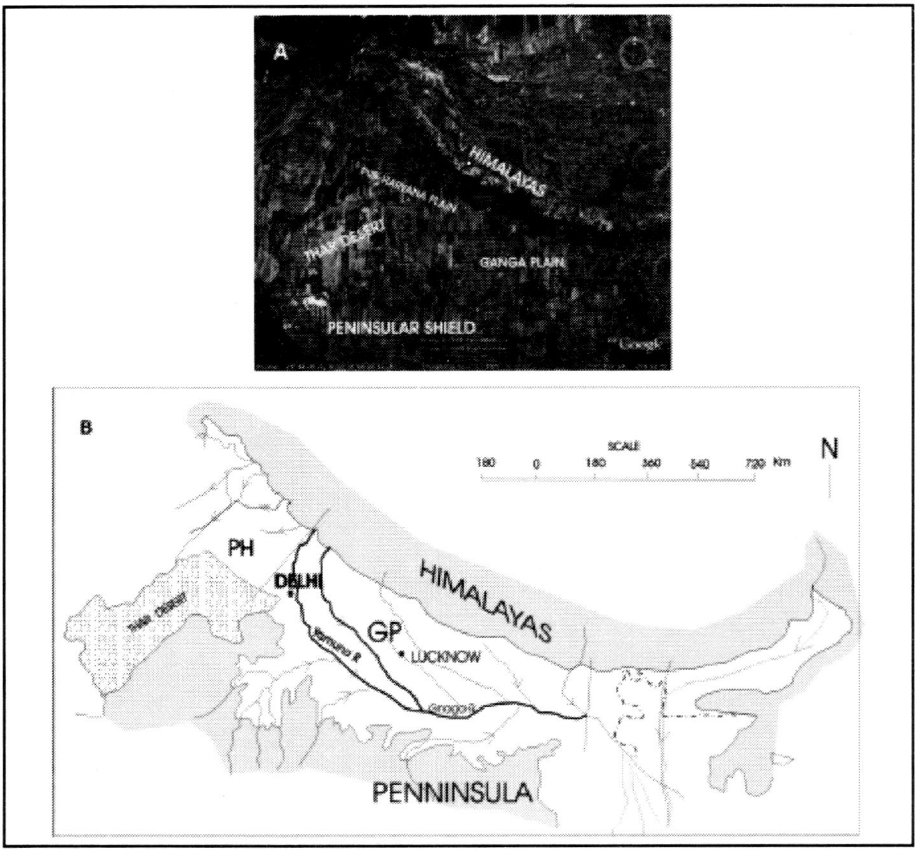

Fig. 1: (A) Regional Goggle image of the northern part of India and its surrounding showing Himalayas, Indo-Gangetic plain and northern part of Peninsula. The white patches in Himalaya are glaciers (B) Figure of the above image showing drainage pattern in Punjab, Haryana and Ganga plains.

The area east of the Yamuna river constitutes the Ganga basin which has a sub-humid to humid climate with 600-1000mm of average rainfall. Unlike Haryana -Punjab plain, it has a well developed drainage of Himalayan-fed, southeasterly flowing perennial system of Ganga and its tributaries like Yamuna, Ramganga, and Kosi. It also has some streams like Gomti, Sai etc which originate in the piedmont and are fed by rain and groundwater. A prominent northerly flowing river system consisting of Betwa, Chambal and Son originating from the northern margin of peninsula join the Yamuna and Ganga rivers in the southern part of the Ganga plain (Kumar *et al.*, 1996). The IGP can be divided into three major geomorphic units namely, piedmont, central alluvial plain and marginal alluvial plain (Singh, 1996; Kar *et al.*, 1997). Some of the important hazards mapped in IGP are shown in Fig.2. Piedmont is a narrow, forested, zone of gravelly-muddy sediments along the Himalayan foothills affected by gully erosion. Rivers passing from the Himalaya shed their coarse

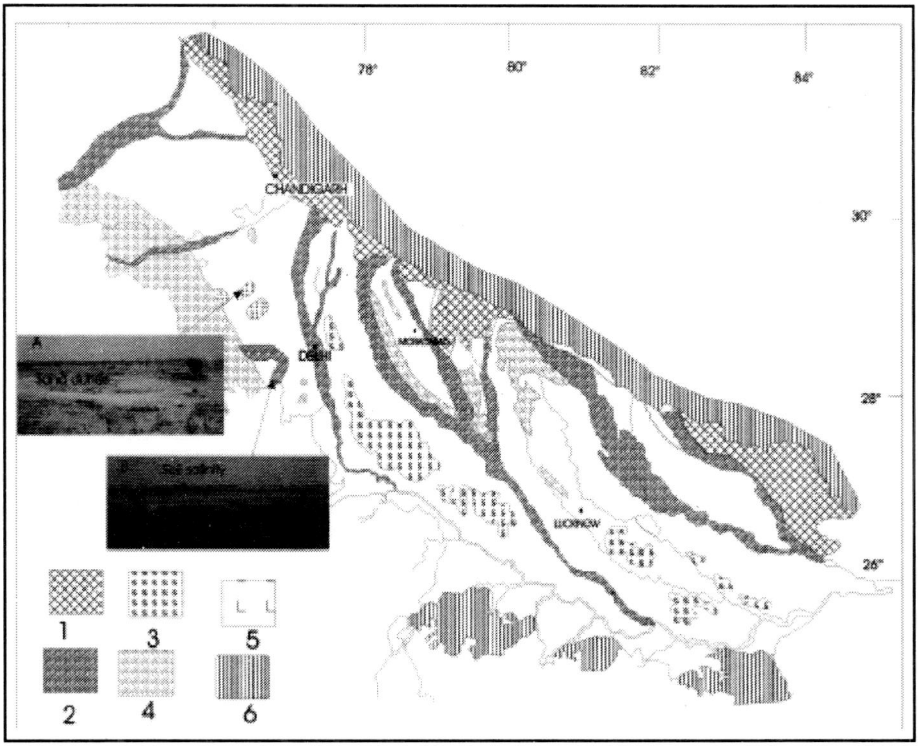

Fig. 2 : Map of parts of Indo-Gangetic plain showing important hazards. (1. Soil erosion, 2. Flooding, 3. Soil salinity, 4. Desertification, 5.water logging, 6. Hard rock exposures)

sediment load in this zone and multiply their channels in braided fashion. The central alluvial plain is 200-400 Km wide stretch with gradual southerly and SE slope in Ganga plain and SW slope in Haryana- Punjab plain. The rivers have 2-8 km wide valleys forming distinct lowlands, about 2-10m lower than the adjoining upland. The lowland valleys have flat terraces (T_1 and T_2) and channel bars. The regional upland surface is the master plain which is extensively cultivated and habited. Sand ridges, lakes, depressions and ponds are main landforms while soil salinity, flooding, riverine erosion are the main hazards in the Ganga plain and desertification and ground water salinity in the Punjab and Haryana Plains (Saini *et al.*, 2001), (Fig. 2). The marginal alluvial plain is a northerly sloping zone which was formed by the deposits of the northerly drainages from the peninsula. In Haryana, this zone is poorly defined due to the overlapping by aeolian deposits.

IGP is Critical

The importance of IGP is due to its high population density and agricultural production. Its climate is controlled by the monsoon rains. Since IGP is strongly connected to the tectonics and climate of Himalaya, any changes in these factors are liable to adversely affect the hydrology, soil fertility, food production and settlement pattern of IGP. Moreover, its proximity to Thar Desert is another cause of concern as the processes of sand movement can force undesirable modifications in the landscape of IGP under varying conditions of temperature and rainfall.

Besides natural threats, the IGP is also overstressed due to over exploitation of the land and water resources. The cultivable land is replaced by urbanization leading to increase in waste production. It can be concluded that IGP is under the twin threat of increasing population and climate change.

Climate Change and Associated Hazards

One of the primary concerns of climate shift is whether precipitation will decrease or increase in a given region? This will decide whether the potential hazards will be due to excessive availability of water or scarcity of water. The traditional model indicates that evaporation and humidity will increase due to warming. However, the warm air will hold the water for long duration that will delay and slow down the precipitation. On the contrary, some recent studies indicate that global warming could increase the precipitation by about 20% (Wentz and Schabel, 2000, Wentz *et al.,* 2007). Assuming that precipitation and temperature will change, we can anticipate rapid as well as slow hazards in the IGP that may nucleate in Himalaya as well as in the IGP. Rapid hazards like flood, erosion and deposition will possibly accompany the increased precipitation while slow hazard like soil -salinization, desertification, drought, ground water subsidence, shrinkage of water bodies and diminished flow in rivers will follow when precipitation is decreased. Both can lead to change in the boundary conditions of landforms, land-water and vegetation. Our concern is how and when the land and water systems in IGP will respond to these variations. Studies have shown that the impacts of climate controlled short and long-term changes on landscape and river systems could be predicted (Moll *et al.,* 2000; Veldkamp and Tebbens, 2001;1 Boggart *et al.,* 2003). In the context of IGP, we need to examine such changes during the last glacial and present interglacial period (*i.e.* 25 ka) to know how landscape, ecology and hydrological systems were affected when the water supply to IGP decreased and increased.

It may be assumed that initially the warming will trigger melting of glaciers and snow in Himalaya and a change in precipitation mode from snowfall to rainfall. It will increase the availability of water leading to accelerate erosion of the Himalayan landforms and production of detritus that may increase the sediment storage in the valley. A close example of this situation in the past can be taken from the hypsithermal event of Holocene, around 8-6 ka BP, when fluvial erosion filled up to 80m of the valleys with regoliths, which were also subsequently transported downstream (Pratt *et al.,* 2002). But it takes few hundred years to complete the cycle because there is a response time for landscape elements and geological agencies to adjust to the new conditions. The IGP and its river valleys will possibly shift towards changed conditions of sediment load, river flow velocity and discharge.

Traditionally, our societal and agricultural needs, conservation and exploitation pattern of water are adapted to the existing monsoonal climate and adjusted with the seasonal behavior of river flow and sub-surface aquifers. After debouching the alluvial plain, the piedmont zone would be the first target of fluvial erosion due to its sloping topography and loose lithology. It is already affected by the gully erosion (Fig. 2) which is likely to be intensified under accelerated flow velocity leading to the extension of badland and loss of forest land. Further downstream, in the central alluvial plain where slope is gentler, deposition of fresh sediment layers in the lowlands may occur. In fact the maximum sedimentary growth of IGP has taken place during the earlier wet interstadials in Late Quaternary as megafans near the Himalayan front and as channel and flood plain deposits in the down stream (Shukla *et al.*, 2001; Kar *et al.*, 1997). The lowlands are few hundred meters to more than 10 kilometer wide tortuous surfaces on either side of the Himalayan mountain-fed rivers. Though less populated, these surfaces are preferred for agriculture on account of fine textured soil, proximity with the perennial water source and fresh quality of shallow ground water. These are stabilized surface in which the cultivation pattern has evolved through centuries of experience. Fresh sedimentation due to change in the hydrodynamics of the rivers would lead to landscape modification in this terrain element and is likely to affect the cultivation style.

In second scenario, when precipitation is decreased, the western part of IGP *i.e.* Haryana -Punjab plains, adjacent to the Thar desert become the most vulnerable sector because it is already a water scarce region with history of drought and desertification. It has undergone a major phase of desertification around 16ka BP when large dunes were deposited up to Delhi (Glennie *et al.*, 2002) followed by several pulses of humid and semi-arid climate (Enzel *et al.*,1999; Saini *et al.*, 2005). These plains are sensitive to even smaller changes in climate and respond much quicker than the desert core. Rainfall at the eastern limit of Thar is around 300mm and 600mm in the latitude of Delhi. If we assume that rainfall decreases by 300mm, it will account for an average of 50% reduction, which is sufficient to transform the area into an arid zone. In the Ganga plain, east of Delhi, where rainfall is between 600-1400mm, a decrease of 300mm rainfall will not change the climatic zone and thus effects will be marginal.

The Haryana -Punjab plain becomes more critical in the light of past events. This area was once a site of the famous Indus valley civilization, when Harappan culture flourished in the Ghagghar-Hakara basin. Hundreds of cultural mounds scattered in northwestern Haryana and southeastern Punjab are testimony that a large population lived on these plains. The exact cause of the downfall of civilization is debatable; however increased aridity, drought, soil salinisation and desertification are cited as the possible causes (Mishra, 1984; Sahai, 1999). The downfall is estimated around 200BC. There is also a growing volume of opinion on the existence of a mighty river named "*Vedic Saraswati*" in this region that flowed from Himalaya to

Arabian sea and later disappeared due to tectonic disturbances and climate change (Wilhelmy, 1999; Oldham, 1999; Roy and Jakhar, 2001). Although a complete chronology of events and precise evidence of the cause of decimation of civilization and the Saraswati river are yet to be established, however, it is evident that the area has witnessed medium to large scale changes in landscape and the climatic variations which appear to be one of the major factors. In the event of a decrease in rainfall in future, there are strong possibilities of the Thar Desert to spread over this region, thereby changing the landscape, and making the land uncultivable.

Even a rise in temperature by 20° C can lead to a fall in farm production between 4 and 34% (IPCC, 2007). The soil salinity, already prevalent in the plains of Haryana, Punjab and western Uttar Pradesh, is bound to extend over the marginally saline areas reducing the availability of agricultural land. This will put additional stress on the already declining water table. The groundwater exploitation is likely to be intensified and consequently the water stressed areas will become water scarce. It can lead to change in river habit from gainer to contributor as shown in Fig.3 of Yamuna river in Faridabad district, U.P.

What can be done?

It is difficult to predict how exactly the warming will effect people in different regions of the globe. In the IGP changing environments have the potential to decrease the productive capabilities. We may have to deal with two kinds of situations: when water resource increases and second when water resource decreases. A possible approach is to model the behavior of IGP under conditions of excess and deficit water. For this we could seek past analogues based on the environmental changes in the last ~ 25ka time slice in the Himalaya and IGP and to analyze past responses of land, water and vegetation to such changes. The task will involve the collection, collation and synthesis of basin wise stratigraphic records, proxies of rainfall and temperature, so that the pattern of monsoon variation in time and space could be reconstructed. This will help in setting up and testing future hypothesis on landscape response to the fluctuations in climatic parameters as well as in identifying the areas according to their vulnerability to different kinds of hazards. Based upon these inputs, management plan(s) for storage, distribution and conservation of water could be attempted; and appropriate modifications in cropping pattern and new techniques for land use and irrigation evolved.

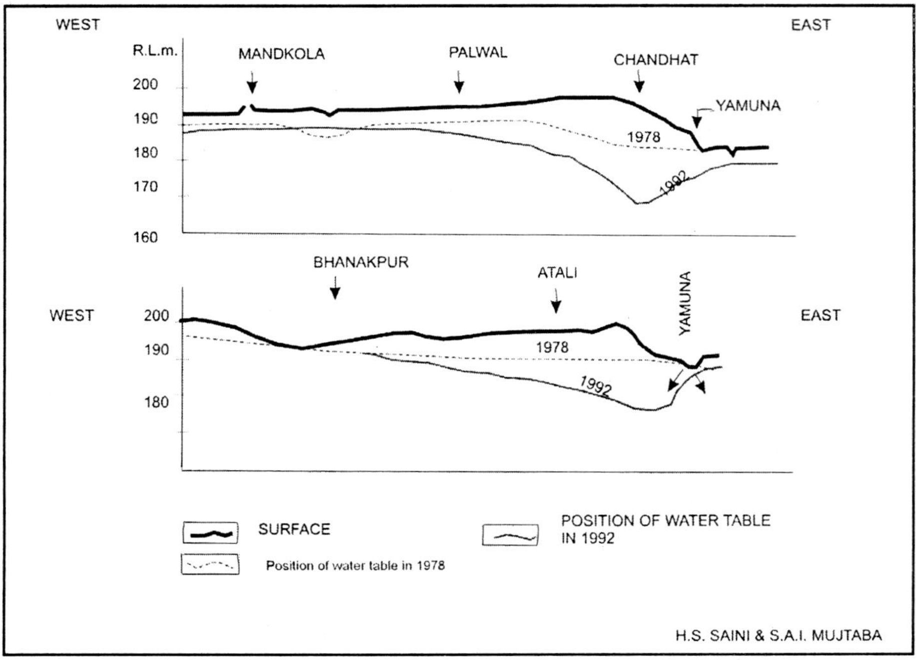

Fig. 3 : The change in the water table in the surrounding of Yamuna river south of Delhi (Faridabad district, U.P.). The excessive withdrawal of ground water has changed the nature of Yamuna River from an affluent to influent river.

There are several conventional approaches like construction of reservoirs, water harvesting, and water recycling which on a limited scale are employed for short term management of water resources. Under global warming scenario, such approaches may require re-assessment.

Today, a large part of the river water flows unutilized to the sea and it may not be practical to store it in reservoirs which have their own adverse impact on the environment. The ideal situation will be to have a regular supply of water through precipitation. Is it possible to 'store' the excess water of Himalayan rivers back in the atmosphere and then facilitate its re-precipitation? It amounts to creating an artificial climatic domain by changing the circulation pattern and enriching the moisture content of air. One possible way is to increase evaporation of the water from water surplus areas through plantation, increase moisture in the atmosphere and receive it back as precipitation. This approach includes identification of water surplus areas *i.e.* pumping zones, afforestation of suitable plant species *i.e.* transpiration surface, and understanding of the atmospheric circulation so as to know whether the evaporated water can be re-precipitated in the areas of interest. In this way a large proportion of water can be retained in the IGP before it goes to the sea. There are good chances that the increased moisture content of air may bring good rains over the IGP. The concept is highly preliminary and speculative but has the potential to address the problem of future water shortages. The need is to make institutionalized efforts with shared expertise and focused approach. Presently there is no such institution, though sporadic and un-coordinated investigations on hydrology, agriculture and geology are being undertaken by various individuals and groups.

Acknowledgements

I wish to record my gratefulness to Prof. S.K.Tandon, University of Delhi who nucleated the thought of preserving the sustenance of IGP as a social surface under changing environments. The idea was refined by several discussions with my colleagues Dr. N. C. Pant, S.A.I Mujtaba and R. K. Khorana of the Geological Survey of India. There could be several deficiencies in the write up as it is just a spark in the hazy horizon of understanding. I also wish to thank Sri D.D. Bhattacharya, whose consistent efforts made me complete the manuscript. The efforts put in by an anonymous reviver are thankfully acknowledged.

References

Boggart, P.W., Van Balen, R.T., Kasse, C. and Vandenberghe, J. (2003) Process-based modeling of fluvial system response to rapid climate change-1: model formulation and generic applications. Quat. Sci. Rev., v. 22, pp. 2007-2095.

Goodbred Jr., S. L. (2003) Response of the Ganges dispersal system to climate change: a source- to- sink view since the last interstade. Sed. Geology, v. 162, pp. 83-104.

Enzel, Y., Ely, S., Mishra, S., Ramesh, R., Amit, R., Lazer, B., Rajaguru, S.N., Baker, U.R. and Sandler, A. (1999) High-resolution Holocene environmental changes in the Thar Desert. Science, v. 284, pp.124-128.

Gibling, M.R., Tandon, S.K., Sinha, R. and Jain, M. (2005) Discontinuity bounded alluvial sequences of the southern Gangetic plain, India: aggradation and degradation in response to monsoonal strength. J. Sed. Research, v. 75, pp. 373-379.

Glennie, K.W., Singhvi, A.K., Lancaster, N. and Teller, J.T. (2002) Quaternary climatic changes over southern Arabia and the Thar desert, India. In: *Clif, P.D., Kroon, D., Gaedicke, C. and Craig, J. (eds.)* The tectonic and climatic Evolution of the Arabian Sea Region. Geological Society of London, sp. Pub. no. 195, pp.301-316.

Inter-Governmental Panel on Climate Change (IPCC). 2007. Fourth assessment Report.

Kar, S.K., Prasad, S. and Kumar, G. (1997) Quaternary sediments of Indo-Gangetic, Brahmputra and adjoining inland basins and the problem of demarcation of Pleistocene -Holocene boundary. Palaeobotanist, v. 46 , pp.196-210.

Kumar, G., Khanna, P.C. and Prasad, S. (1996) Quaternary stratigraphy of the Indo-Gangetic plain and sub-Himalaya, U.P. Contribs. XV Indian Collloq. Micropalaeontology and Stratigraphy, pp. 811-826.

Lamb, H.H. (1984) Some studies of the Little Ice Age of recent centuries. In: *N.A. Morner and W. Karlen (eds.)* Climate Changes on a Yearly to Millenium Basis, pp. 309-330. Reided. Dordreeht.

Mishra, V.N. (1984) Climate, a factor in the rise and fall of Indus civilization-Evidence from Rajasthan and beyond. In: *Lal, B.B. and Gupta, S.P. (eds.),* Frontiers of the Indus Civilisation, Books and Books, New Delhi, pp.461-489.

Mayewski , P. A., Rohling, E. E., Stager, J. C., Karle´n, W., Maasch, K. A., Meeker, L. D., Meyerson, E. A., Gasse, F., Kreveld, S. Van, Holmgren, K., Lee-Thorp, J., Rosqvist, G., Rack, F., Staubwasser, M., Schneider, R. R. and Steig, E. J. (2004) Holocene climatic variability. Quat. Res., v. 62, pp.243-255.

Moll, J., Vandenberghe, J. and Kasse, C. (2000) River response to variations of periglacial climate in mid-latitude Europe. Geomorphology, v. 33, pp.131-148.

Mujtaba, S.A.I. and Saini, H.S. (2001) Ground water environment in semi-arid Hissar and Fatehabad districts, Haryana-Problems and Potential. GSI Spl. Pub. No 65(II), pp 229-236.

Nanson, G.C. and Tooth, S. (1999) Arid zone rivers as indicators of climate. In: *Singhvi, A.K .and Derbyshire, E. (eds.)* Paleoenvironmental Reconstruction in Arid Lands, pp. 175-216. Balkema, Rotterdam.

Oldham, R.D. (1999) On probable changes in the geography of the Punjab and its rivers as historic- geographical studies. Mem. Geol. Soc. India, v. 42, pp. 81-88.

Pratt, B., Burbank, D.W., Hemsath, A. and Ojha, T. (2002) Impulsive alleviation during early Holocene strengthened monsoon, central Nepal Himalayan Geology, v. 30, pp.911-914.

Roy, A.B. and Jakhar, S.R. (2001) Late Quaternary drainage disorganization and migration and extinction of the Vedic Saraswati. Current Sci., v. 81(9), pp.1188-1195.

Raymo, M.E. and Ruddiman, W.F. (1992) Tectonic forcing of late Cenozoic climate. Nature, v. 359, pp. 117-122.

Rowley, D.B. (1996) age of initiation of collusion between India and Eurasia: a review of stratigraphic data. Earth and Planetary Sciences Letters, v. 145, pp. 1-13.

Ruddiman, W.F. (2001) Earth's climate: Past and present. W.H.Freeman and Company, New York.

Sahai, B. (1999) Unravelling the "Lost" Vedic Saraswati. Mem. Geol. Soc. India, v.42, pp.121-141.

Saini,H.S., Mujtaba, S.A.I. and Pant, N.C. (2001) Indo-Gangetic Alluvial Plain – A thrust area for societal geosciences. Geol.Surv. Ind. Special Pub., no. 65(III), pp.123-130.

Saini, H.S., Tandon, S.K., Mujtaba, S.A.I. and Pant, N.C. (2005) Lake Deposits of the northeastern margin of Thar Desert: Holocene palaeoclimatic implication. Curr. Sci., v. 88 (12), pp. 1999-2004.

Sharma,S., Joachimski, M., Sharma, M., Tobschall, H.J., Singh, I.B., Sharma, C., Chauhan, M.S. and Morgenroth, G. (2004) Late glacial and Holocene environmental changes in Ganga plain, Northern India. Quat. Sci. Rev., v. 23, pp.145-159.

Shukla, U.K., Singh, I.B., Sharma, M. and Sharma, S. (2001) A model of alluvial megafan sedimentation: Ganga Magafan. Sedimentary Geology, v. 144, pp. 243-262.

Singh, I.B. (1996) Geological evolution of Ganga Plain- An Overview. J. Palaeontol. Soc. India, v. 41, pp. 99-137.

Srivastava, P., Shukla, U. K., Mishra, M., Sharma, M., Singh, S., Singh,I.B. and Singhvi, A. K. (2000) Luminescence chronology and facies development of Bhur sand in the interflue region of central Ganga plain, India. Curr. Sci., v. 78, pp. 498-503.

Srivastava, P., Singh, I. B., Sharma, M. and Singhvi, A.K. (2003) Luminescence chronology and Late Quaternary geomorphic history of Ganga plain, India. Palaeog. Palaeoclim. Palaeoeco., v.197, pp.15-41.

Thomas, J.V., Kar, A., Kailath, A.J., Juyal, N., Rajaguru, S.N. and Singhvi, A.K. (1999) Late Pleistocen-Holocene history of Aeolian accumulation in the Thar Desert, India. Z. Geomorphologie, Supplt Band, v.116, pp.181-194.

Veldkamp, A. and Tebbens, L.A. (2001) Registration of abrupt climate changes within fluvial system: insight from numerical modeling experiments. Global and Planetry Change, v. 28, pp.128-144.

Wentz, F.J., Ricciardulli, L., Hilburn, K. and Mears, C. (2007) How much more rain will global warming bring? Science, v. 10, pp.1126.

Wentz, F.J. and Schabel, M. (2000) Precise climate monitoring using complementary satellite data sets. Nature, v. 403, pp. 414-416.

Wilhelmy, H. (1999) The ancient River valley on the eastern border of Indus Plain and the Sarswati Problem. Mem. Geol. Soc. India, v.42, pp.95-111.

Wigley, T.M.I., Ingram, M.J. and Farmer, G. (1981) Climate and history: Studies in past climates and their impact on man. Cambridge Univ. Press. Cambridge.

❑❑❑

8

Speleothems from the Himalaya and the Monsoon : A Preliminary Study

V. C. TEWARI
Wadia Institute of Himalayan Geology
33, General Mahadeo Singh Road Dehradun – 248001, Uttarakhand, India
Email: vtewari@wihg.res.in

Abstract

Speleothems (stalagmites, stalactites) have great potential for the records of past climate, environment and monsoon on a global scale particularly in the Asian region. Speleothems are found in all parts of the world and grow continuously through many glacial - interglacial cycles and their stable isotopic ratios and annual growth bands are useful for decadal to seasonal variations in rainfall. Some important speleothems from the Himalayan region such as Sahastradhara, Dehradun; Brahmakhal, Uttarkashi; Bhagirathi valley, Garhwal and Mawasmai caves from the Meghalaya have been studied. In the present paper, the oxygen and carbon isotope ratios from the Sahastradhara, Brahmakhal, Nagni and Pratapnagar speleothems and cave drip water has been recorded for the first time. The significance of speleothems and their relationship with the monsoon is discussed.

Introduction

Speleothems are abundantly found in the Himalayan caves from NW to NE India, where calcium carbonate is precipitated as stalactites and stalagmites (Fig.1 and Fig. 2 a,b,c,d,). The Himalayan speleothems may provide important record of paleomonsoon as they are not subjected to erosion and terrestrial deposits. The Holocene paleomonsoonal studies from the tropical and monsoonal regions of the Indian subcontinent and the southeast Asia has been attempted in recent years (Fleitmann *et al.*, 2003; Cheng *et al.*, 2005, 2006; Wang *et al.*, 2008; Johnson *et al.*, 2006; Sinha *et al.*, 2005; Yadav and Ramesh, 2005). Carbon and oxygen isotopic variations in speleothems especially stalagmite growth laminae are used for interpreting the amount of rainfall (Fleitmann *et al.*, 2003; Johnson *et al.*, 2006; Sinha *et al.*, 2005, 2007; Wang *et al.*, 2008). In the present scientific highlight, three very important speleothem deposits are being described from the Brahmakhal (Prakateshwar), Sahastradhara and Mawsmai caves from NW to NE Himalaya (Fig.1 localities 1, 2 and 3; Fig. 2, a,b,c,d). The sedimentological microfacies of the speleothems, radiaxial fabric of cave calcite and carbon and oxygen isotopic ratios of the stalactites and stalagmites have been discussed for paleoenvironmental and paleoclimatic interpretations. All these caves lie in the high monsoonal region therefore, it is quite significant to study the strength of Indian Summer Monsoon (ISM) and decadal scale seasonal variations. It is proposed to do the high resolution 230 Th dating of stalagmites from these unexplored caves to calculate the amount of rainfall due to west ISM. The Mawsmai cave lies in the Sohra (Cherrapunji) area of the Shillong Plateau, Meghalaya where highest annual rainfall is 11931.7 mm.

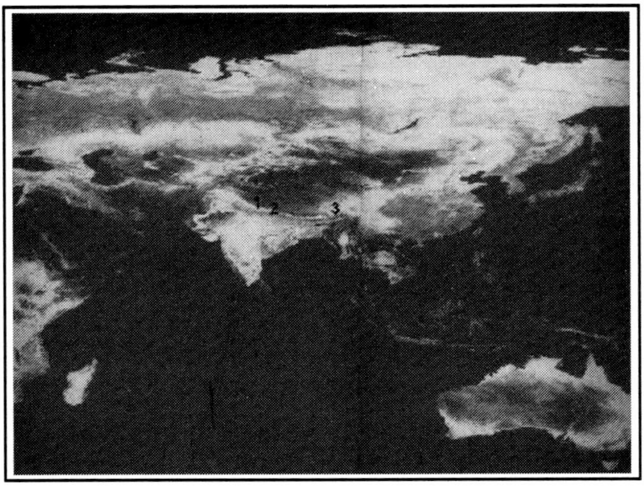

Locality- 1 Brahmakhal (Prakateshwar) cave discovered in 1998 from Uttarkashi district, Garhwal Himalaya, Uttarakhand

Locality- 2 Sahastradhara cave, Dehradun district Garhwal, Himalaya, Uttarakhand.

Locality - 3 Mawsmai cave, 6 km from Cherrapunji, Shillong Plateau, Meghalaya.

Fig. 1 : Map of India (Himalaya in the north), Tibetan Plateau and Indian Ocean showing the locations of caves from where speleothems have been studied

Lesser Himalayan Speleothems from Garhwal Uttarakhand Region

Sahastradhara Cave System

The Sahastradhara cave is situated at a distance of 14 km from the city of Dehradun, State Uttarakhand (Fig. 1, locality 2). The Doon valley is separated by the Lesser Himalayan Neoproterozoic Krol belt carbonate sediments in the north (Tewari, 2007) by Main Boundary Thrust (M.B.T.). The Siwalik foreland basin sediments are found in the south of the Doon valley thrust over the Indo - Gangetic Plains along Himalayan Frontal Thrust (HFT). The GPS location of the Sahastradhara cave is N 30 23 . 145 and E 78 07 .741. The Doon valley is between the river Ganga in the east and Yamuna river in the west. Both these rivers have their origin from two different Himalayan glaciers, namely Gangotri for Ganga and Yamunotri for Yamuna. Sahastradhara in Indian language means *thousand fold springs.*

Williams a British naturalist had described this place in his words "The only other holy spot worthy of special notice is the Suhusra Dhara (pronounced as *Sahastradhara*), a place of the thousand drippings which is a very simple phenomenon has invested with peculiar sanctity in the eyes of the people from the side of a charming valley to the east of Rajpur, oozes a mountain stream, distilling its waters over a precipice thirty feet high, and leaving a crust of lime on everything it touches. Particles, thus accumulating for continuous, have made a projecting ledge forming a sort of cave, from the roof of which falls a perpetual rain that turns every glade of grass coming in contact with it into a petrification. From above hang stalactites innumerable (Fig. 2 a, b). Stalagmite covers the ground beneath. Opposite, there is a sulphur spring also possessing powers of petrification." The seven caves of the Sahastradhara are small in size approximately, 10m long, 2m wide. The smallest one is 2.5m long and 2m wide.

Microfacies of the Speleothems

The Sahastradhara speleothems are made up of stalactites and stalagmites (Fig. 2 a, b). The stalactites show the presence of bacteria to form calcite precipitation. The microfacies of the Sahastradhara stalactites show radiaxial fibrous calcite (Fig. 3c, d). The Sahastradhara caves are located in the Neoproterozoic Krol Dolomite host rock (Tewari, 2007); Mg has triggered the formation of radiaxial fabric. Microbial precipitation of the carbonate is confirmed by the presence of microstromatolites and cellular microbiota (Fig. 3d). The culture experiments have demonstrated the role of microbes (PCR amplification of 16Sr RNA genes; 16Sr DNA) in the stalactite formation (Baskar *et al.*, 2006). The radiating fibrous calcite and finely laminated calcite and organic microbial laminae are also recorded in the Prakateshwar (Brahmakhal) cave speleothem (Fig. 3 a, b). Other carbonate mineral present is aragonite formed in fresh water. Various types of light (carbonate) and dark (organic) laminae are related to the microclimatic decadal scale seasonal variations.

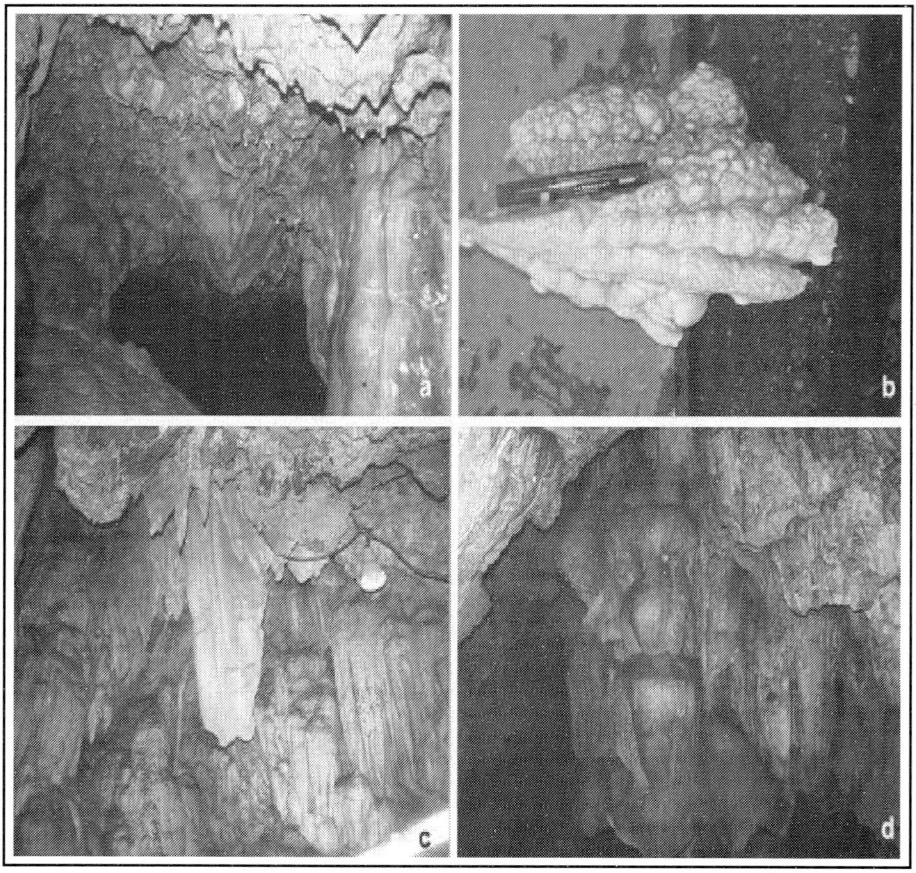

Fig. 2 : (a) Dripping water in the Sahastradhara cave showing stalactites and stalagmites during monsoon studied for oxygen isotope (location 2) (b) Stalactite column from Sahastradhara cave studied for carbonate microfacies and C and O isotope ratios (location 2 in Fig. 1) (c) Speleothems in Brahamakhal (Prakateshwar) cave studied for microfacies and C and O isotope ratios (location 1 in Fig. 1) (d) Speleothems from Mawsmai cave showing pillar like structures near Cherrapunji, Meghalaya, (Location 3 in Fig. 1).

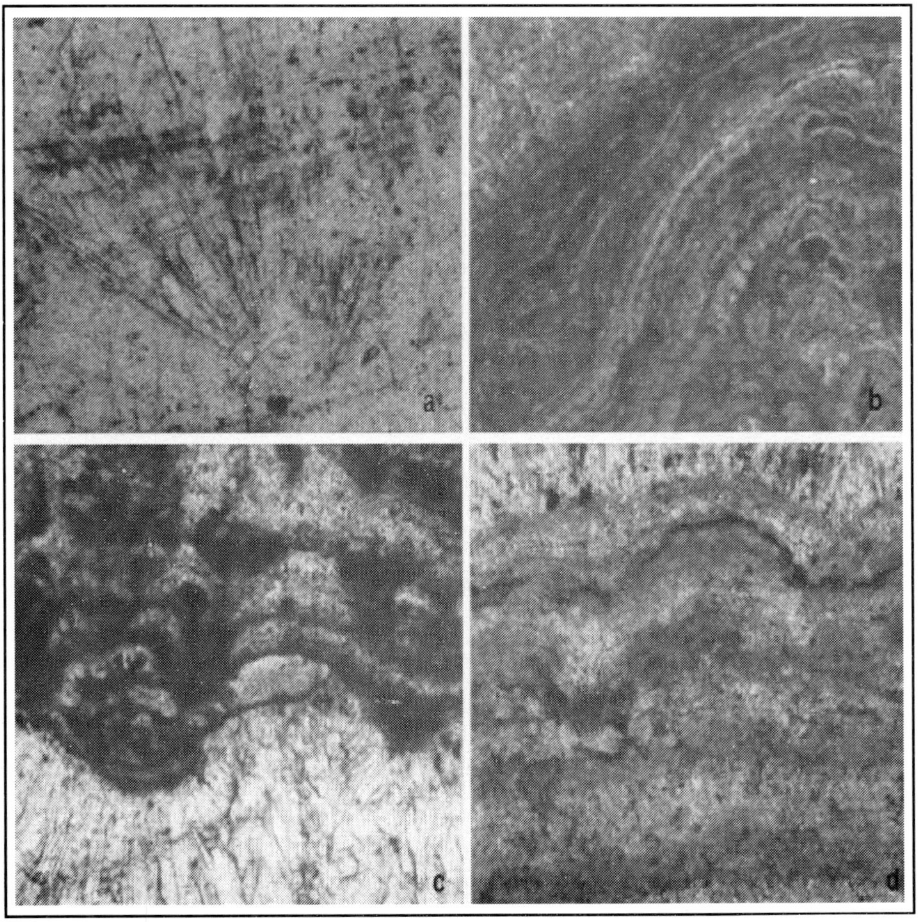

Fig. 3 : Microfacies of the Speleothems from the Himalaya, India (a) photomicrograph of radiating fibrous calcite in Brahmakhal (Prakateshwar) cave, Uttarkashi (b) Finely laminated calcite (white) and dark (organic) laminae in the Brahmakhal (Prakateshwar cave), Uttarkashi (c) photomicrograph of fibrous calcite in Sahastradhara stalactite showing radiaxial fabric (below) and dark (organic/ microbial) microstromatolitic laminae (above) (d) photomicrograph of microstromatolitic light and dark laminae and cellular and filamentous microbiota indicating microbial precipitation of carbonate, Sahastradhara caves, Dehradun.

Stable Isotopes and Palaeomonsoon Reconstruction

The carbon isotopic ratios of the Sahastradhara stalactites range from - 2.47 % (PDB) to -6.06 % (PDB) and oxygen isotope values range from 22.99% (SMOW) to 26.21% (SMOW). In Brahmakhal cave the $\delta^{13}C$ vary from -5.34 % (PDB) to -7.65% (PDB). $\delta^{18}O$ vary from 20.67% (SMOW) to 22.84% (SMOW). The $\delta^{13}C$ value of Pratapnagar modern stalactite calcite varies from -4.85% (PDB) to -9.23% (PDB) and the δ ^{18}O range from 19.50 to 21.57 % (SMOW). The variation of δ ^{18}O in stalagmites is related to the precipitation amount during the monsoon season. The oxygen isotopic composition of speleothem calcite from tropical and monsoon locations are primarily controlled by the δ ^{18}O value of precipitaton. δ ^{18}O values of regional precipitation and that changes in calcite δ ^{18}O over time primarily reflect changes in the amount of monsoonal precipitation. Study of modern cave drip waters and stalagmites demonstrates that stalagmite was deposited in or very near to isotopic equilibrium (Fleitmann *et al.* 2003). The δ ^{18}O isotopic data of drip water from Sahastradhara cave in Dehradun measured during the months of monsoon (August and September , 2007) varies from - 4. 58 % (V- SMOW) to - 5. 14 % (V- SMOW). Cave calcite also contains information about the isotopic composition of meteoric precipitation, is widespread and can be dated with 230^{Th}. Thus, a detailed high resolution O isotope speleothem data from the Himalaya may yield well dated record of paleomonsoon history.

The speleothems show microscopic laminae similar to marine and lacustrine sediments and could be of annual origin and can demonstrate seasonality. In order to establish such a relationship, continuous cave monitoring is essential. Stable C isotope ratios of the light and dark laminae vary in speleothems and may be linked with warm and cold seasons. Table-1 shows the bulk carbon isotope values from different caves. The C- isotope values are mostly negative. The detailed carbon isotopic study of individual light and dark laminae will explain the growth dynamics of the speleothems.

Discussions and Conclusion

Fleitmann *et al.* (2003 and the references therein) have interpreted that the western part of Indian Summer Monsoon (ISM) in Oman and Yemen , the oxygen isotope ratios of stalagmite calcite primarily show variations in the amount of rainfall with more negative δ ^{18}O indicating higher monsoon rainfall . In the present preliminary study also negative δ ^{18}O values have been recorded from the Himalayan speleothems (Brahmakhal and Sahastradhara caves). The δ ^{18}O data is also in agreement with Sinha *et al.* (2005) who have studied Timta cave in the Uttarakhand, NW Himalaya and interpreted that δ ^{18}O reflect varying ISM strength. The Indian Monsoon and East Asian Monsoon systems represent a part of the larger Asian Monsoon System. The East Asian Monsoon system is well documented by oxygen isotope records of the Hulu cave stalagmite (Cheng *et al.*, 2005). However, a detailed study of the known Indian Himalayan cave stalagmites and other unexplored caves in this region is essentially required to further confirm these results and propose a model for paleomonsoonal variation. There are very thick carbonate (karstic) sequences in the entire Himalayan belt from Jammu in the NW to Meghalaya in NE and needs to be studied in detail (Tewari, 2007; 2008).

Table 1: Carbon (permil) of $\delta\,^{13}C_{PDB}$ and Oxygen $\delta\,^{18}O_{PDB\ and\ SMOW}$ isotope ratios from the Lesser Himalayan Speleothems (Brahmakhal cave, Bhagirathi valley, Garhwal, BK1 to BK 3), Nagni (NC2), Pratapnagar, Garhwal (P1 to P6), and Sahastradhara , Dehradun (SD1 to SD8).

Sample Number	$\delta\,^{13}C_{PDB}$	$\delta\,^{18}O_{PDB}$	$\delta\,^{18}O_{SMOW}$
BK 1	-5.37	-9.43	20.67
BK 2	-6.68	-7.49	22.61
BK 3	-7.65	-7.26	22.84
NC 2	1.03	-7.55	22.64
P1	-5.64	-8.31	21.79
P2	-9.23	-9.94	20.16
P3	-4.85	-9.01	21.09
P4	-5.21	-8.53	21.57
P5	-9.22	-10.60	19.50
P6	-6.55	-10.08	20.02
SD 1	-5.97	-6.95	23.15
SD 2	-2.30	-4.12	25.98
SD 3	-3.76	-3.99	26.11
SD 4	-2.47	-3.96	26.14
SD 5	-6.06	-6.98	23.12
SD 6	-3.60	-3.89	26.21
SD 7	-6.07	-6.80	23.30
SD 8	-5.47	-7.11	22.99
Drip Water, Sahastradhara		? D (V -$_{SMOW}$)	$\delta\,^{18}O$ (V-$_{SMOW}$)
SD1	-	- 34.83	- 4.58
SD2	-	- 37.15	- 5.14

There is a plan to explore these caves for boosting geotourism in the fragile Himalayan mountains. Although such cave tourism is very popular in the America and Europe, however, the cave tourism in the Himalaya may be harmful to the environment. There will be tremendous increase in CO_2 in caves and such elevated CO_2 may cause destruction of speleothems within the cave. The speleothem may be at risk from the increased crowd of tourists where the calcium ion concentration of the drip water is low. Therefore, before such cave tourism is finally planned in the Himalayan region, the anthropogenic effects of tourists has to be considered. However, there is a need to explore these caves in detail in India.

Acknowledgements

I am grateful to Padmashree Dr. B.P.Radhakrisna, ex. President, Geological Society of India, Bangalore for inspiration and encouragement and thank Prof. B.R.Arora, Director, W.I.H.G., Dehradun for facilities and permission to publish the article. Professor D. Fleitmann, Switzerland is thanked for general comments on the speleothem photographs from the Himalaya. The anonymous referee is thanked for the critical review of the paper and useful suggestions.

References

Baskar, S., Baskar, R, Mauclaire, L. and McKenzie, J.A. (2006) Microbially induced calcite precipitation in Culture experiments : Possible origin for stalactites in Sahastradhara caves, Dehradun, India. Curr.Sci., v. 90(1), pp. 58- 64.

Cheng, H., Edwards, R.L., Wang, X.F., Wang, Y.J., Kong, Z.G., Yuan, D.X., Zhang, M.Y., Lin, Y.S., Qin, J.M., Tan, M. and Ran, J.C.(2005) Oxygen isotope records of stalagmite from southern China. Quaternary Sciences, v. 25, pp.157-163.

Cheng, H., Edwards, R.L., Wang, Y., Kong, X., Ming, Y., Kelly, M. J., Wang, X., and Gallup, C.D. (2006) A penultimate glacial monsoon record from Hulu Cave and two -phase glacial terminations. Geological Society of America, v.34 (3), pp. 217-220.

Fleitmann, D., Burns, S.J., Mudelcee, M., Neff, U., Kramer, J., Mangini, A. and Matter, A. (2003) Holocene forcing of the Indian monsoon recorded in a stalagmite from Southern Oman. Science, v. 300, pp. 1737-1739.

Johnson, Kathleen , R., Ingram, B., Lynn, Sharp, Warren, D. and Zhang , Ping Zhong (2006) East Asian summer monsoon variability during marine isotope stage 5 based on $\delta^{18}{}_{O}$ records from Wanxiang Cave, Central China. Palaeogeo. Palaeoclim.Palaeoclim. v. 236, pp. 5-19.

Sinha, A., Cannariato, Kevis G., Lowell, D., Stott, Li, Hong -Chun, You, Chen Feng, Cheng, Hai , Edwards, R . Lawrence and Singh, I.B. (2005) Variability of southwest Indian summer monsoon precipitation during the Bolling - Allerod. Geol. Soc. America, v. 33 (10), pp.813-816.

Sinha, A., Cannariato , K.G., Stott, L.D., Cheng, H., Edwards, R.L., Yadava, M.G., Ramesh, R. and Singh , I.B. (2007) A 900 year (600 to 1500) record of the Indian Summer monsoon precipitation from the core monsoon zone of India. Geophysical Research Letters, v. 34 , L 16707. doi : 1029 / 2007 GL 030431.

Tewari, V.C. (2007) Rise and decline of the Ediacaran biota : Palaeobiological and stable isotopic evidence from the NW and NE Lesser Himalaya, India. Special Publication, Geological Society of London, v. 286, pp.77- 102.

Tewari, V.C. (2008) Speleothem : Monsoon and Climate Change (In Hindi), Ashmika, June, Wadia Institute of Hmalayan Geology, Dehradun.

Wang, Y., Cheng, H., Edwards, R.L., Kong, X., Shao, X., Chen, S., Wu., J., Jiang, J., Wang , X. and An, Z. (2008) Millennial - and orbital -scale changes in the East Asia monsoon over the past 2,24,000 years. Nature, v. 451, pp. 1090- 1093.

Yadava, M.G. and Ramesh, R. (2005) Monsoon reconstruction from radiocarbon dated tropical Indian speleothems. The Holocene, v. 15 (1) , pp. 48- 59.

❑❑❑

9

Sediment Characteristics and Heavy Mineral Distribution in Tamiraparani Estuary and off Tuticorin, Tamil Nadu-SEM Studies

M. SURESH GANDHI, A. SOLAI,
K. CHANDRASEKARAN and V. RAMMOHAN
Department of Geology, University of Madras,
Guindy Campus, Chennai - 600 025, India
Email: surgan@yahoo.co.uk; geosolai@gmail.com

Abstract

The study area Tamiraparani located between latitudes 8° 25′ and 9° 13′ N, and longitudes 77° 10′ to 78° 10′ E along the south- east coast of India. All samples were collected between Pre-monsoon July, 2003 and Post-monsoon February, 2004. The heavy mineral assemblage of the study region is governed by the distribution of different type of minerals. However, the assemblage is restricted to the dominance of few selective minerals like garnet (colourless and pink), zircon, rutile, chlorite, etc. From the SEM studies, it is observed that the grains are sub-angular with various surface solution features, rounded crescent like pits, straight net like sutures; v-shaped pits are noticed in pre- and post-monsoon samples. The coarser nature of the grains may be due to the influence of river activities. The weathered surface futures also noticed in few samples. The sediment texture, mineral and SEM studies, clearly indicate that in pre-monsoon the erosional activities are predominant than the post-monsoon. In both the periods, sediment transport was from river and estuary towards the beach and marine. In pre-monsoon, the deposited sediments were transported and shifted due to longshore current action. But, in the post-monsoon period the sediments deposited due to the multi-source like riverine and marine influence is observed.

Introduction

Scanning electron microscopy (SEM) plays an important role in geology as a tool for understanding the nature of surface morphometric features in sediment samples. During last 15 years quartz grain surfaces have been studied extensively (Kuenen and Perdok, 1962; Krinsley and Donahue, 1968; Margolis, 1968; and Margolis and Krinsley, 1974). These studies have shown that fine surface textures indicate depositional environment and mode of transport. Heiken (1972, 1974) and Walker and Croasdale (1972) studied ash using the SEM and related ash morphology to magma composition and eruption type. Honnorez and Kirst (1976) combined SEM along with optical microscopy to develop a system of morphometric quantification. Huang and Watkins (1976) and Huang *et al.*, (1980) have analyzed angular surface pits on deep sea ash particles in order to estimate the frequency of impacts in volcanic eruption columns of silicic composition.

For beach sediments, heavy mineral studies Scanning Electron Microscope (SEM) is used to verify the general morphological characterizes (Robson, 1984; Mallik, 1986). However, the majority of studies focus on the examination of surface features. Krinsley and Doornkamp (1973) have studied the surface textures of quartz grains for over a decade. Tamiraparani basin is one of the major river basins in Tamil Nadu. Even though studies have been carried out on the different aspects of grain size and beach placer distribution in the study area (Cherian, 2003; Angusamy and Rajamanickam, 2007), but no systematic study has been taken up to understand the sedimentological and heavy mineral variation using SEM studies in this region.

Study Area

The study area Tamiraparani is located between latitudes 8° 25′ and 9° 13′ N and longitudes 77° 10′ to 78° 10′ E along the south-east coast of India. Study area and sampling locations of Tamiraparani Estuary off south-east coast of India are shown in the Fig.1. Geographic locations of sampling stations are shown in the Table-1 covering Tirunelveli and Tuticorin districts of Tamil Nadu. Tuticorin is geographically located in the Gulf of Mannar. The coastal belt between Vembar to Thiruchenduris is represented by raised beaches with sand bars parallel to the present coastline. The sand bars trend in north-south direction. In the coastal area between Thiruchendur to Manappad, there are sand dune and terri dune complexes. The Tamiraparani and its tributaries originate in the eastern slopes of Western Ghats, the hilly catchment area of which is influenced by both southwest and northeast monsoons.

Table 1: Geographical locations of Tamiraparani River, Estuary and Marine sediment samples.

Location	Longitude	Latitude	Location	Longitude	Latitude
R1	78° 00' 07.17"	08° 37' 16.94"	M1	78° 08' 27.44"	08° 40' 03.23"
R2	78° 01' 41.40"	08° 37' 16.94"	M2	78° 09' 29.80"	08° 40' 15.17"
R3	78° 03' 01.78"	08° 37' 15.55"	M3	78° 10' 28.01"	08° 40' 08.78"
R4	78° 04' 36.01"	78° 04' 36.01"	M4	78° 11' 37.30"	08° 40' 03.23"
R5	78° 05' 07.89"	08° 38' 15.14"	M5	78° 12' 49.36"	08° 40' 17.09"
R6	78° 07' 41.71"	08° 35' 52.41"	M6	78° 14' 12.50"	08° 40' 10.16"
R7	78° 07' 12.61"	08° 35' 59.33"	M7	78° 15' 30.11"	08° 40' 03.23"
R8	78° 07' 07.07"	08° 36' 29.82"	M8	78° 16' 51.59"	08° 40' 10.16"
R9	78° 07' 22.31"	08° 36' 51.99"	M9	78° 18' 11.97"	08° 40' 04.62"
R10	78° 07' 19.54"	08° 37' 23.87	M10	78° 18' 13.35"	08° 35' 51.02"
R11	78° 07' 20.92"	08° 38' 04.06"	M11	78° 16' 41.89"	08° 35' 42.71"
R12	78° 06' 49.05"	08° 37' 39.11	M12	78° 15' 18.74"	08° 35' 28.85"
R13	78° 06' 04.71"	08° 38' 06.83"	M13	78° 14' 13.61"	08° 35' 53.79"
R14	78° 07' 32.01"	08° 38' 38.70"	M14	78° 13' 15.41"	08° 35' 33.00"
R15	78° 06' 50.44"	08° 38' 33.16"	M15	78° 12' 04.73"	08° 35' 48.25"
R16	78° 05' 56.39"	08° 38' 53.94"	M16	78° 10' 38.81"	08° 35' 55.18"
R17	78° 07' 02.91"	08° 39' 09.19"	M17	78° 09' 40.61"	08° 36' 07.65"
R18	78° 06' 51.82"	08° 39' 41.06"	M18	78° 08' 35.48"	08° 35' 44.09"

Methodology

Samples were collected from selective localities like river, estuary and marine for SEM studies. Three size fractions were chosen from sieved samples to represent each sample. Sizes used were fine, medium and coarse in order to study size dependency of textures. The micro-features of grain size for sediment characteristics and heavy minerals like illmenite, garnet, etc. from the study area are attempted using Scanning Electron Microscope (Model; JEOL JSM 6360). After washing, samples were mounted on stainless steel stubs using double-stick tape. Approximately (5 grams) 20 grains from each of the three size categories were mounted on a single stub pasted with double side adhesive tape.

Samples were coated with platinum-palladium (JFC 1600) in order to counteract grain surface charging while scanning with the electron beam. Both whole grain and detailed surface photographs were taken using the secondary electron mode at 20 keV. For heavy mineral SEM studies, selective minerals from both the monsoons were picked and photographed.

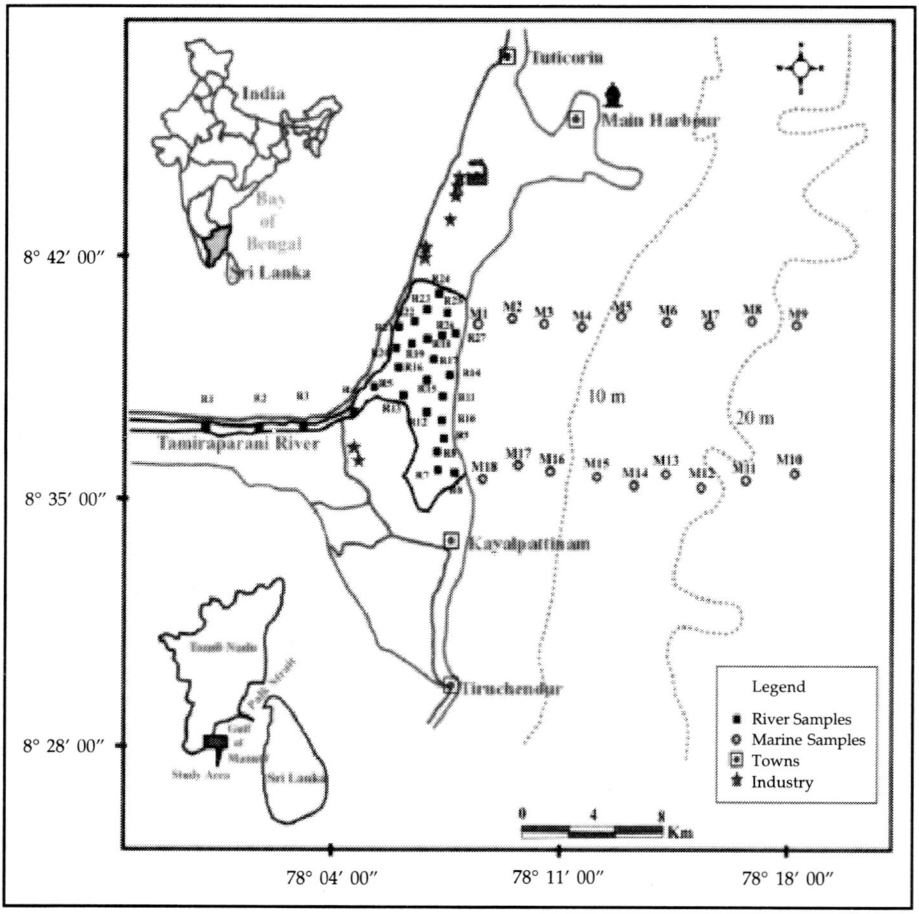

Fig.1: Map showing sampling locations in the study area

Result

Table 2 shows the mechanical and chemical features of the selected grains with respect to river, estuary and marine samples.

Granulometric Surface Feature Studies

Pre-Monsoon Samples

The grains are sub-angular with various surface solution features is observed in pre-monsoon river samples. Rounded crescent like pit, straight net like sutures, V- shaped pits are noticed. The coarser nature of the grains may be due to the influence of river activities. The weathered surface features also noticed in few samples. The estuary grains shows more or less similar pattern as of river samples. This indicates that the sediments were transported and deposited in the estuaries. The quartz grains are irregular in shape and show conchoidal fractures. In pre-monsoon marine samples, the grains are again sub-angular to irregular in nature. Few crack like features are also noticed. This may be due to mechanical activities of waves and currents. Further, due to erosional activities, the grains surfaces are smoothless in nature (Plate I-II).

Table 2 : Morphological features for the grains in river, estuary and marine sediments in pre and post monsoon- (Approximately for 0.25 grams)

	Surface Features	River Pre-M.	Estuary Pre-M.	Marine Pre-M.	River Post-M.	Estuary Post-M.	Marine Post-M.
MECHANICAL FEATURES	V- shaped pits	5	4	2	4	3	3
	Rounded crescent Shaped pits	11	12	4	2	8	5
	Straight net like Sutures	21	0	2	8	0	2
	Conchoidal fracture (<10 micron	18	4	4	12	6	18
	Conchoidal fracture (>10 micron	21	3	2	5	4	14
	Arcuate step like Furrows	2	0	0	3	0	2
	Dish shaped Depressions	6	4	2	2	4	8
	Upturned plates	0	0	0	0	0	0
	Crevasses	2	4	12	4	4	18
	Etch V's	4	0	0	2	0	4
CHEMICAL FEATURES	Crystalline	0	0	0	0	0	0
	Silica precipitation	2	4	0	0	0	0
	Silica flowers	0	0	0	0	0	0
	Silica globules	0	0	0	4	0	0
	Solution pits	12	14	4	2	12	84
	Surface solution Features	0	2	4	0	11	96
	Triangular oriented Pits	0	0	2	2	4	4
	Block removal	0	0	0	0	0	0

Post-Monsoon Samples

The river samples show sub-angular, angular and irregular grains. Numerous pits are observed in the surface of the few grains which may be due to chemical and mechanical disintegration. In comparison to pre-monsoon samples, the post-monsoon samples are less disturbed by mechanical activities. In estuary, the grains are angular to sub-angular in nature. Numerous pits and weathered samples are noticed which may be due to the depositional activities. In marine samples fossil like preservations were also observed. The preservations like v-shaped pits, grooves are noticed in these samples. Few rounded grains are also observed which may be due to long transportation (Plate III-IV). It is clearly indicated that the depositional activities are predominant in post-monsoon compared to pre-monsoon.

Heavy Mineral SEM Studies

Fig. 2 shows the distribution of total percentage of heavy mineral in Tamiraparani River and off Tuticorin. From the study of heavy mineral assemblages, it is observed that the non- opaque percentage is 3.26 to 5.47% in marine samples and opaque ranging from 3.67 to 5.72%. Overall the higher percentage is noticed in the river samples may be due to the fast moving action of wind and water. In estuary, the heavy mineral distribution is lesser (0.8% to 11.6%) due to the movement of sediments from river and the erosional activities. Plate-V shows the SEM photographs of selective minerals.

Ilmenite

The Ilmenite grains show varied features. Most of the grains are sub-angular to rounded with moderate relief. Under the higher magnification a number of oriented crescentic pits are noticed from solution cavity (Plate-V). A long residence time allowed the grain to develop this type of features. Some grains shows development of undulatory wavy surfaces probably because of mechanical reaction and removal of large blocks. Impact 'V' marks are also noticed which are sometimes modified by etching. Spit like features also observed in some minerals. This indicated that the grains must have longer residential time in the basin. According to Mallik (1986) the most important features that reflect the sub-aqueous environment are the V- shaped pits. This feature is observed in the Ilmenite near Tuticorin zones.

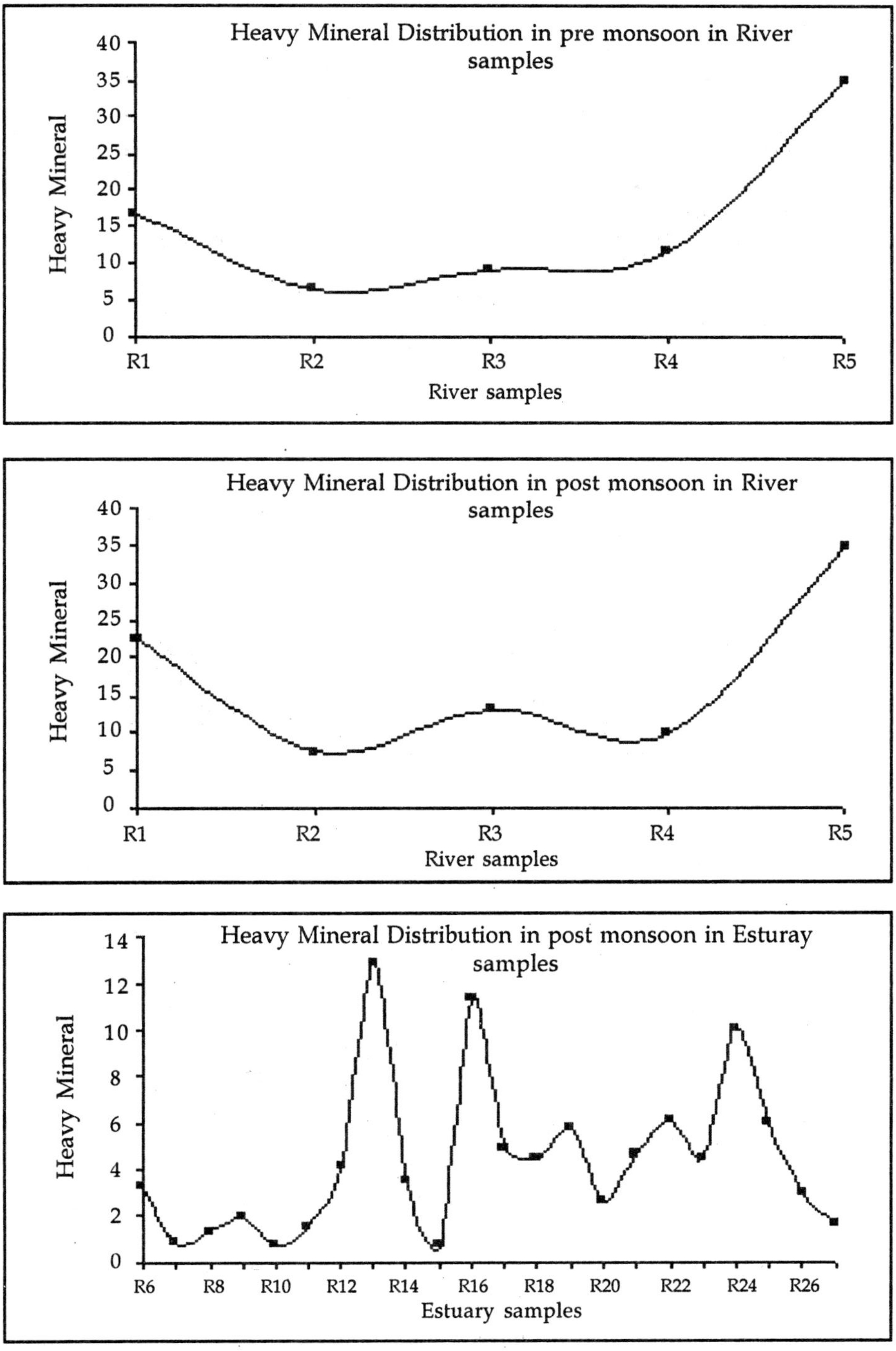

Fig. 2 : Heavy Mineral Distribution of pre-monsoon and post-monsoon of Tamiraparani River, estuary and marine samples.

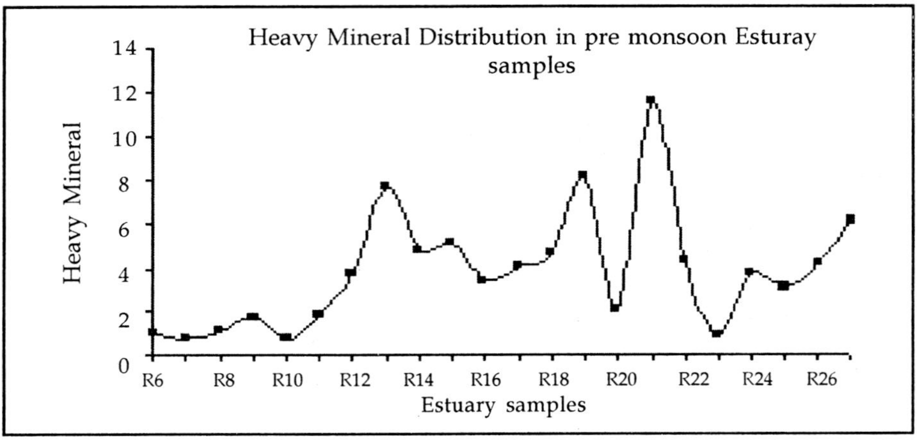

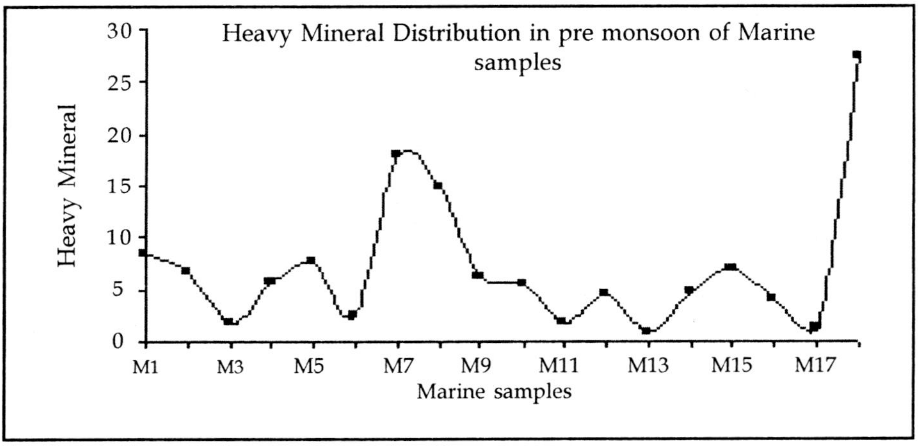

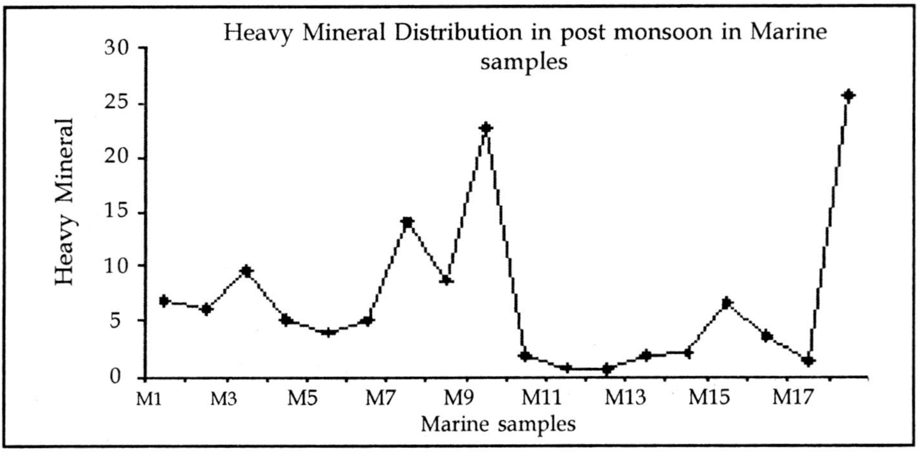

Fig. 2 : (Contd...) Heavy Mineral Distribution of pre-monsoon and post-monsoon of Tamiraparani River, estuary and marine samples.

The rounded edges of Ilmentite and mechanical abrasion suggest their prolonged transportation and the solution pits indicate their longer residential time in the basin. However, the Ilmenite of this area shows the features produced due to mechanical breaking. These are with irregular or regular breaking along or across cleavage plane, abrasional marks, mechanical 'V' marks, etc. Most of the Ilmenite grains are sub-angular to sub-rounded in shape. These mechanical features indicate Ilmenties are being transported in aqueous medium and they were not subjected to chemical processes in this region. Krinsley and Doornkamp (1973) have also reported these characteristic features for the paleo-sediments. Similarly, Gujar (1996), Mallik (1986) and Cherian (2003) have reported the same features off South Konkan coast, Kerala beach and Valinokkam - Tuticorin sector respectively.

Garnet

Garnet grains are angular to sub-angular with moderately high relief. Some angular grains show conchoidal fracture with embayments formed due to solution effect. As a result of etching, large depressions have formed within which precipitation structures are identified.

In most of the grains, impact features and etch marks prevail which designate considerable physical activity in the nearshore zone. The rounding has been caused mainly by transportation that, in turn, indicates polycyclic nature of sediments and change in energy conditions.

Zircon

The minerals are euhedral, broken, rounded in nature. Few varieties show well-developed crystal faces. A deep and shallow depression also noticed in zircon, may be due to various transport mechanism.

Sillimanite

The minerals are elongated in nature. Well developed crystal faces are noticed. Triangular pits are also observed. This indicates the mechanical actions taken place due to transportation.

Rutile

The grains are sub-rounded in nature. Due to long transportation some grains obtain rounding. It is frequently distributed along the investigated area.

Light Mineral SEM Studies

Quartz and feldspar are the principal light minerals present, among which quartz occupies the major part followed by orthoclase and plagioclase feldspar.

The SEM studies of quartz grains are applied to understand the post-depositional or digenetic history of the sediments (Plate-III & IV). During the processes of transportation and deposition, various micro-features are developed by mechanical and chemical processes and are influenced by physical and chemical properties of grains such as hardness, cleavage, solubility, tenacity, etc.

Based on the mode of formation, Rahman and Ahmed (1996) have classified various micro-textures into four groups. They are,

(i) Mechanical formation of conchoidal fracture, pitted surface, etc.,

(ii) Mechanical/ chemical process- these are responsible for textures like angular outline, relief features, V form etc.,

(iii) Dissolution chemical process- generates cavities, solution pits, etc.,

(iv) Precipitational chemical process- forms silica, precipitation and crystalline outgrowths.

The common features observed are conchoidal fractures with cavities and solution pitting, chemical etching marks. In the present study, the quartz grains of the marine and beaches show both conchoidal fractures and etched marks indicating high- energy environment as well as the longer stay of sediments in the depositional basin.

Grain Morphology

Few textural features were chosen to describe surfaces of the grain. These features constitute the most common surface morphologies of different mineral grains determined after reconnaissance observations and considerations of the extensive work done with grain surfaces by

Margolis and Krinsley (1974). The essential characteristics identifiable with each micro-feature are briefly discussed. Heiken (1972) and Honnorez and Kirst (1976) described the concept of overall grain morphology with respect to basaltic ash. For our study, selected grains were studied and photographed.

Conchoidal Fractures

These are distinguishing characteristics of glassy materials and quartz, and result from brittle deformation due to compressive contact between two surfaces. In the beach environment, this situation occurs during transport from different environments. Conchoidal fracture patterns vary from regular dish shapes (uncommon) to irregular elongate fan-like or trough-like depressions. The common parallel step-like fractures that curve around the conchoidal depression are thought to be expressions of planes of weakness in the glass similar to cleavage planes in crystals (Margolis and Krinsley, 1974). These featherlike fractures that produce a rippled surface texture may also be produced by acoustic wave phenomena (Kragelskii, 1965). Elongate conchoidal fractures are most evident on grain edges. The conchoidal fractures are shown in pre-monsoon samples (Plate I & II) and few in the post-monsoon samples in our study region.

V-Shaped Depressions

These are micropits which vary from triangular depressions to elongate grooves that widen in one direction. Chemical effects as well as abrasion quickly obscure these features. Many of the V-shaped depressions may be only several microns in maximum dimension. These are only evident under very high magnification. Two processes can account for the V-shaped depressions: 1) tangential impacts with sliding of one grain over another (Lawn and Wilshaw, 1975), and 2) chemical solution in areas of localized order or microlite development. Huang *et al.* (1980) considered them to be the result of glancing impacts. They appear to form on all types of tephra particles. In our study, V-shaped depressions are noticed more in post- monsoon samples (Plate III & IV).

Upturned Plates

These features are visible only at magnifications above about 1,000 X arrow. Plates resemble sub-parallel ridges that are slightly raised above the general surface. They form ridge like structures that are generally less than 10 μm in long dimension. Plate edges are smooth and generally of equal height due to solution/precipitation effects. Plates are found near edges of grains, especially on exposed surfaces as opposed to semi-protected hollows (Plate-I Fig. 8. depressions with sharp edges). Krinsley and Smalley (1973) have suggested that upturned plates are oriented along traces of cleavage planes in quartz and are likely cleavage plates. Their formation is due to failure along planes of weakness due to mechanical stress applied during impact or crushing.

Grooves

Grooves include elongate scratches and troughs that may be slightly curved. These grooves are oriented in a preferred direction, occur with conchoidal fracture, and appear in sets. Grooves are, however, relatively uncommon on the samples studied.

Cracks

Cracks are mainly due to mechanical process, and are straight or slightly curved. Separation along cracks generally is less than 10 μm. Cracks are best developed on vesicle surfaces and may radiate from equal angles in groups of two or four. Overall, these features appear similar to mud cracks and, when they intersect, form polygonal plates on surfaces; they appear to form both before and after conchoidal fracture. Cracks that project through a grain may be due to the thermal stress of quick cooling or by impact with another grain or surface. Cracks in the grain skin could be due to grain expansion after formation of a brittle skin or contraction of the hydrated skin often followed by accumulation of alteration materials. The grooves and depression are more in pre- monsoon samples in our study region than the post-monsoon (Plate- I, Fig. 12-14).

Chemical Alteration

Grains are sugary in appearance when totally altered, however, they retain their vesicular morphology. Solution and precipitation occur together on the same grain. Resulting textures include pitted or scalloped surfaces on a micron scale, rounding of upturned plates or other sharp features, and development of a frosted, light diffused surface as compared to the vitreous surface of fresh glass. Solution pits and silica precipitation are more in our samples (Plate-III, Fig. 17.

Solution pits due to chemical activities) in all the monsoons; this may be due to chemical alteration.

Discussions

The study mainly focused on the nature of sediments and its characteristics with respect to grain size and mineral distributions using SEM studies in pre- monsoon and post- monsoon of Tamiraparani and adjoining regions. In the Tuticorin samples, the sediments are medium grained; nevertheless, they are better sorted as a result of the prevailing high wave energy conditions there. At Tuticorin itself, however, the sediments are poorly sorted, even though their mean size is finer. The growth of the spit at Tuticorin impedes the movement of sediment-laden littoral currents; as a result, lateral updrift causes sediment deposition, especially in the northern part of the Tuticorin sector. In this process, the materials removed from the Tamiraparani River mouth are immediately deposited in the Tuticorin region, and no further sorting is possible. Besides the above, the deposition of fine sediments from the adjacent reefs may be a further factor causing the poor sorting of sediments. The better sorting at the remaining stations is probably due to the prevalence of wave convergence throughout the year and the finer size of the sediments.

The post-monsoon period of near by Tuticorin region has been identified as zone of erosion. Due to more wave energy updrift alongshore current whereas, in Tuticorin, the sediments in this marine region are medium to fine sand but only medium sand in river and estuary samples. The beach slope is steep and the width is less. The long shore current direction is NE/ SW direction. But, in Tuticorin, the zone of accumulation is taking place. Here the wave energy is decreased and suspended sediment transport become predominant. Due to the accumulation of sediments, it is identified a leaned zone of heavy mineral concentration. In river and estuary samples the sediments are medium sand and moderately sorted in nature.

In the marine samples, the nature of sediment is medium-to-fine, moderately-to-well sorted and positive symmetrically skewed sediments indicate, probably as a result of the influence of palaeo-sediments deposited by rivers from inland as well as by waves and currents from offshore. This fact is supported by Angusamy and Rajamanikam (2007). Further, in the post-monsoon the estuary sediments are medium to poorly sorted with positive symmetrical and fine skewed. Further, it indicates the deposits of near source region.

The mean size of the post-monsoon sediments on the marine region indicates the predominance of very fine sediments with an admixture of medium-grained sands. The preponderance of such fine sediments is probably due to the deposition and indicating the influence of aeolian activities in transporting fine sediments in suspension and of saltation from the adjoining coastal landforms as well as from estuaries.

In pre-monsoon, the sediments are medium to fine, moderately sorted and fine skewed indicates the erosional activities. Similar type of observations has been made by Cherien (2003). The erosion may be due to onshore and offshore movement of materials. In Tuticorin zone, the erosion may be due to the northward movement of sediments by longshore current. The beach morphology shows that during pre-monsoon the beach width is only 16m whereas in post-monsoon it is 26 m. The beach slope is 4° and 7° in post- and pre- monsoon respectively.

The distribution of heavy minerals is higher in river and marine sediments than in estuary in pre-monsoon compared to the post-monsoon. This may be due to erosional activities and high winnowing actions in pre-monsoon, the heavy mineral enrichment is favorable in marine sediments and in depositional periods of post- monsoon, the distribution of heavy minerals are lesser in amount may be due to long shore current actions the deposition of heavy are transported to northern side.

The predominance of garnet populations in beach, dune marine and river sands from Tuticorin, southern India, which almost exclusively comprise low grossular high-pyrope garnets. These are derived from the high-grade (granulite facies) metasedimentary and charnockitic rocks that form the basement in this area. During the monsoon periods, the lesser percentage of topaz, glaucophane, actinolite, sillimanite and kyanite helps to infer that the Tamiraparani river sediments may not be reaching these beaches in significant proportion (Loveson, 1994). Hence, in the estuary the distribution is lesser in amount in both the monsoons. Similar studies on the sediments of Vaippar River (Udayaganesan, 1993) as well as the beaches of northern Tamilnadu (Muthukrishnan, 1993; Mohan, 1995) have shown the negligible contribution of heavy minerals by the rivers to the respective beaches. The presence of strongly etched chlorites and garnet in this zone suggest either a longer distance transportation of sediments or a longer stay of sediments in the depositional basin.

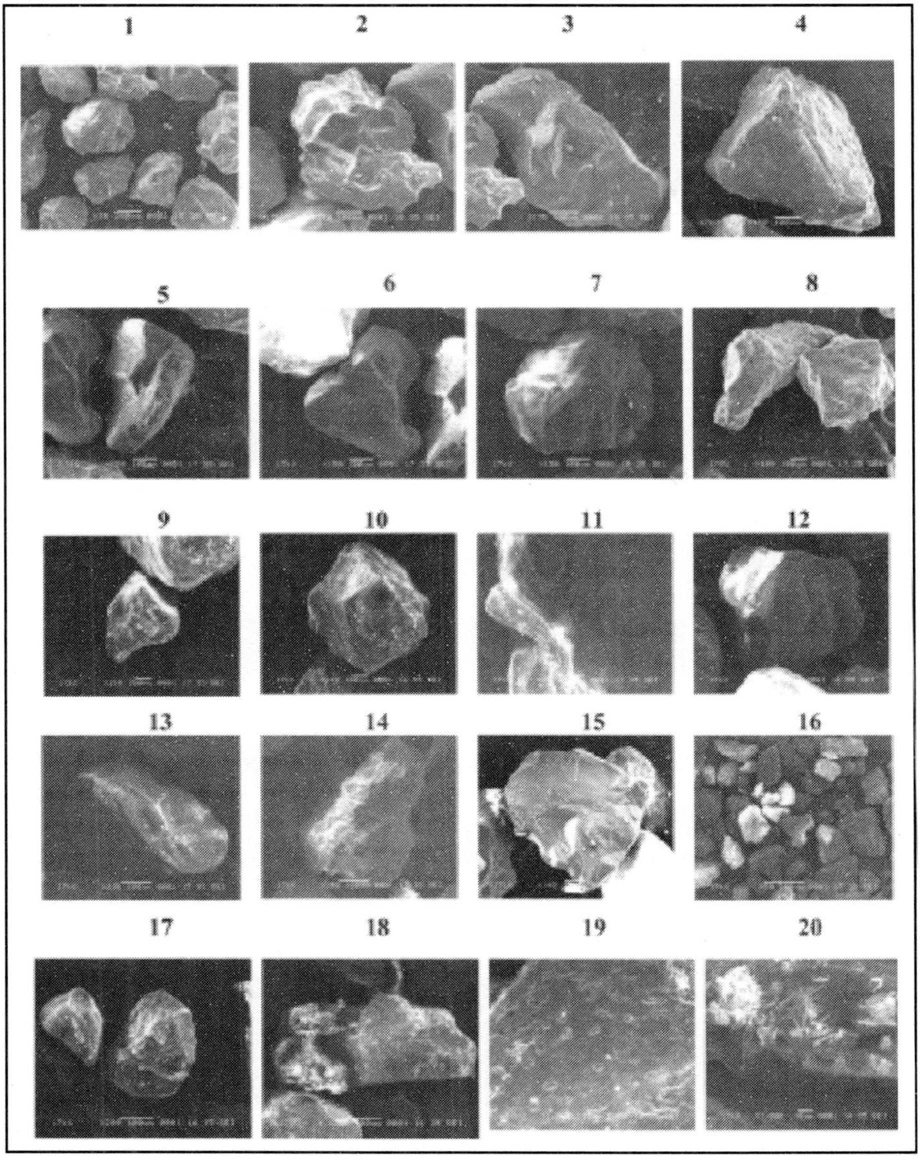

Plate I : Fig.1: Angular to sub-angular with mixed grains Fig.2: Conchoidal fractures like quartz Fig.3:V-marks and pitting at edges Fig.4: Solution pits and minor cavities Fig.5-6: Mechanical breakages at the edges Fig.7: Sub- rounded with small pits Fig. 8: Depressions with sharp edges Fig.9 -10: Precipitations with pits Fig.11: Sharp edges Fig.12: Cracks and fractures Fig.13-14: Cracks with depressions Fig.15: Triangular pits Fig.16: Sub-angular mixed grains Fig.17: Solution pits and precipitation with sub angular grains Fig.18: Precipitations structures inside the depressions Fig.19: Rectangular solution pits Fig.20: Fibrous needle with cracks.

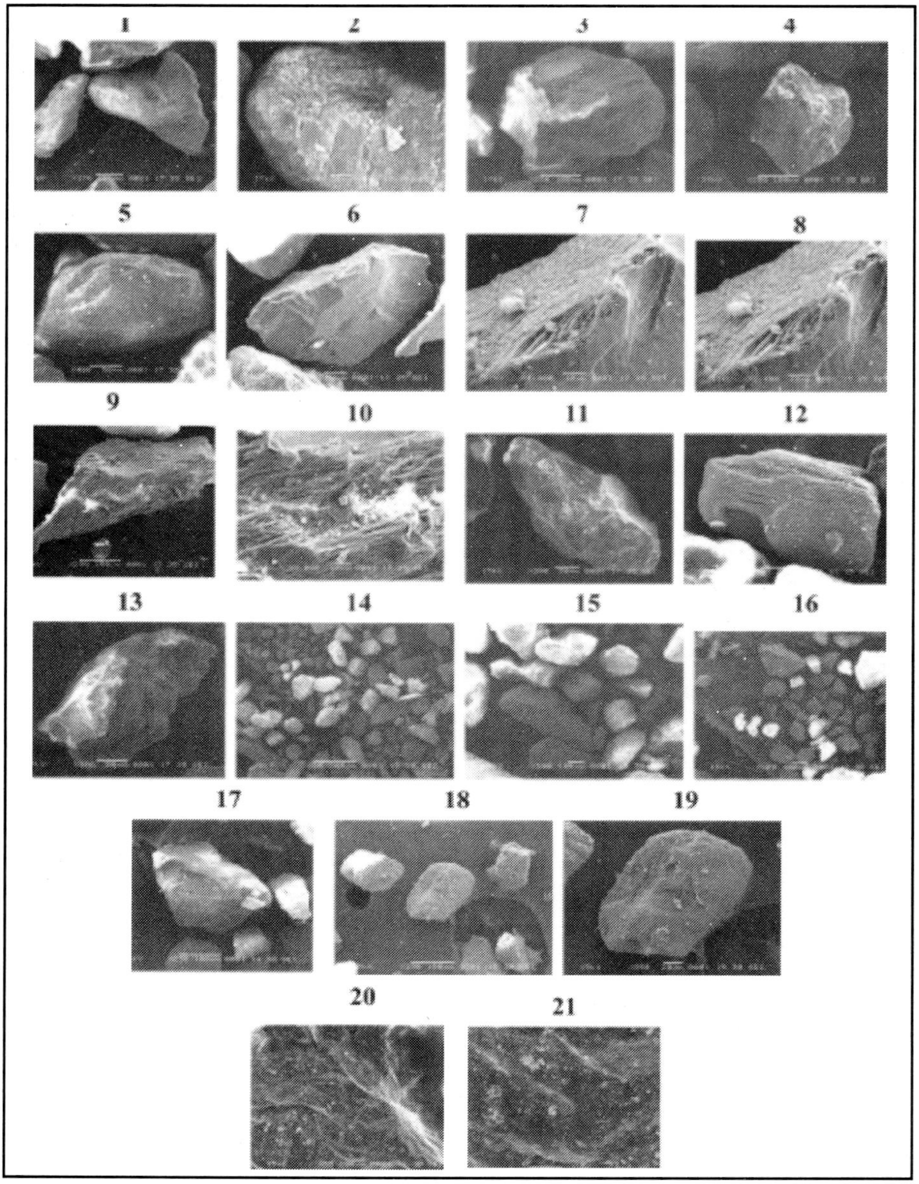

Plate II : Fig.1: Angular with overgrowth grains Fig.2: Solution pits and depressions due to chemical activities Fig.3: Surrounded grains Fig.4: Silica precipitates with sub angular Fig. 5: Sub-angular with shallow depressions Fig.6: Conchoidal fracture with sharp edges Fig.7-8: Suture surface with evenly cut cleavages Fig.9 -10: Long scratch with elongated cleavages Fig.11: Deep depressions Fig.12: Fibrous like straight cleavages Fig. 13:Conchoidal fracture Fig. 14-16: Uneven and sub angular to angular grains Fig. 17-19: Sub-angular grains Fig. 20-21: Solution pits and precipitation, enlarged view.

Plate III : Fig.1: Sub-angular grains Fig.2: Depressions at the surface Fig.3: Sharp angular grains with pits Fig.4: Solution pits Fig. 5-6: Over growth and elongated grains Fig. 7: Sharp and angular grains Fig.8-9: Conchoidal fractures Fig.10: V-shaped depression and fractures Fig.11: Sub-angular grains Fig.12: Depressed edges Fig.13: Precipitations and cracks Fig.14: Straight factures, Fig. 15: Sub- rounded grains Fig. 16: Small depression at the edges, Fig. 17: Solution pits due to chemical activities Fig. 18: Large Depressions due to solution effects Fig. 19: V-shaped depressions Fig.20: V-shaped depression with precipitations Fig.21: Dish shaped cavities Fig.22: Straight fractures with depressions Fig.23: Straight fractures Fig.24: Depression filled with precipitations.

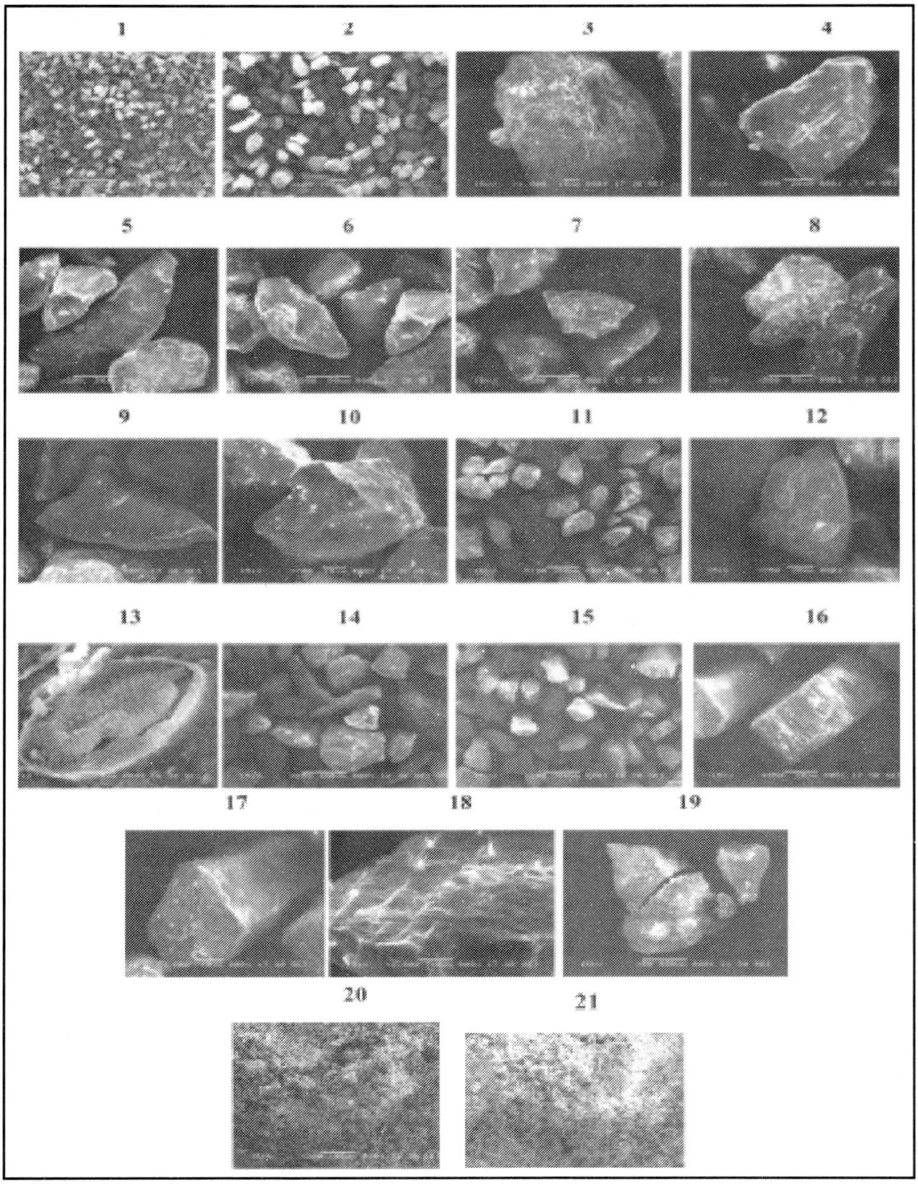

Plate IV : Fig.1: Sub-angular to sub rounded Fig. 3: Small pits due to solution Fig. 4: Craks with sub angular grains Fig. 5-7: Conchoidal fracture with solution pits Fig. 8: Depressions with solution pits Fig.9-10: Sub-angular grains Fig.11: Sub-angular to surrounded grains Fig.12: Angular with fossil like impression Fig.13: Diatoms like impression? Fig. 14-15: Sub- rounded to sub- angular grains Fig. 16: Disc shaped with vertical fractures Fig. 17: Elongated triangular grains Fig. 18: Step like furrows Fig.19: Sub- angular grains Fig.20-21: Solution pits due to chemical activities.

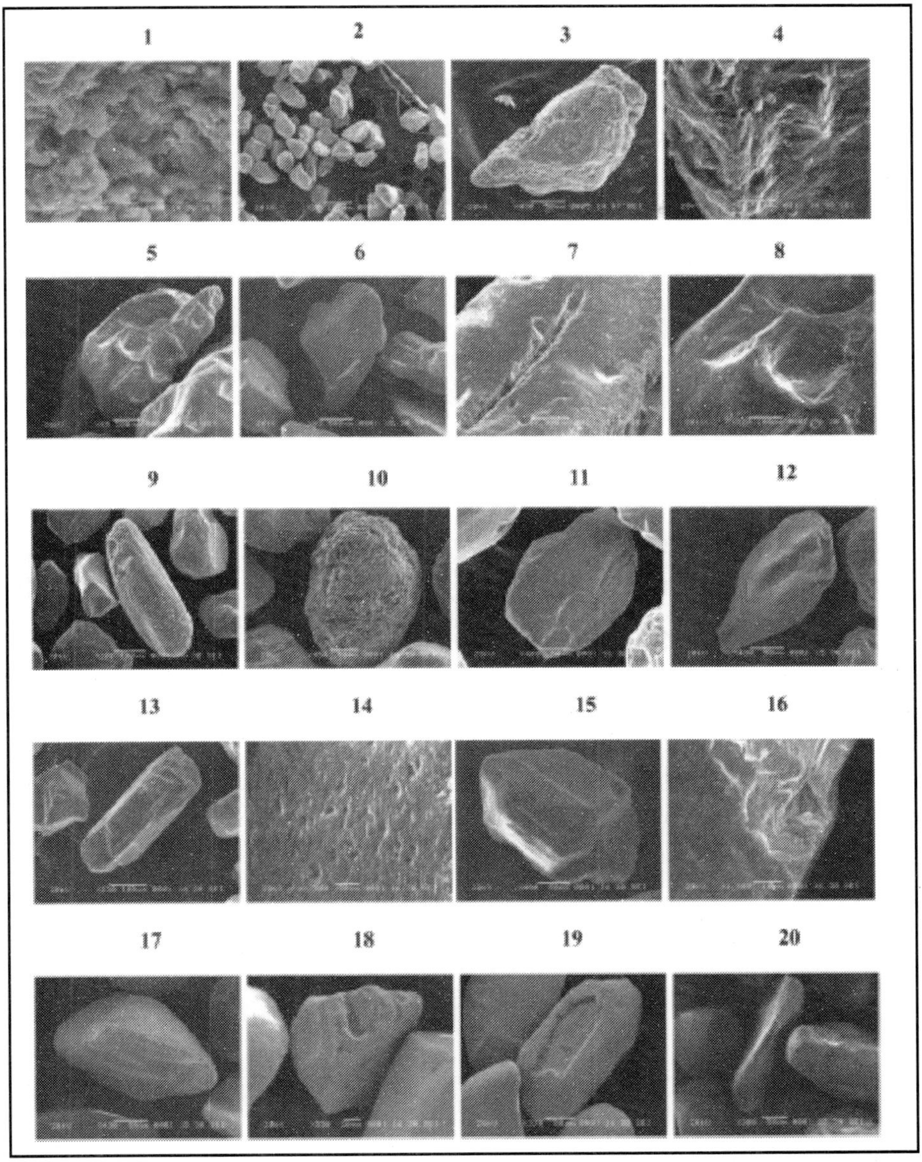

Plate V : Fig.1: Illmenite filled with precipitations Fig.2: Sub-rounded illmenite Fig.3: Illmenite with small pits Fig.4: Illmenite with surface cracks Fig. 5-6: Garnet with conchoidal fracture Fig.7: Garnet with cracks Fig. 8: Garnet with deep pits and V- shaped pits Fig. 9: Rutile with subangular to rounded Fig.10: Sub-rounded rutile with solution activities Fig. 11-12: Elongated rutile Fig.13: Sillimanite with crystal faces Fig.14: Sillimanite with solution pits Fig.15: Broken sillimanite with crystal faces Fig. 16: Sillimanite with triangular pits at the edges Fig.17: Sub-rounded zircon Fig.18: Zircon with surface depressions Fig.19: Elongated zircon with deeper depression by chemical action Fig. 20: Elongated zircon.

Conclusion

The SEM studies of placer minerals indicate that a number of micro-textures are developed by mechanical and chemical processes influenced by physical properties mainly cleavage. The SEM studies support that the sources of sediments are from crystalline rocks either from igneous - metamorphic rocks or paleosediments, that too from mixed sources from littoral currents and riverine and beach. The SEM study shows that in pre- monsoon, the morphological changes like pits, grooves, depressions etc. are lesser than the post- monsoon grain samples. The river and estuary grains show more or less similar patterns in both the monsoons. In the present study the quartz grains of the marine and beaches show both concordant fractures and etched marks indicating high - energy environment as well as the longer stay of sediments in the depositional basin.

Acknowledgement

We are thankful to Professor S.P. Mohan, Head, Department of Geology, University of Madras for providing SEM facilities and constant encouragement during this work. The first author wishes to express his gratefulness to Prof. V. Rajamanickam, Dean, Department of Disaster Management, SASTRA University, Thanjavur & former Head, Department of Earth Sciences, Tamil University, Thanjavur for interpretation in the Heavy mineral studies.

References

Angusamy, N. and Rajamanickam G.V. (2007) Coastal processes of Central Tamil Nadu, India: clues from grain size studies. Oceanologia, v.49 (1), pp. 41–57.

Cherian, A. (2003) Sedimentological studies in the beaches between Valinokkam and Tuticorin, Tamil Nadu. Ph.D. Thesis (Published), Tamil University, Thanjavur, pp120-189.

Gujar, A. R. (1996) Heavy mineral placers in the near shore areas of south Konkan Maharashtra: Their nature, distribution, origin and economic evaluation. Ph.D. Thesis published in Tamil University, Thanjavur, 234 p.

Heiken, G. (1972) Morphology and petrography of volcanic ashes. Geological Society of America Bulletin, v. 83, pp.1961-1988.

Heiken, G. (1974) An atlas of volcanic ash. Smithsonian Contributions to Earth Science, v.12, pp.1-101.

Honnorez, J. and Kirst, P. (1976) Submarine basaltic volcanism: Morphometric parameters for discriminating hyaloclastites from hyalotuffs. Bulletin of Volcanology, v. 34, pp.1-25.

Huang, T. C. and Watkins, N. D. (1976) Volcanic dust in deep-sea sediments: Relationship of microfeatures to explosivity estimates. Science, v.193, pp.576-579.

Huang, T. C., Varner, J. R. and Wilson, L. (1980) Micropits of volcanic glass shards: Laboratory simulation and possible origin. J. Volcanology and Geothermal Research, v. 8, pp.59-68.

Kragelskii, I. V. (1965) Friction and Wear. Butterworths, London, 225 p.

Krinsley, D. H. and Donahue, J. (1968) Environmental interpretation of sand grain surface textures by electron microscopy. Geological Society of America Bulletin, v. 79, pp.743-748.

Krinsley, D. H. and Doornkamp, J. (1973) Atlas of Quartz Sand Grain Textures: Cambridge University Press, Cambridge, England, 91 p.

Krinsley, D. H. and Smalley, I. J. (1973) Shape and nature of small sedimentary quartz particles. Science, v. 180, pp.1277-1279.

Kuenen, Ph. H., and Perdok, W. G. (1962) Frosting and defrosting of quartz sand grains. J. of Geology, v. 70, pp. 648-658.

Lawn, B. R. and Wilshaw, T. R. (1975) Review indentation fracture: principles and applications. J. of Material Science, v. 10, pp. 1049-1081.

Loveson, V.J. (1994) Geological and Geomorphological investigations related to sea level variation and heavy mineral accumulation along the Southern Tamilnadu Beaches, India, Ph.D. Thesis, Madurai Kamaraj University, Madurai, 223 p.

Mallik, T.K. (1986) Micromorphology of some placer minerals from Kerala beach, India. Marine Geology, v.71, pp.371-381.

Margolis, S. V. (1968) Electron microscopy of chemical solution and mechanical abrasion features on quartz sand grains. Sedimentary Geology, v. 2, pp. 243-256.

Margolis, S. V and Krinsley, D. H. (1974) Processes of formation and environmental occurrence of microfeatures of detrital quartz grains. American Journal of Science, v. 274, pp. 449-464.

Mohan, P.M. (1995) Distribution of heavy minerals in Parangipettai (Porto Novo) beach, Tamilnadu. J.Geol.Soc. India, v.46, pp.401-408.

Muthukrishnan, N. (1993) A study of detrital minerals from the sediments of Gadilam river basin, Tamilnadu. Unpublished M.Phil. Thesis, Tamil University, Thanjavur, 193 p.

Rahman, M.H. and Ahmed, F. (1996) Scanning electron Microscopy of quartz grain surface texture of Gondwana sediments Barapukyric Dinajpur, Bangladesh. J. Geol. Soc. India, v.47, pp.207-214.

Robson, D.A. (1984) Scanning electron micrographs of some common heavy minerals from the Sandstones of the Northumberland Trough. Trans. Nat. History. Soc., Northumbaria, v.52, pp.27-34.

Udayaganesan, P. (1993) A study of detrital minerals from the sediments of Vaippar basin, Tamilnadu. Unpub. M.Phil., Thesis, Tamil University, Thanjavur, 118 p.

Walker, G.P.L. and Croasdale, R. (1972) Characteristics of some basaltic pyroclastics: Bulletin Volcanologique, v. 35, pp. 305-317.

❑❑❑

10

New Trends in Gondwana Palaeobotany

A.K. SRIVASTAVA
Birbal Sahni Institute of Palaeobotany
53, University Road, Lucknow - 226007 (U.P.), India
E-mail : ashwinisrivastava@hotmail.com

Abstract

Discoveries of insect wings, insect activities in fossil flora and microbial association with plant fossils have helped to understand the interrelationship of plants and insects and role of mico-organisms in Gondwana flora.

Introduction

Taxonomy, systematics, origin, evolution and distribution of plants and their significance in stratigraphy, palaeoecology, palaeoclimatology and palaeogeography have traditionally been carried out by many workers in Gondwana sediments from different basins of India (Lakhanpal, *et al.* 1976; Chandra and Tewari, 1991). Recently, findings of insect wings, insect damaged plant fossils and evidences of fungal and bacterial assault in the flora have revealed the existence of well organized insect-plant relationship and mutualistic, parasitic and saprophytic role of microorganisms in Gondwana flora of India. The study provides a new trend in understanding the ecosystem dynamism, as evident in the extant flora.

Insect-Plant Association

Earliest evidence of insect-plant association in fossil flora is known from the Lower Devonian plant assemblages of Rhynie Chert (Smart and Hughes, 1973; Kevan *et al.*, 1975). It has been discussed that insects coevolved with the development of plant communities (Strong *et al.*, 1984) and systematic development of insect fauna and insect activities are evident in the subsequent flora of Carboniferous (Scott and Taylor, 1983; Scott *et al.*, 1992).

Feistmantel (1880) while describing the Lower Gondwana flora of Raniganj Coalfield doubted the presence of insect wing in the Gondwana sediments of India. Due to its uncertain nature he described the specimen as insect wing-like fragment of leaf (Feistmantel, 1880; Pl. 16, Figs.7, 7a).

Handlirsch (1906-1908) for the first time reported the well preserved and complete hind wing of cockroach, *Gondwanoblatta reticulata* from Gondwana equivalent sediments of extra-peninsular regions of Kashmir. Later Bana (1964), Verma (1967) reported insect wings from different Gondwana localities. Recent investigations carried out by Dutt (1977), Srivastava (1988a, b, 1996), Pant and Srivastava (1995) have revealed the presence of insect wings in association with plant fossils. On the basis of venation pattern wings are classified with insects belonging to Blattoidea, Homoptera, Mecoptera and Coleoptera groups of insect.

Plant fossils showing insect activities in the form of eaten/chewed leaves, mining activity, gall impressions, egg pouches over the surface of leaves, burrowing behaviour in stems, seeds (Srivastava, 1988a, b, 1996, 1998; Pant and Srivastava, 1995) and trace fossils of faecal pelletes with plant material and horizontal/vertical burrows possibly representing the locomotary behaviour of insect (De, 1990; Maheshwari and Bajpai, 1990; Chandra and Singh, 1996; Srivastava *et al.*, 1996) indicate the existence of mutualistic relationship of insect-plant in Gondwana flora of India. Insects used plants for feeding, shelter and reproductive purposes and in turn plants benefitted themselves in dispersal and propagation mechanisms.

The informations in hand suggest that there is enough scope to examine the insect-plant relationship in Gondwana flora of India. To study animal-plant interaction in flora there is a need to examine each and every fossil specimen. Lack of data in this discipline is probably due to identification of such features in fossils rather than their absence, for example leaves with severely chewed or eaten margin might well have been discarded because they were considered to have been badly preserved (Fig. 1).

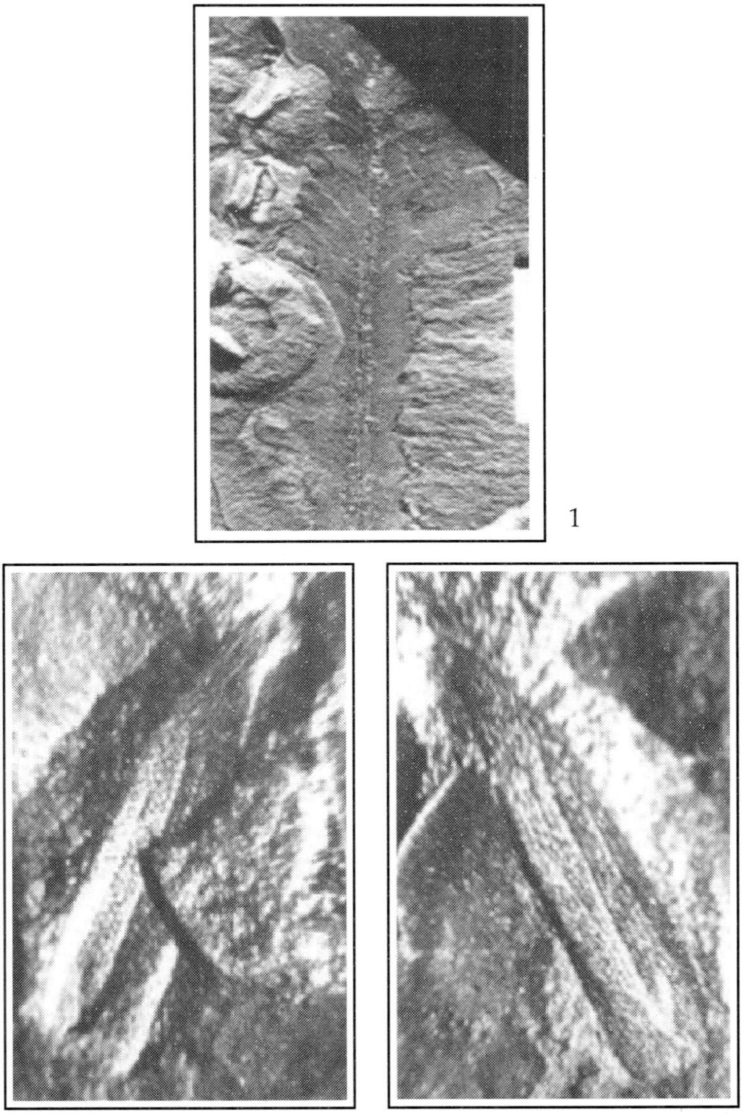

1 2 3

Fig. 1-3 : 1.*Glossopteris* leaf showing insect chewed/eaten margin. BSIP No. 36229 x 1 (Srivastava, 1988b). 2&3. Part and counterpart of coleopteran insect wing comparable with *Kaltanicupes* sp. BSIP No. 37041 Ca 7 (Srivastava, 1996).

Fungal Association

Investigations of fungal association in fossil floras have increased the interest of palaeo-mycologists in understanding sedimentology, palaeoecology, palaeontology and evolutionary dimensions of plants and environment (Pirozynsky, 1981; Pirozynsky and Malloch, 1975; Malloch *et al.*, 1980; Stubblefield and Taylor 1988). Fungal diversity, interaction and role of fungi in plant fossils as heterotrophs, decomposers, saprophytes, parasites/pathogens and in palaeoecological interpretations are considerably examined in the flora of Europe and America (Boullard and Lemoigne, 1971; Lyons, 1991).

Gondwana plants from Permian and Triassic deposits of Antarctica show good deal of fungal interaction (Osborn and Taylor, 1989). In India fungal hyphae and fruiting bodies are mostly known from post-Gondwana sequence of Tertiary age (Jain, 1974; Rao and Ramanujam, 1976; Kar and Saxena, 1976; Saxena and Khare, 1992). Petrographic analysis of Gondwana coals has revealed the presence of fungal spores (Pareek, 1958). Biradar and Bonde (1976) have described *Rhizoctonia-a* deuteromycetaceous fungi in pith region of gymnospermous wood from Upper Permian beds of Maharashtra. The occurrence of fungi over the cuticular surface of *Glossopteris* and *Thinnfeldia* is known but apparently interference with the host plant is not known.

Firm evidence of fungal interaction with host plant showing biological response has been observed in the leaf specimens of *Glossopteris* and *Noeggerathiopsis* (Fig. 4& 5). The leaves are preserved as coalfield compression and externally bear small irregular shaped spots over the surface. Scanning Electron Microscopic (SEM) study of infected portion of leaves show number of fungal hyphae and fruiting bodies in association with degraded and damaged cellular tissues of plant (Srivastava, 1993). The specimens signify the direct bearing of fungi with host plant and possibly suggest the parasitic or saprophytic nature of fungi. The example proposes to critically evaluate the Gondwana flora in terms of palaeomycology because plants ought to be affected by fungi for various types of biological and geological processes.

Bacterial Association

Bacteria play an active role in the formation and evolution of the bisophere and bacterial activity results into the genesis of different types of ores. Biodegradation process associated with bacteria help in decomposition of organic matter to release fixed carbon as a requisite step in the carbon cycle.

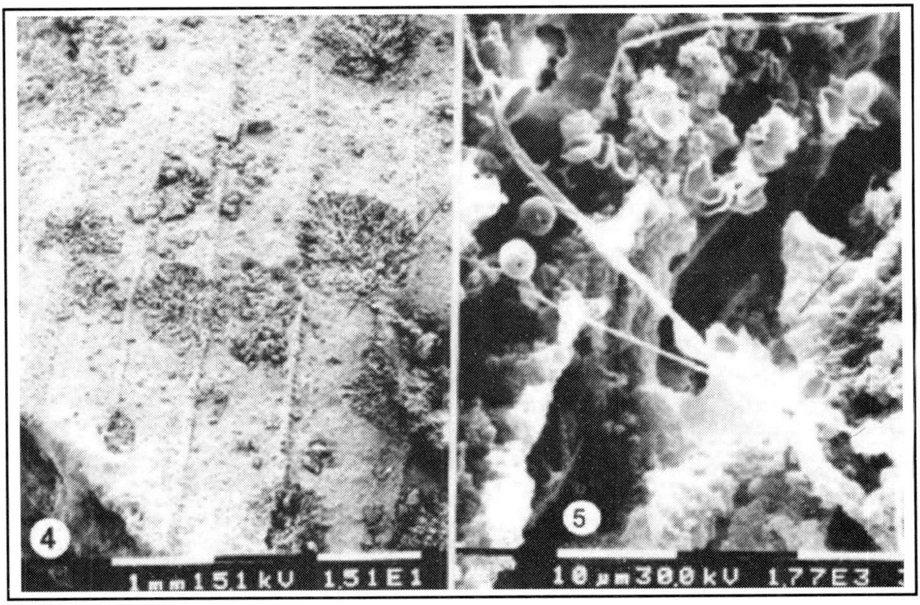

Fig. 4-5 : 4. *Neoggerathiopsis* leaf under SEM to show irregular spots over the surface (Srivastava, 1993) x 250. 5. SEM photograph of infected leaf showing fungal hyphae, spoes and degraded cellular tissues (Srivastava, 1993) x 2500.

It is difficult to identify bacterial association with plants in fossils because of their small size and difficulty in identifying them from other smaller organisms like cynophytes and fungi. Sometimes artefacts formed during the preparation of sample also abduct their identification. However, bacterial colonies are known to exist in sediments ranging from Pre-Cambrian to Recent.

The decay of plant material by bacteria and fungi is well acclaimed in Indian Gondwana (Bobde, 1979) but study of bacteria or sign of bacterial degradational product has not been attempted positively in the Gondwana flora. Lower Gondwana plant fossils investigated from Jharia Coalfield, Bihar show their faithfulness towards an understanding of bacterial association with plants. Cuticular pieces recovered from *Scutum*- type glossopterid fructification, under SEM study indicates the presence of bacterial colony and degraded cellular tissues (Fig. 6&10). Quite often marred surface of seed cuticle with plaque/pit-like structures demonstrate association of rod shaped bacteria (Srivastava and Tewari, 1994). The specimen shows the testimony of bacterial association and bacterial degradation in Gondwana flora of India. The information derived from such specimens suggests that to examine bacterial association in Gondwana plants the structural features especially the cuticles of leaf and fructifications be studied at higher resolutions preferably under Scanning Electron Microscope.

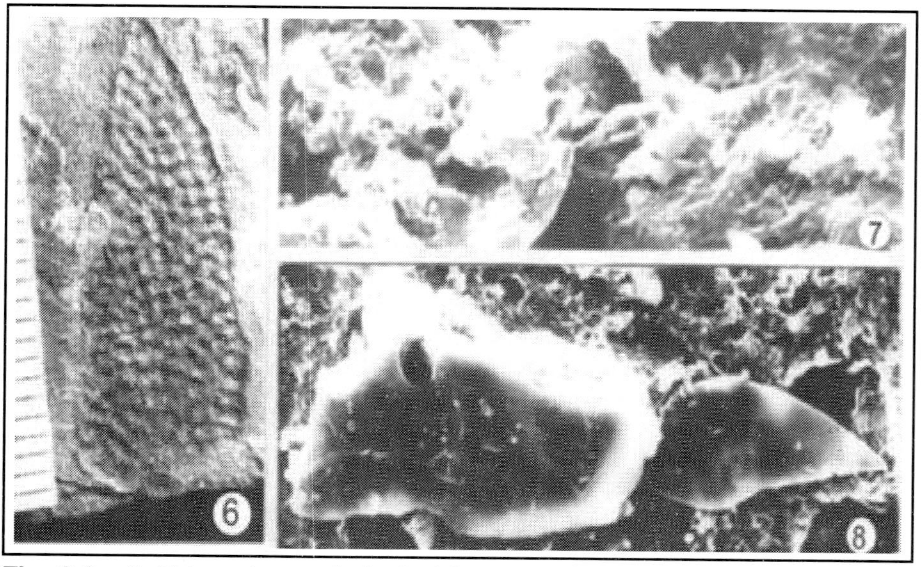

Fig. 6-8 : 6. External morphological features of *Scutum* – type glossopterid fructification x 2 (Srivastava and Tewari, 1994). 7. SEM photograph of fructification cuticle showing damaged and degraded cellular tissues x 1500 (Srivastava and Tewari, 1994). 8. SEM photograph showing seed/ovule possessing plaque, pit-like structures with rod shaped bacterial colony x 1000 (Srivastava and Tewari,1994).

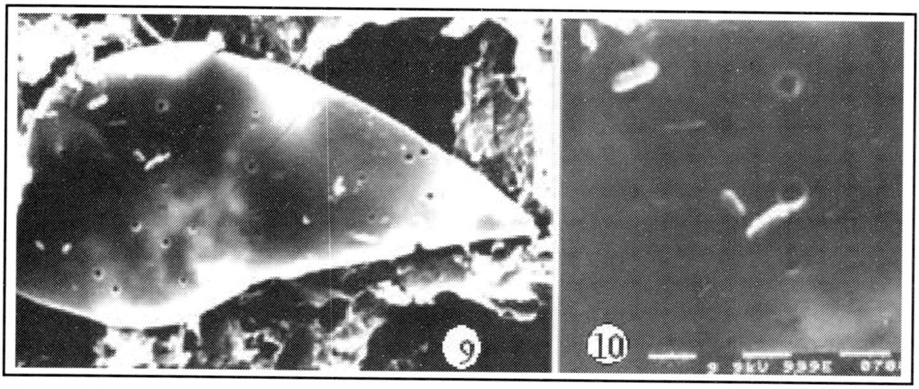

Fig. 9-10 : 9. Enlargement of photo of fig. 8 to show the details of seed cuticle and bacteria x 12000. 10. Further enlargement to show the nature of rod shaped bacteria x 2500 (Srivastava and Tewari, 1994).

The knowledge of Indian Gondwana flora has greatly advanced with the established practice of morphology and taxonomy. Nevertheless, it is imperative to make sincere approach to examine the flora to find out the insect activities, trace fossils, association of micro-organisms and degradational phenomenon of the flora. The study will help in evaluating the interaction of plants with animals, microorganisms and in discussing the evolutionary and environmental stress during Gondwana in India.

Acknowledgement

Thanks are due to Dr N.C.Mehrotra, Director, Birbal Sahni Institute of Palaeobotany, Lucknow for permission to send the paper for publication.

References

Bana, H. R. (1964) *Prognoblattina collumbiana* Schudder from Permo-Carboniferous beds near Srinagar, Kashmir, India. Int. geol. Cong. 22nd Session, New Delhi, Pt vii, pp. 274-281.

Biradar, N.V. and Bonde, S.D. (1976) Maharashtra Vigyan Mandir Patrika, v. 11, pp. 210-213.

Bobde, S.P. (1979) Environmental conditions of deposition during the Barakar Period in India-with special reference to paucity of animal fossils. Proc. IV th Gond. Symp. Geol Surv. Ind., Calcutta 1977, v.1, pp. 327-335.

Boullard, B. and Lemoigne, Y. (1971) les champignons endophytes du Rhynia gwynnevaughanii. *K. et al.* Le Botaniste, v.54, pp. 49-89.

Chandra, S. and Singh, K. J. (1996) Plant fossils from the type locality of Talchir Formation and evidence of earliest evidence of plant-animal activity in Gondwana of India. Gondwana Nine, v.1, pp. 397-414.

Chandra, S. and Tewari, R. (1991) A catalogue of Indian Fossil Plants - Part 2. Palaeozoic and Mesozoic megafossils, Birbal Sahni Institute of Palaeobotany, Lucknow.

De, C. (1990) Upper Barakar lebensspurren from Hazaribagh, India. J. Geol. Soc. Ind., v.36, pp. 430-438.

Dutt, A.B. (1977) *Rajharablatta laskarii* a new Lower Gondwana fossil insect from Daltonganj Coalfield. Bihar. Rec. Geol Surv. India, v.108, pp. 167-169.

Feistmantel, O. (1880) The fossil flora of the Gondwana System: Flora of Damuda-Panchet divisions. Mem. Geol. Surv. India Palaeont. Indica, Ser 123, pp 1-77.

Handlirsch, A. (1906-1908) Die fossilen Insekten und Phylogenie der Rezenten Formen. W. Engelman. Leipzig, pp. 351-352.

Jain, K.P. (1974) Fossil fungi: Aspects and appraisal of Indian Palaeobotany. In: *Surange, K.R., Lakhanpal, R.N. and Bharadwaj, D.C. (eds.)*: 38-46 Birbal Sahni Institute of Palaeobotany, Lucknow.

Kar, R. K. and Saxena, R. K. (1976) Algal and fungal microfossils from Matanomadh Formation (Palaeocene), Kutch, India. Palaeobotanist, v. 23, pp. 1-15.

Kevan, P.G., Chaloner, W.G. and Savile, D.B.O. (1975) Interrelationships of early arthropods and plants. Palaeontology, v. 18, pp. 391-417.

Lakhanpal, R.N., Maheshwari, H.K. and Awasthi, N (1976) A Catalogue of Indian Fossil Plants. Birbal Sahni Institute of Palaeobotany, Lucknow.

Lyons, P. (1991) Bacteria-like bodies in coalfield Carboniferous xylem enigmatic microspheroids or possible evidence of microbial saprophytes in a vitrinite precursor? Int. J. Coal Geol., V.18, pp. 293-303.

Malloch, D.W., Pirozynsky, K.A. and Raven, P.H. (1980) Ecological and evolutionary significance of mycorrhizal sybioses in vascular plants (a review). Proc. Nat. Acad. Sci., USA, v. 77, pp 2113-2114.

Maheshwari, H.K. and Bajpai, U. (1990) Trace fossils from the Permian Gondwana of Rajmahal Hills. Geophytology, v. 20, pp. 45-47.

Osborn, J.M. and Taylor, T.N. (1989) *Palaeolibulus* gen. nov A clamp bearing fungus from the Triassic of Antarctica. Mycologia, v. 81, pp. 622-626.

Pareek HS (1958) Fungal bodies in Talchir coals. J. palaeont. Soc. India, v. 3, pp. 45-50.

Pant, D.D. and Srivastava, P.C. (1995) Lower Gondwana insect remains and evidence of Insect Plant interaction. Proc. Int. Conf. Global Environment and Diversification of plants through Geological Time. Allahabad, pp. 317-326.

Pirozynsky, K.A. (1981) Interactions between fungi and plants through the ages. Can. J. Bot., v.54, pp. 1824-1827.

Pirozynsky, K.A. and Malloch, D.W. (1975) The origin of land plants: a matter of mycotrophism. Bio Systems, v.6, pp. 153-164.

Rao, C.N. (1959) Fossil insects from the Gondwana the Gondwanas of India. Indian Minerals, v. 13, pp. 3-5.

Rao, K.P. and Ramanujam, C.G.K. (1976) A further record of microthyraceous fungi from the Neogene deposits of Kerala in South India. Geophytology, v. 6, pp. 98-104.

Saxena, R.K. and Khare, S. (1992) Fungal remains from the Neyveli Formation of Tiruchirapalli District, Tamil Nadu, India. Geophytology, v. 21, pp. 37-43.

Scott, A.C. and Taylor, T.N. (1983) Plant/Animal interactions during the Upper Carboniferous. Bot. Rev., v. 49, pp. 259-307.

Scott, A.C., Stephensen, J. and Chaloner, W.G. (1992) Interaction and coevolution of plants and arthropods during the Palaeozoic and Mesozoic. Phil. Trans. R. Soc. Lond., v. B335, pp.129-163.

Smart, J. and Hughes, N.F. (1973) The insect and the plant: Progressive palaeoecological integration. In: *van Emden HF (ed.)* Insect plant relationship. Symp. Roy. Ent. Soc. London, 6. Blackwell, London, England, pp.143-155.

Srivastava, A.K. (1988a) An insect wing from the Lower Gondwana of India. J. Paleont., v. 62, pp. 827-828.

Srivastava, A.K. (1988b) Lower Baraker flora of Raniganj Coalfield and Insect/ Plant relationship. Palaeobotanist, v. 36, pp. 138-142.

Srivastava, A.K. (1993) Evidence of fungal parasitism in the Glossopteris flora of India. Comte Rendu XII ICCP Buenos Aires, pp. 141-146.

Srivastava, A.K. (1996) Plant / Animal relationship in the Lower Gondwanas of India. Ninth International Gondwana Symposium, Hyderabad, India, January 1994. Gondwana Nine, v.1, pp.549-555.

Srivastava, A.K. (1998) Fossil records of Insects and Insects related plant damage in India. Zoos' Print, v. 13, pp. 5-9.

Srivastava, A.K., Chandra, S. and Singh, K.J. (1996) Trace fossils from Talchir Formation, Talchir Coalfield, Orissa. Geophytology, v.24, pp. 35-40.

Srivastava, A.K. and Tewari, R. (1994) Possible evidence of bacterial degradation in Glossopteris flora of India. Palaeobotanist, v. 35, pp. 174-177.

Stubblefield, S.P. and Taylor, T.M. (1988) Tansley Review No. 12. Recent advances in palaeomycology. New Phytol., v. 108, pp. 3-25.

Strong, D.R., Lawton, J.H. and South, Wood S.R. (1984) Insects on plants: Community pattern and mechanisms. Harvard University Press, Cambridge, Massachusetts.

Verma, K.K. (1967) A new fossil insect from the Lower Gondwana of Kashmir. Curr. Sci., v.36, pp. 338-340.

□□□

11

Brissid Echinoid *Eupatagus* L. Agassiz, 1847, from the Khuiala Formation, Jaisalmer District, Rajasthan, India

D. K. SRIVASTAVA[1] and HUKAM SINGH[2]
[1]Centre of Advanced Study in Geology, University of Lucknow, Lucknow – 226 007, India
[2] Birbal Sahni Institute of Palaeobotany, Lucknow – 226 007, India
Email:sirdkdr@gmail.com

Abstract

The spatangoid genus *Eupatagus* L. Agassiz, 1847 is recorded, for the first time, from the sediments of the Khuiala Formation (Early Eocene) exposed near Habur, Jaisalmer district, Rajasthan, India. The specimens are placed in a new species characterised by its large, oval test with mild frontal sinus; convex aboral surface; flat oral surface; anteriorly eccentric ethmolytic apical system; anteriorly eccentric kidney- shaped peristome; long, petaloid ambulacral petals; longitudinally oval periproct at the posterior truncation and imperforate, non crenulated tubercles.

Introduction

The paper records, for the first time, a brissid echinoid *Eupatagus* L. Agassiz, 1847 (Fischer,1966 and Smith, 2008) from the sediments of the Khuiala Formation (Early Eocene) exposed at about 625 m west of

Habur, Jaisalmer district, Rajasthan, India (Fig.1) as *Eupatagus rajasthanensis* n. sp. The field work was carried out by the second author (HS) jointly with Dr. J. S. Guleria, Birbal Sahni Institute of Palaeobotany, Lucknow, while the palaeontological work has been carried out by the first author (DKS). Earlier, the three echinoid genera recorded from Rajasthan (Sahni, 1955; Sahni and Bhatnagar, 1958; Srivastava and Mathur, 1996 and Srivastava *et al.*, 2008) are *Recrosalenia* (*R. jaisalmerensis* from the Jurassic sediments), *Echinocyamus* (*E. jaisalmerensis* from the Middle Eocene sediments) and *Megapneustes* (*M. jaisalmerensis* from the Early Eocene). However, the genus *Eupatagus* L. Agassiz, 1847 has so far been reported in India from the Tertiary rocks of Kachchh (Duncan and Sladen, 1883; Srivastava, 1981; Srivastava, 2004) and Assam (Spengler, 1923) only.

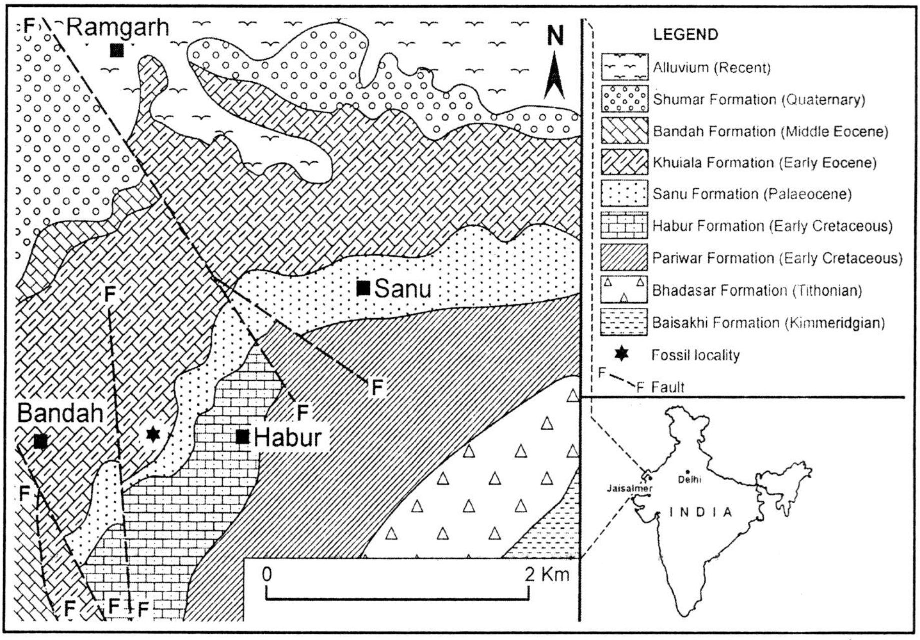

Fig. 1: Geological map of the area showing the fossil locality (after Singh, 2007)

Geology of The Area

The isolated outcrops of the fossiliferous Mesozoic and the Cenozoic formations in the Jaisalmer Basin are interspersed in the southeastern part of the basin within the desert sands (Singh, 2006). The Tertiary sequence in the Jaisalmer Basin (Das Gupta, 1974; Pareek, 1984; Singh, 1996; Singh, 2003; Sinha *et al.*, 1998; Bhandari, 1999 and Singh, 2007) is represented by the Sanu, Khuiala, Bandha and Shumar formations in ascending order. The stratigraphic succession of the studied area is given in Table 1.

Table 1: Stratigraphic succession of the Palaeogene sediments in Jaisalmer area, Rajasthan, India

Age	Formation	Member (surface)	Member (subsurface)
Pleistocene To sub - Recent	Shumar		
Middle Eocene	Bandah	Bakhri tibba Batrewala Limestone	Bakhri tibba Batrewala Limestone
Early Eocene	Khuiala	Khinsar	Upper khinsar Sirhera Lower khinsar
		Tetakkar Limestone	Tetakkar Limestone
Palaeocene	Sanu		Kharatar
		Mohammaed Dhani	Mohammaed Dhani
Cretaceous	Abur/Pariwar		

The age of echinoid-bearing horizon of the Khuiala Formation (calcareous sandy marl) is constrained by the presence of index larger foraminifer *Assilina lacunata* Cizancourt (Srivastava *et al.*, 2008) which indicates an Early Eocene (Ypresian) age. The ostracode, bivalve and other biotic elements previously described from the Khuiala Formation (Khosla, 1972; Das Gupta, 1974; Pareek, 1984; Ghosh, 1987; Bhandari, 1995; Singh, 1996; Singh, 1997 and Singh, 2003) are also consistent with this age assignment.

Systematic Palaeontology

(Fischer, 1966)

Order:	Spatangoida Clauss, 1876
Family:	Brissidae Gray, 1855
Genus:	*Eupatagus* L. Agassiz, 1847
Subgenus:	*Eupatagus* L. Agassiz, 1847

Eupatagus (*Eupatagus*) *rajasthanensis* sp. nov.

(Figure 2. a-g)

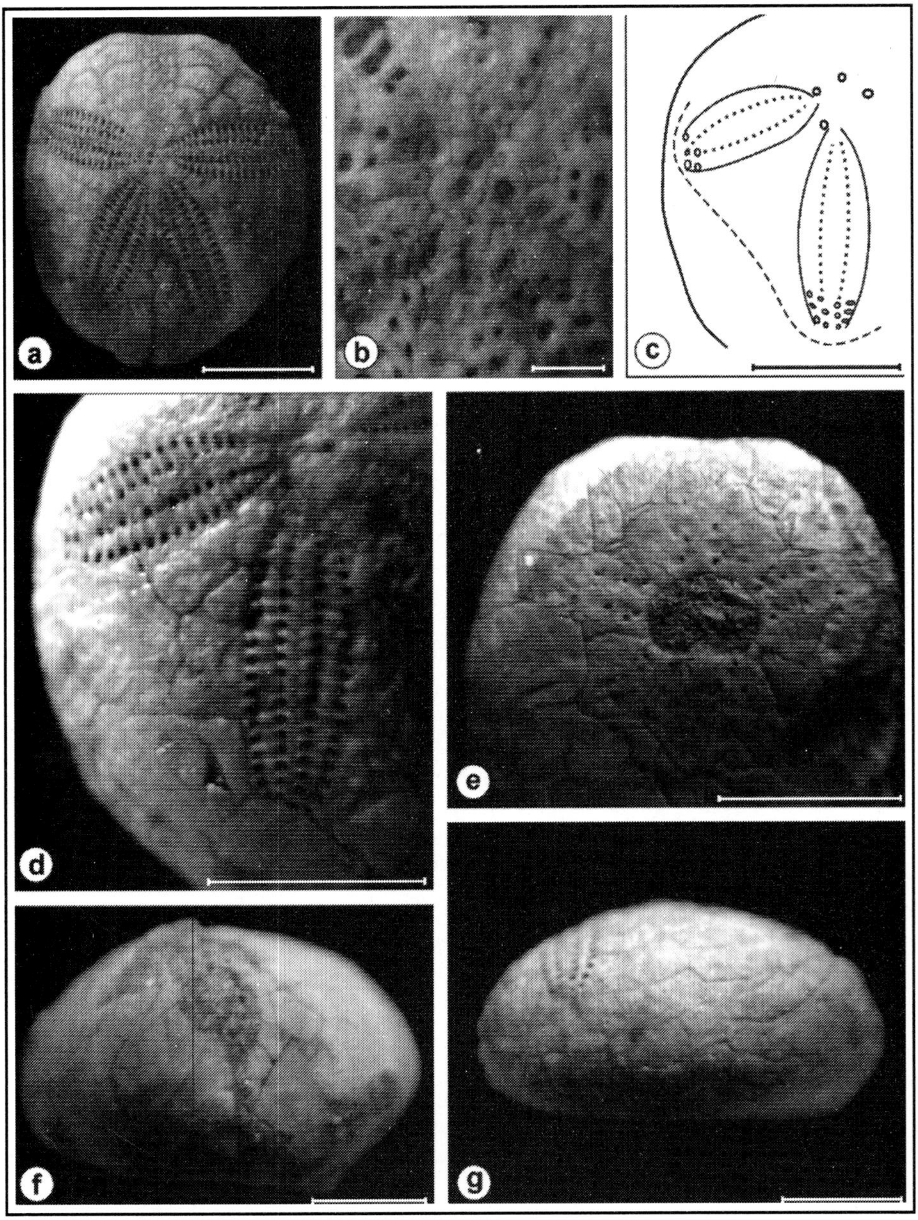

Fig. 2 : a-g *Eupatagus* (*Eupatagus*) *rajasthanensis* n. sp. (Scale bars = 5.0 mm otherwise as stated) (a) Aboral view [7097/BSIP 39528 (Holotype)]; (b) Apical disc (bar = 1.0 mm) [7097/BSIP 39530 (Paratype)]; (c) Line drawing showing peripetalous fasciole; (d) Aboral view showing ambulacral plates in petals IV & V [7097/BSIP 39529 (Paratype)]; (e) Oral view showing plates around peristome [7097/BSIP 39531(Paratype)]; (f) Posterior view showing periproct [7097/BSIP 39532 (Paratype)]; (g) Lateral view [7097/BSIP 39528 (Holotype)]

Material: 20 specimens, few broken and incomplete but preservation is fairly good; (Holotype: 7097/BSIP/39528; Paratypes: 7097/BSIP/39529 to 7097/BSIP/39547).

Derivation of name: The species has been named after Rajasthan State (area of its occurrence), India.

Diagnosis: The species is characterized by its large, oval test with mild frontal sinus; convex aboral surface; flat oral surface; anteriorly eccentric ethmolytic apical system; anteriorly eccentric kidney-shaped peristome; long, petaloid ambulacral petals; longitudinally oval periproct at the posterior truncation and imperforate, non crenulated tubercles.

Description: The specimens are characterized by a large, oval test with mild frontal sinus, convex aboral surface, flat oral surface, highest point of the test at the centre of the test, margin tumid and well rounded. The test is longer than wide and wider than high, truncated towards posterior side and moderately keeled. Apical system anteriorly eccentric, ethmolytic with four small and circular genital pores, the two anterior genital pores are closer than the posterior ones. Paired ambulacra petaloid and flush with the test. Petal III indistinct, petals I & V longer than the petals II & IV. Poriferous zones slightly depressed and consists of inner circular and oval to slit like outer pores which are conjugated by a deep groove. Periproct longitudinally oval, lies on the vertical posterior truncation and not visible from the aboral side. Peristome on oral surface, anteriorly eccentric and kidney-shaped. The peripetalous fasciole passes around the petals and do not indent in the interambulacral areas. The subanal fasciole, however, is not visible. The test is ornamented with imperforate, non-crenulated tubercles which are sunken in small scrobicules. These are large, scarce and circumscribed by the peripetalous fasciole on the aboral side while orally these are dense and small.

Dimensions (in mm)

Specimen No.	Length (L)	Breadth (B)	Height (H)	Ratio between L, B & H		
7097/BSIP 39528 (Holotype)	17.82	15.42	09.15	1.0	0.86	0.51
7097/BSIP 39529 (Paratype)	20.64	17.73	09.52	1.0	0.85	0.41

Remarks: The new species is close to *E. alatus* Arnold and Clark described from the Eocene sediments of Jamaica (Arnold and Clark,

1927) but it differs in having longer petals I & V, posteriorly protruded test and lesser number of tubercles. It also differs from *Eupatagus* cf. *antillarum* (Cotteau) reported from the mid-middle to low-late Eocene sediments of Clarimont Formation, White Limestone Group, Jamaica (Donovan and Rowe, 2000) in having less broad oval test which is broadest in the middle and more anteriorly eccentric apical system. *Eupatagus* (*Eupatagus*) *rajasthanensis* n. sp. differs from *Eupatagus* (*Eupatagus*) *rostratus* (d'Archiac) reported from the Eocene and Oligocene (Srivastava, 2004) and Oligocene and Miocene rocks of Kachchh (Duncan and Sladen, 1883) in not having flexuous petals II & IV. It also differs from *Eupatagus* (*Eupatagus*) *singhi* Srivastava described from the Middle Oligocene sediments of Kachchh (Srivastava, 1981) in having moderate frontal sinus.

Type locality: About 625 m west of Habur, Jaisalmer district, Rajasthan, India.

Type horizon: Khuiala Formation (early Eocene).

Repository

All described, undescribed and photographed specimens of fossil echinoids (Holotype: 7097/BSIP/39528; Paratypes: 7097/BSIP/39529 to 7097/BSIP/39547) have been deposited in the Museum, Birbal Sahni Institute of Palaeobotany, Lucknow, India.

Acknowledgements

The authors express a deep sense of gratitude to Prof. A. K. Jauhri, Head, Centre of Advanced Study in Geology, University of Lucknow, Lucknow and to Dr. N. C. Mehrotra, Director, Birbal Sahni Institute of Palaeobotany, Lucknow for providing facilities and encouragement. Dr. A. B. Smith (Natural History Museum, London) is thanked for critically reading through the original manuscript and offering helpful comments. They are obliged to Prof. Ashok Sahni, Panjab University, Chandigarh for suggestions. The financial assistance granted to the first author (DKS) by the Department of Science and Technology, New Delhi for the laboratory work is thankfully acknowledged.

References

Arnold, B. W. and Clark, H. L. (1927) Jamaican fossil echini, with descriptions of new species of Cainozoic echinoidea by H.L.Hawkins. Memoirs of the Museum of Comparative Zoology (Harvard), v. 50(1), pp. 63.

Bhandari, A. (1995) Early Eocene ostracodes from the subsurface of Jaisalmer Basin, Rajasthan. Geosci. Jour., v. 15, pp. 73-99.

Bhandari, A. (1999) Phanerozoic Stratigraphy of Western Rajasthan: A review. In: *Kataria, P. (editor)*, Geology of Rajasthan-Status of Prespective, Proceedings of a Seminar (A.B. Roy Felicitation volume), pp. 126-174.

Das Gupta, S. K. (1974) Stratigraphy of western Rajasthan Shelf. Proceedings IV Colloquium Indian Micropaleontology and Stratigraphy, Dehradun, pp. 219-233.

Donovan, S. K. and Rowe, C. D. (2000) Spatangoid echinoids from Eocene of Jamaica. J. Plaeont., v. 74(4), pp. 654-661.

Duncan, P. M. and Sladen, W. P. (1883) The fossil Echinoidea of Kutch and Kattyawar. Palaeontologia Indica, v. 14, 104p, 13 pls.

Fischer, A. G. (1966) Spatangoids. In: *Moore, R. C. et al. (eds.)*, Treatise on Invertebrate Palaeontology, Part U(3), Echinodermata 3, Echinoidea, Geologial Society of America and University of Kansas Press, pp. U543-U628.

Ghosh, C. C. (1987) Report of serpulid framestone from Khuiala Formation (lower Eocene) Jaisalmer Basin, Rajasthan. Curr. Sci., v. 56, pp. 414-415.

Khosla, S. C. (1972) Ostracodes from the Eocene beds of Rajasthan. Microplaeontology, v. 18(4), pp. 476-507.

Pareek, H. S. (1984) Pre-quaternary geology and mineral resource of northwestern Rajasthan. Mem. Geol. Surv. India, v. 115, pp. 1-96.

Sahni, M. R. (1955) Recent researches in the Palaeontological Division, Geological Survey of India. Curr. Sci., v. 25(6), pp. 187.

Sahni, M. R. and Bhatnagar, N. C. (1958) New fossils from the Jurassic rocks of Jaisalmer, Rajasthan. Rec. Geol. Surv. India, v. 87(2), pp. 418-436.

Singh, N. P. (1996) Mesozoic-Tertiary biostratigraphy and biogeochronological datum planes in Jaisalmer Basin, Rajasthan, In: *Azami, et al., 1998 (eds.)*, Proceedings XVth Colloquium Indian Micropaleontology and Stratigraphy, pp. 63-89.

Singh, N.P. (2003) Contribution of biostratigraphic studies in stratigraphic evaluation of west Rajasthan shelf. Gondwana Geology Magazine, Spcial vol. no. 6, pp.79-104.

Singh, N. P. (2006) Mesozoic lithostratigraphy of the Jaisalmer Basin, Rajasthan. J. Pal. Soc. India, v. 51(2), pp. 53-56.

Singh, N. P. (2007) Cenozoic lithostratigraphy of the Jaisalmer Basin, Rajasthan. J. Pal. Soc. India, v. 52(2), pp. 129-154.

Singh, P. (1997) Ostracodes from the subsurface Khuiala Formation (lower Eocene) of Manhera Tibba well-1, Jaisalmer, Rajasthan, India. Geosci. Jour., v. 18, pp. 149-233.

Sinha Roy, S., Malhotra, G. and Mohanty, M. (1998) Tertiary. In: Geology of Rajasthan. Geol. Soc. India, pp. 219–228.

Smith, A. B. (2008) The echinoid directory [Electronic Publication]. http://www.nhm.ac.uk/palaeontology/echinoid, (accessed 19th January, 2008).

Spengler, E. (1923) Contributions to palaeontology of Assam. Palaeontologia Indica, N.S., v. 8(1), pp. 1-74, 4 pls.

Srivastava, D. K. (1981) Echinoid genus Eupatagus in Tertiary rocks of Kutch, India. J. Pal. Soc. India, v. 25, pp. 38–41.

Srivastava, D. K. (2004) Fossil spatangoids (Echinodermata) from the Indian subcontinent - a review. J. Pal. Soc. India, v. 49, pp. 125–149.

Srivastava, D. K. and Mathur, U. B. (1996) New fibulariid echinoid from the Middle Eocene rocks of Rajasthan, India. J. Pal. Soc. India, v. 41, pp. 53–56.

Srivastava, D. K., Rana, R. S. and Singh, H. (2008) Record of *Megapneustes* Gauthier (Brissid echinoid) from the Khuiala Formation, Jaisalmer district, Rajasthan, India. J. Pal. Soc. India, v. 53 (1), pp. 31-36.

❑❑❑

12

Recently Studied Sedimentary Records from the Eastern Arabian Sea : Implications to Holocene Monsoonal Variability

RAJESH AGNIHOTRI and SIBY KURIAN
National Institute of Oceanography, Dona Paula, Goa - 403 004, India
Email: ragnihotri@nio.org; siby@nio.org

Abstract

Deep insight into exact mechanism(s) controlling past monsoonal variability on inter-annual to millennial time scales is a pre-requisite for developing predictive capability of monsoon on timescales relevant to human life. Reliable monsoon predictions developed by climate modelers play a crucial role for making country's future economic programme. Basic data input to generally applied monsoon models is instrumental rainfall data, beyond which search for monsoon variability relies upon proxy records stored in various natural repositories such as tree rings, corals, cave deposits, lake and marine sediments. While tree rings, cave deposits and corals are capable of yielding paleo-monsoonal information on extremely high temporal resolution, continental margin sediments depositing in the Arabian Sea have provided relatively coarser but longer proxy records of monsoonal variability. Here we review some of such recently studied sedimentary records from precipitation dominated eastern Arabian Sea in an attempt to comprehend past monsoonal variability in Indian as well as global context. We also focus upon different proxies used so far in the Arabian Sea, their fidelity, limitations and future scope of using novel proxies for a better understanding of past monsoonal variability especially in the anthropocene epoch.

Introduction

Knowledge of past monsoon variability has been an immensely studied research area attracting inter-disciplinary scientists since long, owing to its socio-economic significance for monsoon affected and densely populated countries on Indian sub-continent *e.g.* India, Pakistan, Sri Lanka and Bangladesh. Instrumentally measured All India Summer Monsoon Rainfall (AISMR) data spanning last 130 years (Parthasarathy *et al.*, 1995) have been available to scientific community; however, to develop realistic and adaptable monsoon models, natural variability of south Asian monsoon (or southwest monsoon) has to be known beyond the anthropogenic era. This requires a wide-ranging search for monsoon variability stored in various natural repositories such as tree rings, corals, cave deposits, lake and marine sediments in the form of different biological, geological and chemical proxies. Quality and applicability of paleo-monsoonal research beyond the instrumental era is strongly dependent on choice of repository, location, sampling resolution and fidelity of the used 'proxy'. Though some of the continental and shallow marine repositories (cave deposits, tree rings, corals, for example) do provide extremely high temporal resolution varying from weekly to annual time scale, these proxy records at times are strongly influenced by 'local' rather than 'regional climate' and in turn possess significant 'noise'. These proxy based monsoonal reconstructions are also often limited in total time-stretch requiring a substantial patch work (time series alignment) to yield an integrated monsoonal variability of a particular region. In contrast, well dated sedimentary records preserved in oxygen deficient waters (oxygen minimum zone; OMZ) of the Arabian Sea have shown their capability to catalogue the global monsoonal variations on annual to orbital time scales (Schulz *et al.*, 1998; von Rad *et al.*, 1999; Suthhof *et al.*, 2001; Lückge *et al.*, 2001; Altabet *et al.*, 2002; Agnihotri *et al.*, 2002; Anderson *et al.*, 2002; Gupta *et al.*, 2003) relatable to global climate. It is important to state here itself that, depending on the locale and context, usage of the monsoonal term is also adapted. For example, the broad south Asian monsoon term signifies itself as southwest (SW; henceforth) monsoon or northeast (NE; henceforth) monsoon for all the paleoceanographic studies implicating to past monsoonal variability in the Arabian Sea.

Sediments depositing beneath OMZ of the western Arabian Sea have been extensively exploited for paleo-monsoon reconstructions owing to the following facts (i) surface biological productivity and associated biogeochemistry of the region substantially responds to seasonal reversal of monsoonal winds (Fig. 1) (ii) higher sediment accumulation rates enable sediments to record short term (decadal to centennial scale) climatic oscillations in conventionally analyzable mm to cm slices and (iii) presence of thickest OMZ in water column extending from ~200 to 1200m (Naqvi, 1987) probably help better preservation of various biogenic (organic and inorganic) proxies with minimal particle reworking through bioturbation.

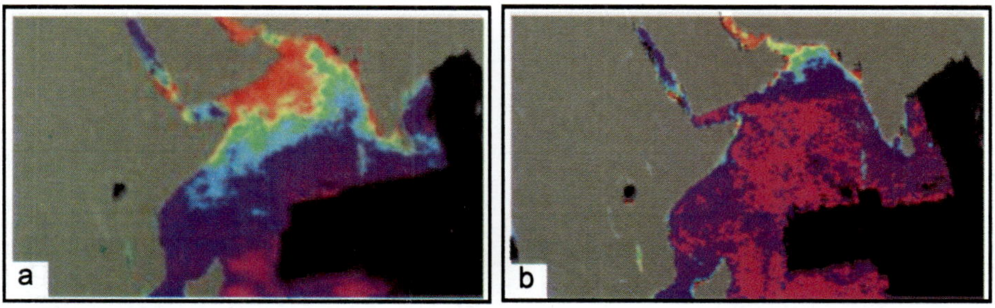

Fig. 1 : Satellite pictures of the Arabian Sea revealing seasonal contrast of surface biological productivity during (a) SW monsoon (left) and (b) Non-monsoon season (right). Red & green colors indicate regions of high surface productivity, while blue and purple colors represent low productivity conditions (pictures are taken from http://daac.gsfc.nasa.gov).

Despite voluminous temporally well-resolved proxy data available from sediments of the Arabian Sea especially from western and northwestern side, there is neither a uniform picture of past monsoonal variability emerged out, even for the 'traditionally thought' stable climatic period *i.e.* the Holocene (last ~10,000 years), nor a unanimously preferred 'proxy' for reconstructing paleo-monsoon variability has been suggested. Moreover, as the case for the modern climate, paleoceanography of the western and northwestern sides of the Arabian Sea is dominated by intense monsoonal winds and associated upwelling resulting in high surface biological productivity (Prell, 1984; Sirocko *et al.*, 1993, 1996, 2000; Reichart, 1997; Anderson *et al.*, 2002; Gupta *et al.*, 2003) rather than precipitation. Therefore, it is always a pertinent question about utility of these paleo-monsoonal reconstructions in terms of their applicability in models dealing with monsoonal precipitation on Indian sub-continent. On the other hand, too few well-dated proxy records are ironically available from the precipitation dominated eastern Arabian Sea (Sarkar *et al.*, 2000; Agnihotri *et al.*, 2002, 2003 a & b; 2008a; Thamban *et al.*, 2001 & 2007; Kurian *et al.*, 2008) to provide desired basic data input to monsoon models. In contrast to western Arabian Sea, where biogenic and eolian proxies have been quite successfully employed (Sirocko *et al.*, 2000) and paleo-monsoon variability in terms of monsoonal winds have been reconstructed ranging from last few centuries to more than ~200 ka BP (Reichart *et al.*, 1998; Anderson *et al.*, 2002), the eastern Arabian Sea, in general, receives very less dust input from Indian sub-continent and upwelling induced surface productivity is also limited to mainly coastal regions off Pakistan and western coast of India. Nonetheless, eastern part of the Arabian Sea is the only locale in the Arabian Sea where heavy precipitation takes places on all along the west coast (~300-400 cm.a^{-1}) during SW monsoon (Fig. 2).

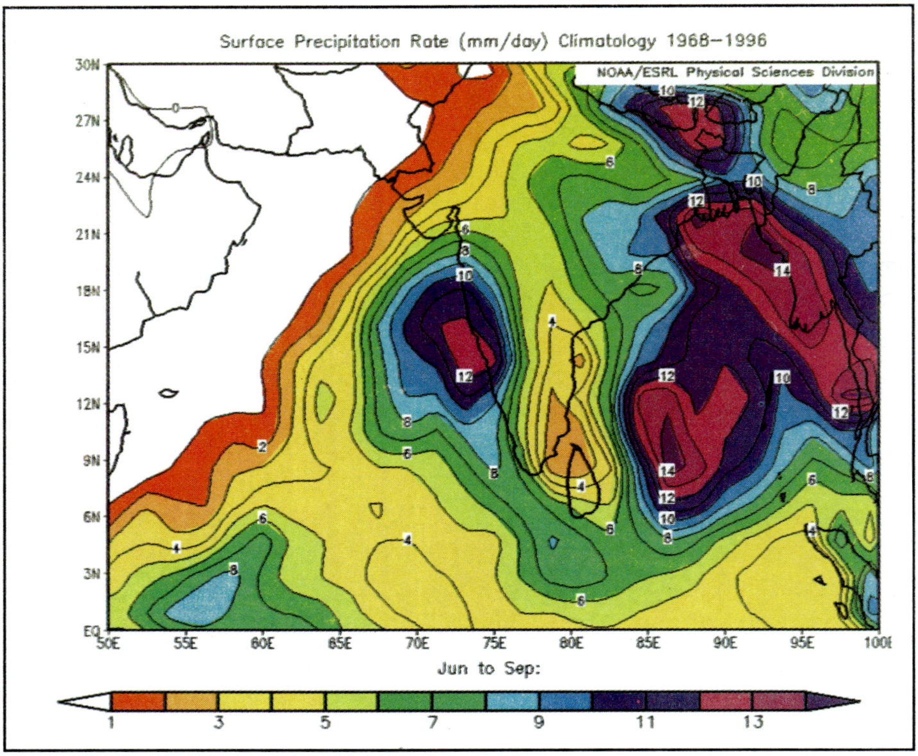

Fig. 2 : Surface precipitation map generated by satellite measurements clearly demonstrates eastern Arabian Sea as a precipitation dominated region. Area off Konkan-Goa coast and Karnataka receives high precipitation during SW monsoon season.

This results in large amount of continental runoff influencing coastal hydrography significantly traceable on a seasonal basis. Precipitation on west coast of India systematically increases from Gujarat to Kerala coast by ~350 mm. (degree latitude)$^{-1}$ (Sarkar *et al.*, 2000). In addition to precipitation, conspicuous surface biological productivity can also be seen on all along the west coast of India during SW monsoon (Fig.1a). Besides these natural manifestations of SW monsoon, increasing anthropogenic activity in the urban and relatively developed west coast of India has been proposed to be capable of perturbing coastal ecology/ biogeochemistry of the region (Naqvi *et al.*, 2000; 2006 a,b). Interestingly and probably unique in the world, anthropogenic activity is also manifested here (the eastern Arabian Sea) via a dominant natural forcing *i.e.* the SW monsoon. Thus, eastern Arabian Sea presents itself as a unique locale for paleo-monsoonal reconstructions ideally suited for delineating anthropogenic versus natural (SW monsoonal) forcing by studying temporally well-constrained and well resolved sedimentary records with suitable proxies and their inter-comparison with other contemporaneous global climate data archives. Fidelity of various proxies can also be tested by comparing proxy time series with available modern instrumentally measured climate parameters such as rainfall, sea surface temperatures etc.

Despite these great potentials, sedimentary archives stored in the eastern Arabian Sea have not been extensively explored yet. Nonetheless, available quality monsoonal reconstructions from this region on varying timescales (decadal to millennial) have provided paleo-monsoonal history in terms of changes in hydrological cycle, which is conspicuously different than that obtained by wind intensity records (especially in Holocene), underscoring importance of the eastern Arabian Sea for paleo-monsoonal research. We review here some of recently studied such proxy records from the eastern Arabian Sea and also explore possibility of new proxies in conjunction with usually employed ones for monsoonal reconstructions in this region. We mainly focus here on high-resolution (decadal to centennial) paleo-monsoonal records spanning Holocene with special emphasis on the last millennium (last ~1000 years) and the Anthropocene epoch (the last ~100-150 years) in order to characterize most dominant natural forcing on SW monsoon variability just before the advent of greenhouse era. We first briefly discuss hydrography of the eastern Arabian Sea in response to SW monsoon and applied major proxies in the paleo-monsoonal research carried out so far.

Hydrography of the Eastern Arabian Sea

Coastal circulation along the west coast of India as elsewhere in the North Indian Ocean, reverses annually during April and October associated with the two monsoons (Shetye and Gouveia, 1998; Naqvi *et al.*, 2006a,c). In winter (NE monsoon), the West India Coastal Current (WICC) flows poleward, causing downwelling and well-oxygenated conditions in the water column over the shelf. In contrast, circulation in summer (SW monsoon) is typical of an eastern boundary -an equator-ward flowing WICC, a pole-ward undercurrent and coastal upwelling. In addition, the entire west coast receives large precipitation as high as ~400 cm during just four summer months between June and September. The freshwater inputs result in very strong but shallow thermohaline stratification; yet, the high nutrient concentrations in the freshly upwelled water that reaches the euphotic zone ensure high surface biological productivity (Fig. 1a) (Naqvi *et al.*, 2006c). In fact, fresh water input to coastal areas off Karnataka (Mangalore) becomes so high that a portion of this part of the Arabian Sea becomes a sink for atmospheric CO_2, a typically unique feature for the Arabian Sea which, as a whole, acts as a global source of CO_2 (Kumar *et al.*, 1996). The ensuing high oxygen demand for respiration of the copious organic matter produced by phytoplankton leads to the development of very intense oxygen deficiency just a few meters beneath the surface. This seasonally-occurring natural suboxic zone occupies an area of about 200,000 km^2 at its peak, and is the largest of its kind in the world (Naqvi *et al.*, 2000, 2006a). As the biogeochemical processes (for example subsurface denitrification) over the western Indian shelf are strongly forced by the SW monsoon, ideally high-resolution paleo-monsoon variability should be traceable through their signatures in form of proxy records stored in sedimentary archives. Important to mention here, seasonal suboxic zone off west coast of India is not contiguous with perennial OMZ of the Arabian Sea which occupies significant portion of slope sediments of the eastern Arabian Sea. Sedimentary records lying under seasonal suboxic zone and perennial OMZ have not been studied in tandem to explore fidelity of conventional geochemical proxies. Another noteworthy point about the eastern Arabian Sea is that, its northern part (off Pakistan and Gujarat) is significantly influenced by winter convective mixing (Madhupratap *et al.*, 1996) and thereby experiences considerable biological productivity in response to NE monsoon. Whereas the Konkan-Goa (KG) coast and areas southward are affected by upwelling and intense precipitation (100-400 $cm.a^{-1}$) during the SW monsoon but do not experience winter convective mixing and hence any significant surface biological productivity during NE monsoon. Detailed hydrography of the eastern Arabian Sea comprising seasonally reversing surface/ subsurface circulation, development of seasonal suboxia and its plausible influence on regional biogeochemistry has been discussed in several detailed reports such as Naqvi *et al.* (2000; 2006a,b), Agnihotri *et al.* (2008a) and Kurian *et al.* (2008).

As said earlier SW monsoon not only manifests itself heavy continental precipitation on the coastal belt of west coast of India (Fig. 2), significant increase in surface biological productivity is also seen (Fig. 1b). Majority of monsoonal proxies employed in this region thereby are actually traces of these monsoonal manifestations *viz.* upwelling induced surface biological productivity, continental runoff in terms of its influence on SST and SSS (sea surface temperature and salinity, respectively). We give brief introductions of all major proxies in following sub-sections. For the first time, we compare subsurface denitrification variability in the eastern Arabian Sea in relation with SW monsoonal intensity variations in both perennial and seasonal suboxic zones of the eastern Arabian Sea. While doing so, we emphasize that volume of chronologically constrained data in both sub-oxic zones is just indicative of inferences made; leaving several questions unanswered demanding many more quality datasets from both oxygen deficient zones.

The other important point which has to be addressed here itself is intricacies involved in achievable temporal resolution in a particular sediment core. Characteristic temporal resolution in a sediment core is mainly governed by (i) Sediment deposition rate and uncertainties involved in the chronologies/ age-depth model used for the sediment core (ii) Sampling interval in commonly employed sampling schemes (cm scale) necessary to yield material enough to analyze different biogeochemical proxies and finally (iii) Analytical uncertainty involved in the measurement of a particular proxy. Some aspects of uncertainties involved in the usually employed ^{14}C chronologies are discussed in another report by K. Dutta of this very issue. Unfortunately not any significant improvement has come recently to overcome chronological constraints by usually employed methods. However, there have been remarkable advancements in *in-situ* analytical methods. For example, advent of scanning XRF has provided means of determination of sediment core elemental composition at sub-millimeter scale with reasonable accuracy and precision (Agnihotri *et al.*, 2008b).

An Overview of Proxy Indicators Used in the Eastern Arabian Sea to Infer Past Monsoonal Variability

Surface Biological Productivity

SW monsoonal upwelling brings subsurface nutrients to euphotic zone of the eastern Arabian Sea and thereby induces copious amount of surface biological productivity (Fig. 1a). Remains of bulk productivity carriers *viz.* C_{org}, N, $CaCO_3$ are expected to be well preserved in sediments depositing beneath both perennial as well as seasonal OMZ due to faster sedimentation rates, better preservation under reducing conditions and minimal bioturbation. The sedimentary C_{org} and N have thus been used quite successfully as proxies of SW monsoon intensity in sediments of the northeastern and eastern Arabian Sea (Schulz *et al.*, 1998; Agnihotri *et al.*, 2002, 2008a). Likewise, sedimentary $CaCO_3$ also has been used as a productivity proxy in the southeastern Arabian Sea (Bhushan *et al.*, 2001; Agnihotri *et al.*, 2003a, b; 2008a). However, following limitations need to be noted with bulk carbonate content usage such as (i) significant contribution can come from relict carbonate deposits and fossil coral reefs in the eastern Arabian Sea (Rao *et al.*, 2003) and/or (ii) post depositional dissolution of $CaCO_3$ under highly suboxic conditions within the sediments might cast doubts on fidelity of this traditional proxy (Bhushan *et al.*, 2001). Availability of biogenic silica (BSi) data, the third major bulk carrier of productivity in the eastern Arabian Sea is too limited to discuss here.

In using remnants of these aforementioned bulk carriers of surface bio-productivity, an inherent limitation is always present *i.e.* what is the relative proportion of material produced *in-situ* and transported by numerous rivers/ streams by precipitation runoff. Traditionally, source characterization of organic matter has been done using C isotopes ($\delta^{13}C_{org}$) and C/N wt. ratios. Fontugne and Duplessy (1986) reported typical $\delta^{13}C_{org}$ of terrestrial and marine type organic matter of the Arabian Sea as -26 and -20‰ (with respect to PDB). Likewise, mean C/N wt. ratio ranges ~8.0± 2 for typical marine organic matter in this region (Calvert *et al.*, 1995; Bhushan *et al.*, 2001). However, $\delta^{13}C_{org}$ of bed-load sediments of the Mandovi estuary of Goa range from -30.6 and -26.3‰, with an average of -29±1.71‰ (*n*=6) (Rajesh Agnihotri, unpublished data). Several studies carried out in the eastern Arabian Sea even in very shallow marine environment revealed majority of organic matter is typically 'marine' (Bhushan *et al.*, 2001; Agnihotri *et al.*, 2002, 2003, 2008a) despite heavy monsoonal precipitation runoff on the west coast of India.

Subsurface Denitrification Variations in the Eastern Arabian Sea

Development of intense SW monsoon induced biological productivity leads to consumption of almost all dissolved oxygen resulting in seasonal suboxia (Naqvi *et al.*, 2000). Though water column along the shelf off west coast of India experiences only seasonal suboxia, sediment depositing beneath these waters appear to be strongly reducing throughout at least during the last ~700 years (Agnihotri *et al.*, 2008a). During this period, though surface productivity appears to be varying in tandem with SW monsoonal intensity but subsurface denitrification intensity does not appear to follow same rhythm. Alternatively, sedimentary $\delta^{15}N$, the most reliable and probably only means to measure the extent of past subsurface denitrification in most of the world denitrification zones may not be working in this biogeochemically complex marine realm (Agnihotri *et al.*, 2008a). Contrary to these inferences, in the perennial OMZ region of the eastern Arabian Sea, subsurface denitrification intensity (as measured by sedimentary $\delta^{15}N$) appeared to be varying in concert with surface productivity and hence SW monsoonal intensity on millennial time scale during the Holocene (core 3268G5; Figure 3; Agnihotri *et al.*, 2003a) and same was the case, reported during the late-Quaternary (core MD76-131; Fig. 3; Ganeshram *et al.*, 2000).

Molecular Biomarkers

Molecular biomarkers have been proposed as more source-specific than bulk proxies *viz.* C_{org}, $CaCO_3$ and biogenic silica. Sterols such as dinosterol and brassicasterol can help in identifying the organisms contributing to overall surface bio-production and its past variability (Volkman *et al.*, 1998; Schulte *et al.*, 1999). Sedimentary concentration of dinosterol, a well established biomarker of dinoflagellates (Boon *et al.*, 1979), and brassicasterol, a marker for diatoms have been linked to primary productivity in the open ocean region of the Arabian Sea (Schubert *et al.*, 1998, Schulte, *et al*, 1999). β-sitosterol and stigmasterol are major constituents of terrestrial plants (Volkman, 1986; Saliot *et al.*, 1991), and have been used to track the terrestrial organic matter supply in coastal marine environments (Jaffé *et al.*, 2006). However, since they are also produced by marine phytoplankton (Volkman, 1986), the validity of these two sterols as terrestrial markers is not well ascertained (Mudge and Norris, 1997). n-alkanes are also used as a marker of terrestrial organic carbon input to the marine environment as they present in epicuticular waxes synthesized by higher plants

(Prahl and Muehlhausen, 1989). Likewise, alkenones (C37 to C39) are produced in the pelagic ocean mainly by two species of organisms (*Emiliania huxleyi* and *Gephyrocapsa oceanica*) belonging to the family Haptophyceae (Volkman *et al.*, 1980). The unsaturation index ($U_{37}^{K'}$) calculated from the concentration of alkenones is used for the reconstruction of paleo- temperature in all over the oceans (*e.g.* Brassell *et al.*, 1986; Prahl and Wakeham, 1987). To check the fidelity of biomarkers, Prahl *et al.* (2000) analyzed sediment trap samples from the central Arabian Sea and showed that they display distinct maxima at the start and stop of the NE and SW monsoons due to rapid change in surface mixing conditions which alter light and nutrient availability, thereby triggering these biomarker signals of export production. Biomarkers have been used to reconstruct changes in paleo-productivity in the eastern and equatorial regions of the Arabian Sea for the last ~300 ka BP and showed enhanced (decreased) productivity during glacial stages (interstadials) (Schubert *et al.*, 1998, Schulte *et al.*, 1999). However, certain studies have addressed the issue of degradation of biomarkers during post burial diagenesis, which might affect the reconstruction of past productivity changes (Prahl *et al.*, 2000). Likewise, Sinninghe Damstè *et al.* (2002) showed effect of oxygen exposure time on the preservation of biomarker signals stored in the sedimentary records.

Despite acknowledging the fact that aforementioned biomarkers of productivity also might have some complications/ limitations as studies carried out in other regions, we state that fidelity and potential of these biomarkers in shallow marine environment of eastern Arabian Sea have to be explored in detail, especially during Anthropocene epoch and recent past (Fig. 5). In fact, our recent work Kurian *et al.* (2008) demonstrated first ever application of specific biomarkers (sterols) to reconstruct surface productivity of coastal eastern Arabian Sea for the last 700 years, in conjunction with traditional proxies. Steep increases observed in all the analyzed biomarkers during the Anthropocene (especially last ~50 years) indicate coastal productivity is at its all time maxima (Kurian *et al.*, 2008). Delineation of role of anthropogenic forcing in the underlying natural forcing is still needs to be explored with more suitable proxies. Specific biomarkers *viz.* C35/C31 n-alkane ratio and lycopane/ C31 n-alkane ratio could be used as very potential proxies of redox changes (Schulte *et al.*, 1999; Sinninghe Damstè *et al.*, 2003), if used in conjunction with other relatively established inorganic proxies.

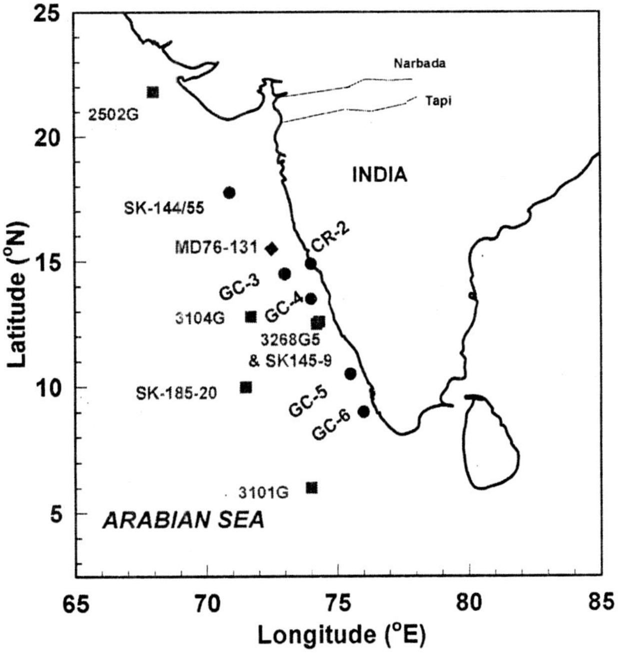

Fig. 3 : Sediment core locations from the eastern Arabian Sea for reconstructing paleo-monsoonal history on decadal to centennial time scale in different studies carried out in the region so far. Sarkar *et al.*, 1990 (core #185-20); Sarkar *et al.*, 2000 (core #3268G5); Ganeshram *et al.*,2000 (core #MD76-131); Agnihotri *et al.*, 2002 (core #2502G);2003b (3104G & 3101G); 2008a (core #CR2); Thamban *et al.*, 2001 (core #GC-5), 2002 (core #GC-3 & GC-5; 2007(core #148/55); Thamban and Rao, 2001 (GC-4 & GC-6); Tiwari *et al.*, 2005a (core #SK145-9); Kurian *et al.*, 2008 (core #CR2).

Major and Trace Element Abundance

As employed anywhere else in marine realms, several paleo-studies carried out in the eastern Arabian Sea have attempted a suite of major and trace elements for tracking monsoon related processes. For instance, sedimentary abundance of characteristic major and minor elements such as Al, Mg, Ti and also Fe track changes in lithic fraction either by changes in terrestrial runoff/ aeolian inputs (supply) or by just dilution by other biogenic components or both. Fe, although is a redox sensitive element, amount of Fe taking part in redox-changes in continental margin regions is very small compared to its total abundance. Redox conditions in bottom waters/ sedimentary column strongly influence the distribution of some of trace elements such as Mn, Cr, V and U. In sediment pore waters, depending on their oxidation states, these elements have different solubilities. For example, Mn is found to be enriched in sediments from oxic zone, whereas in reducing environment it gets mobilized to pore waters, therefore, gets depleted in solid phase (Bonatti *et al.* 1971; Somayajulu *et al.* 1994; Yadav, 1996). In contrast, Cr, V and U are found to be depleted in oxic zone and enriched in the reducing strata (Dean *et al.* 1997). Thus, an anti-correlation is expected in the depth-profiles between Mn/Al and Cr/Al or V/Al (wt. ratios) during diagenetic changes in sedimentary column.

Sediment cores lying outside the OMZ of the eastern Arabian Sea were found to have much lower abundance of C_{org} and N contents due to much lower sedimentation rates and thereby longer oxygen exposure time at the sediment-water interface, leading to a probable poor preservation of these traditional surface productivity proxies (Agnihotri *et al.*, 2003b). However, elemental ratio of Ba/Al or biogenic Ba content of the sediment has demonstrated its potential to be reliable indicator of surface productivity proxy (Dymond *et al.*, 1992) in these settings (Agnihotri, 2001; Agnihotri *et al.*, 2003b).

C and O Isotopes of Foraminiferal Tests

In addition to aforementioned proxies, most preferred paleoceanographic proxies in the eastern Arabian Sea have been C and O isotopes of surface dwelling foraminifera ($\delta^{13}C$, $\delta^{18}O$) in order to reconstruct evolution of past surface and bottom water hydrographic conditions (Duplessy *et al.*, 1981; Berger *et al.*, 1985; Rostek *et al.*, 1997; Sarkar *et al.*, 1990; Sarkar *et al.*, 2000; Thamban *et al.*, 2001). The $\delta^{18}O$ of foraminiferal calcite is a function of temperature and salinity of the waters in which they grow. Therefore, either a logical way or independent proxy (alkenone derived temperature, for instance) can be used to correct the effect of minor influence from the major one, from the total signal. For example, $\delta^{18}O$ showed a significant positive relationship with surface salinity in the eastern Arabian Sea and hence can be used as proxy indicator to decipher past surface hydrographic conditions, if appropriate temperature corrections are used suitably (Sarkar *et al.* 2000). On the contrary to $\delta^{18}O$, $\delta^{13}C$ of foraminifera reflects productivity of surface waters. Bio-productivity, in general, consume ^{12}C preferentially compared to ^{13}C leaving its signature in form of enriched $\delta^{13}C$ in foraminiferal tests, thereby enabling them to act as proxy for surface productivity. However, upwelling influences on surface waters have to be considered while interpreting this proxy in terms of surface bio-production.

Using these proxies, Duplessy (1982) showed that the SW monsoon was much weaker during the last glacial maximum (LGM, ~18 ka BP) than present, whereas NE monsoon was much stronger at that time. Sarkar *et al.* (1990) corroborated these findings using oxygen isotope foraminiferal record of sediment core SK-185-20 (Fig. 3) from the southeastern Arabian Sea. Tiwari *et al.* (2005a), however, further constrained the timing of strengthening of NE monsoon by AMS ^{14}C dating as ~19-17 ka BP, based on a sediment core analyzed from equatorial Indian Ocean. Likewise, Thamban *et al.* (2001) derived paleo-monsoonal history from the LGM to present, by measuring oxygen isotope of surface dwelling foraminifera (*G. rubber* and *G. sacculifer*), geochemical and other sedimentological parameters in a sediment core (GC-5; Fig. 3) from the shelf region of the eastern Arabian Sea. They also applied clay mineral composition and grain size parameters in order to reconstruct the variations in the intensity of summer monsoon precipitation during the late-Quaternary using sediment cores (GC-3 and GC-5; Fig. 3) from the eastern Arabian Sea (Thamban *et al.*, 2002).

Though $\delta^{18}O$ of surface dwelling foraminifera has been regarded as most natural choice for tracking past monsoonal variability, various different external effects such as global ice volume, SST and contribution of local runoff to sea surface salinity require appropriate corrections to retrieve monsoonal (precipitation) signal (Sarkar *et al.*, 2000). Therefore, there is a strong need to go some steps forward involving coupled trace element analyses of foraminiferal tests in addition to oxygen isotopic composition. Trace metals incorporated in the calcium carbonate of foraminiferal shells are becoming increasingly popular for the reconstruction of past environments. Past sea water temperature, carbonate chemistry, nutrient levels and possibly salinity are all reflected by differences in trace metals to calcium ratios. Although it is generally assumed that the incorporation of these trace metals follow thermodynamic relations known from inorganic calcium carbonate precipitation experiments, large offsets have been observed in case of foraminiferal bio-mineralization. The mechanisms responsible for these offsets are, however, poorly known. Exploration of these coupled proxies in search of paleo-monsoon from the eastern Arabian Sea has not yet taken off.

Holocene Monsoonal Variability from the Eastern Arabian Sea on Millennial to Decadal Timescales

We discuss Holocene monsoonal variability here in three time windows (i) millennial scale changes during the entire Holocene (*i.e.* last ~10,000 years), (ii) focusing on centennial to decadal scale variability during the last millennium, in order to understand underlying forcing mechanism(s) and (iii) Anthropocene (last ~100-150 years). Depending on water column hydrography, distance and type of particular continental terrain and most importantly sedimentation rate, suitable locales are chosen for collection of a sediment core for studying a particular time window. For example, outer shelf /slope regions of the eastern Arabian Sea are suitable for studying Holocene epoch on millennial scale as sedimentation rates are typically few cm per thousand years; while inner shelf and locales near to river mouths are capable to yield sedimentation rates typically in range of mm per year. Comparisons were also made between paleo-monsoonal evolution inferred from the eastern Arabian Sea with inferences obtained from northwestern and western Arabian Sea.

Millennial Scale Holocene Monsoonal Variability

First chronologically well-constrained and temporally well resolved Holocene monsoonal reconstruction was obtained using a sediment core 3268G5 (Fig. 3) raised from heavy precipitation zone of the eastern Arabian Sea (off Mangalore, Figure 2) (Sarkar *et al.*, 2000). Using oxygen isotopes of surface dwelling foraminifera, the parameter P-E index which is the net balance of precipitation over evaporation was used as an index of monsoonal precipitation in the region. Based on its downcore variations, monsoonal intensity was found to have steadily increased from ~10 to ~2 ka BP (Sarkar et *al.* 2000). Surface productivity and sub-surface denitrification intensity, the other manifestations of SW monsoon were also found to be in unison with SW monsoonal intensity during the Holocene (Agnihotri *et al.*, 2003a). Spectral analyses of proxy data from the core 3268G5 also revealed presence of common and significant ~1400 and ~700 year periodicites in the Holocene monsoonal climate coinciding with famous Bond cycles (1400±300 years) observed in north Atlantic proxy records at high latitude (Bond *et al.*, 1997) and the 775-year periodicity observed in the South Asian (Chinese) monsoon (Wang *et al.*, 1999). In a contemporaneous study, Thamban *et al.*, (2001) reconstructed paleo-monsoonal variability for a period since the Last Glacial Maximum

(~18-20 ka BP) to present; using oxygen isotope data and other sedimentological parameters from a sediment core from the eastern shelf region of the Arabian Sea. They too found significant oscillations in $\delta^{18}O$ of surface dwelling foraminifera during deglaciation and the Holocene periods suggesting large variations in sea surface hydrography related to monsoonal precipitation, similar to those reported by Sarkar *et al.* (2000). Surface productivity at their core site (shelf region of the southeastern Arabian Sea) appeared to have reduced between ~13 to 6 ka BP, whereas it increased substantially during the last glacial period (~18 to 15 ka BP). They attributed this to enhanced strength of NE monsoon and subsequent convective mixing and nutrient injection (Madhupratap *et al.*, 1996), leading to high surface biological productivity. Ganeshram *et al.* (2000) studied one sediment core (MD76-131; Fig. 3) from the eastern Arabian Sea for tracking past surface productivity and sub-surface denitrification variability during the late-Quaternary, but due to very coarser temporal resolution during Holocene, any significant inference can not be made in the present context. In general, not many high resolution paleo-monsoonal records could be generated till then from the slope/ outer shelf region of eastern Arabian Sea owing to relatively slower sedimentation rate resulting in poor temporal resolution and probable bioturbation. Nonetheless, since eastern Arabian Sea is the only locale where from SW monsoon could be reconstructed in terms of precipitation (Fig. 2), with modern advances in new analytical techniques such as scanning XRF in capable of achieving much better temporal resolution than conventional ones (Agnihotri *et al.*, 2008b), it is possible to unravel many *hitherto* aspects of past SW monsoon in Indian context.

In contrast to the precipitation dominated eastern Arabian Sea, upwelling induced surface productivity is supposed to be the best choice to reconstruct past monsoonal variability in the wind dominated western Arabian Sea. Numerous studies from the western Arabian Sea have indeed demonstrated this fact on orbital to millennial timescales. In the present context of Holocene, Gupta *et al.* (2003) measured abundance of *G. bulloides* foraminifera in the core 723A and found a typical decreasing pattern in monsoonal intensity closely following solar (July) insolation at 65° N with conspicuous maxima at ~9-10 ka BP. The observed Holocene monsoonal wind intensity pattern was found to be well supported by continental high resolution stalagmite record (Fleitmann *et al.*, 2003). These studies also find consistency with several other paleomonsoonal reconstructions from the western Arabian Sea which suggested early Holocene period ~8-10 ka BP as most humid interval when monsoon intensity was at its maximum (Sirocko *et al.*, 1996, 2000). Latest Holocene monsoonal record of highest temporal resolution from stalagmite record of Dongee Cave (Wang *et al.*, 2005) also supports inferences made by aforementioned studies in the western Arabian Sea.

Therefore, proxy records from the precipitation dominated eastern Arabian Sea (Sarkar *et al.*, 2000; Thamban *et al.*, 2001) appear to be in an obvious contrast with those from western Arabian Sea. Thamban *et al.* (2007) corroborated this fact in their syntheses of results from the entire Arabian Sea where it appears that as a whole, there is no uniform pattern of paleomonsoonal variability during the Holocene in the Arabian Sea. The contrasting Holocene monsoonal patterns between western and eastern Arabian Sea can be possibly understood in terms of gradual southward shift in mean latitudinal position of ITCZ during the Holocene (Fleitmann *et al.*, 2007). A typical pattern of Holocene monsoonal variability inferred from eastern Arabian Sea sediments has been shown schematically in Fig. 4.

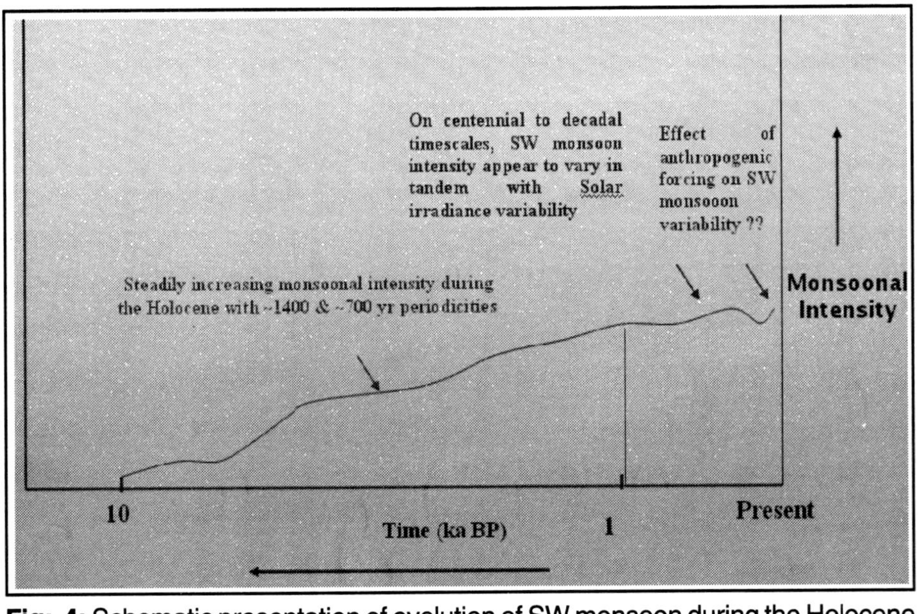

Fig: 4: Schematic presentation of evolution of SW monsoon during the Holocene as inferred from proxy records of the eastern Arabian Sea.

It is worth mentioning here that sub-decadally resolved 5000 yr long monsoonal reconstruction has been carried out using varve thickness and other proxy data on a core duplet from the northeastern Arabian Sea (von Rad *et al.*, 1999; Lückge *et al.*, 2001). However, the observed pattern is neither consistent with eastern or the western Arabian Sea trend. As this region is significantly influenced by NW monsoon in addition to SW monsoon, retrieval of the SW monsoonal signal is complicated. This might be a plausible cause responsible for the observed trend of monsoonal intensity in the northeastern part of the Arabian Sea.

Monsoonal Variability During the Last Millennium

Nature and magnitude of climate variability during the last few millennia with sufficient temporal resolution has received immense interest in paleo-scientists and modelers (Agnihotri *et al.*, 2002, 2008b; Hu *et al.*, 2003; Mangini *et al.*, 2005; Mann, 2007) in order to determine dominant forcing factors, mutual interactions among different climatic components. Since temperature reconstructions for the last two millennia by various workers demonstrated quasi- cyclicity in global temperatures, *i.e.* arrival of warm and cold phases such as medieval warm period (MWP; ~800-1300 AD) and little ice age (LIA; ~1500 to 1900 AD), attempts have been made to understand plausible forcing(s) responsible for this. Various forcing mechanisms were suggested to have influenced global climate on decadal to centennial scale over the last millennium, in which solar irradiance forcing is one of them (Agnihotri *et al.*, 2002). Then, one school of thought believes that extra-tropical changes in north Atlantic were most likely drivers of global climate change, others presume climatic changes occurring at the tropics are likely to be principal drivers of global climate. SW monsoon is one of the major tropical climatic phenomena intimately linked with seasonal migration of inter-tropical convergence zone (ITCZ) across the globe and thereby affecting significant proportion of the world population on inter-annual to decadal timescales. SW monsoon not only impacts billions of human life on Indian sub-continent but also influences significantly hydrography, biology and biogeochemistry of the entire northern Indian Ocean (Arabian Sea and Bay of Bengal). Therefore, a natural thinking developed *i.e.* to study past SW monsoonal history in terms of past changes in hydro-biogeochemistry stored in sedimentary records in form of different proxies arresting signatures of these biochemical changes (Suthhof *et al.*, 2001; Altabet *et al.*, 2002; Agnihotri *et al.*, 2003a; Agnihotri *et al.*, 2008a) and comparing inferences with those made from high latitude climatic changes mainly in north Atlantic (Bond *et al.*, 2001). Several studies

(Schulz *et al.*, 1998; Altabet *et al.*, 2002; Gupta *et al.*, 2003; Suthhof *et al.*, 2001) have revealed close connection of monsoonal variability recorded in the Arabian Sea with north Atlantic climate.

Which climatic component responds first to the external forcing and spreads signal across globe is, however, not still known with certainty due to poor age controls. Careful examination of last millennium climate can probably help to a greater extent if generated time series are better chronologically-constrained. Therefore, to investigate synchrony of SW monsoon with global climate various continental and marine archives have to be studied from diverse geographic locations. We discuss here some records from the eastern Arabian Sea spanning last few millennia and compare inferences made with those made from the western Arabian Sea. Agnihotri *et al.* (2002) and Kurian *et al.* (2008) studied sediment cores 2502G and CR-2 from the eastern Arabian Sea, off Saurasthtra (Gujarat) and Goa spanning ~1200 and 700 years respectively (see Fig. 3 for core locations). Outcome of both the studies is common *i.e.* upwelling induced surface productivity varied in tandem with a common external forcing *i.e.* solar irradiance variability on decadal to centennial timescales during the last millennium. These studies along with some contemporaneous ones from the western Arabian Sea (Gupta *et al.*, 2005) revealed the fact that solar intrinsic forcing appears to dominate significantly upwelling strength in both western and eastern Arabian Sea and hence monsoonal intensity on decadal to centennial timescale. In a very recent paleoceanographic reconstruction carried out in the Peru margin area (Southern hemisphere) by Agnihotri *et al.* (2008b), it appears that coastal upwelling strength off Peru also shows signatures of a plausible solar irradiance forcing at least during the last two millennia (Agnihotri *et al.*, 2008b) and thus reinforce already made inferences from the eastern and western Arabian Sea. Solar irradiance not only influences coastal upwelling strength in the tropics but also appears to control continental precipitation as well during the instrumental period (Bhattacharyya and Narasimha, 2005). As far as historical background of solar activity on Indian rainfall is concerned, possibly the earliest references on solar forcing of Asian monsoon are Meldrum (1873) and Lockyer (1900).

Since total change in magnitude of solar irradiance forcing on decadal timescale is very minute (0.06-1.0%), it requires amplifying mechanism(s) within the climate system to produce traceable effect. Various physical mechanisms have been proposed to produce this amplification, including alteration of the atmospheric circulation pattern as a result of changes in solar ultraviolet radiation during the solar cycle (Shindell *et al.*, 1999; Haigh, 2001). The exact physical mechanism involved in the observed coupling between solar irradiance changes at the top of the atmosphere and upwelling stimulated surface productivity in the tropics is still not yet clear. However, sensitivity of the mean latitudinal position of ITCZ to solar irradiance changes may be one pathway (Agnihotri *et al.*, 2002 & 2008b; Agnihotri and Dutta, 2003). Stratospheric modulation of upwelling in the equatorial troposphere, which produces a north-south seesaw of convective activity during seasonal migration of ITCZ from northern to southern hemisphere, may be a more acceptable pathway (Kodera, 2004).

Tiwari *et al.* (2005b) measured oxygen isotopes of surface dwelling foraminifera *G. ruber* in a core SK-145-9 (off Mangalore; Fig. 3) and demonstrated that estimated precipitation proxy (P-E) also seem to resonate with solar variability during the last millennium. Thus, by and large, all the high-resolution paleoceanographic studies carried out in the eastern Arabian Sea emphasize a significant role of external solar irradiance forcing on SW monsoon. Upwelling indices broadly vary positively in unison with solar irradiance variability during the last millennium, *i.e.* enhanced surface productivity during all solar maxima and vice versa. Same is the case for coastal rainfall on the west coast of India *i.e.* higher solar activity produces enhanced rainfall and vice versa. From the western Arabian Sea, Anderson *et al.*, (2002) reconstructed monsoonal wind intensity variations using high-resolution upwelling record based on the foraminiferal abundance of *G. bulloides* for the last four centuries. This record is also in general agreement with inferences made from the eastern Arabian Sea *i.e.* suppressed upwelling strength during the LIA and prominent intensification thereafter.

Kurian *et al.* (2008) employed specific biomarkers in the eastern Arabian Sea region to track past surface productivity in conjunction with other conventional proxies like sedimentary C_{org}, N and $CaCO_3$ contents. Among the four measured sterols, Dinosterol, an indicator of dinoflagellate productivity appears to be relatively more sensitive than other proxies to the exerted forcing (Kurian *et al.*, 2008). It is interesting to know that subsurface denitrification intensity in the seasonal suboxic zone off Goa do not appear to be dominantly controlled by surface productivity during the last 700 years (Agnihotri *et al.*, 2008a). This is in an apparent disagreement with concurrent relationship observed between the two during the Holocene in the eastern Arabian Sea, but in the perennial OMZ (off Mangalore) (Agnihotri *et al.,* 2003a).

Monsoonal Variability During Anthropocene Era

Whether the anthropogenic (greenhouse) forcing has any noticeable influence on the intensity of SW monsoon or not? This is a still vital and open question, not only of scientific importance but also of considerable socioeconomic relevance. Conventionally, it was assumed that with elevated emissions of CO_2 in the atmosphere and increasing temperatures would help plant photosynthesis on land and thereby would cause increased rainfall intensity. In contrast, Goes *et al.* (2005) suggested with increasing global temperatures in response to global warming would reduce high latitude Eurasian ice extent. This would probably reduce land-ocean thermal contrast between Eurasia and Arabian Sea and in turn SW monsoonal strength during the modern Anthropocene epoch. Instrumental (sub-divisional) rainfall data for India do not corroborate suggestions made by Goes *et al.* (2005). However, Goswami *et al.* (2006), based on their detailed analyses of sub-divisional rainfall data suggested that, though there is no discernable influence of global warming/greenhouse forcing on the average rainfall over India, but frequency of extreme events have tremendously increased in recent years. A quantitative delineation between the natural and anthropogenic forcing factors on monsoonal intensity is essential and for.that so called 'ultra-high resolution techniques' can play vital role in paleo-monsoonal research. Shallow coastal marine realm of the eastern Arabian Sea again appears to be an in ideal locale for delineation between anthropogenic versus natural forcing factors as precipitation on Konkan-Goa coast showed maximum coherence with Sunspot activity in the instrumental era; and upwelling induced surface productivity indices *e.g.* dinosterol concentration in sediment core CR-2 (Fig. 3) showed appealing covariance with external natural forcing *i.e.* Sunspot acivity (or solar forcing) together with regional continental precipitation in the recent past (Fig. 5).

In the sediment core CR-2 from coastal Goa (Fig. 3, for core-site), steep increases were observed in all the analyzed biomarkers during the Anthropocene (especially last ~50 years) indicating coastal productivity is at its all time maxima (Kurian *et al.,* 2008). This unprecedented increase in coastal surface bio-productivity during the last few decades is natural or anthropogenic, could not be characterized with enough certainty. However, this unique increase is not likely to be a preservation artifact as redox sensitive elements distribution in the same core indicates bottom waters/ sediment water interface have remained strongly reducing throughout the deposition period (Agnihotri *et al.* 2008a). Nonetheless, preservation versus supply issue is an important one for applying these new proxies in paleo-monsoonal research and hence should be studied in detail with many more sediment cores in same marine locale with varying sedimentation rates. Both anthropogenic (enhanced supply of fertilizer derived nutrients in coastal waters) and natural (increased Sun's activity in recent decades; Solanki *et al.,* 2004) forcing factors might have resulted in recent increase in surface productivity observed in the coastal zone of the eastern Arabian Sea but their relative role(s) have to be quantified using multi-core-multi-proxy approach.

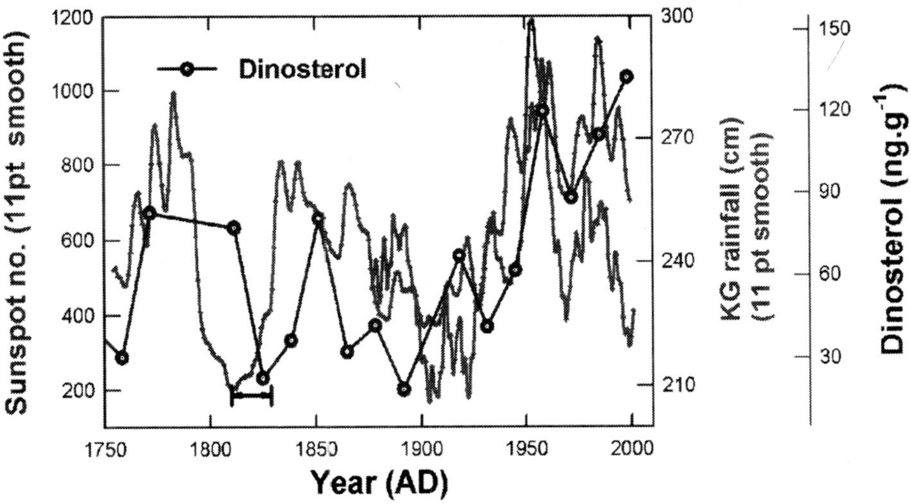

Fig. 5 : Covariance observed between monsoonal precipitation on Konkan-Goa coast and surface productivity as mimicked by dinosterol concentration ($ng.g^{-1}$) in sediment core CR-2 in response to external solar forcing (Sunspot activity). Precipitation data for Konkan-Goa coast and Sunspot no. are adapted from websites of Indian Institute of Tropical Meterology (Pune, India) and paleoclimatology site of National climate Data Center (NOAA) respectively. This figure is adapted from Kurian *et al.* (2008).

Concluding remarks

We have presented here a comprehensive scenario of Holocene monsoonal history on millennial to decadal time scales revealed by recently studied paleo-records obtained from the precipitation dominated eastern Arabian Sea. Inferences made were also compared with those obtained from the wind dominated western Arabian Sea. Paleo-monsoonal variability in the eastern Arabian Sea exhibits a spatial concordance, however, overall monsoonal variability inferred from the eastern Arabian Sea appears to show a contrasting pattern with respect to that obtained from the western Arabian Sea. A plausible southward shift in mean latitudinal position of ITCZ during the Holocene within the Arabian Sea is a likely interpretation for this anomaly (Fleitmann *et al.,* 2007). However, this anomaly underlines the importance of the eastern Arabian Sea region as a suitable paleoceanographic locale for reconstruction of past monsoon in terms of precipitation.

In contrast to millennial scale variability of monsoon during the Holocene, past monsoonal variability on centennial to decadal time scale at least during the last few millennia, appears to be significantly influenced by external solar irradiance forcing in the entire Arabian Sea. Therefore, solar forcing can be treated as the most dominant natural forcing for the variability of SW monsoon in the pre-anthropogenic era. The eastern Arabian Sea also presents itself as an ideal locale for calibration of various monsoonal proxies with instrumental rainfall data in the rapidly depositing shelf sediments under seasonally suboxic waters. Chronologies of sediments can be improvised by compound specific dating schemes using new accelerator mass spectrometer (AMS) technique available in India (Institute of Physics, Bhubaneswar). Future paleoceanographic research should focus on a quantitative assessment of anthropogenic forcing which is most likely riding over natural forcing component of monsoonal variability.

Acknowledgements

We thank Director, NIO, for the facilities provided, and to Council of Scientific and Industrial Research for financial support. This manuscript got significantly improved by constructive comments made by Dr. M. Dileep Kumar from NIO and one anonymous reviewer. Dr. K. Dutta is accordingly thanked for supplying surface precipitation map of the eastern Arabian Sea (Figure 2). This is NIO contribution no 4444.

References

Agnihotri, R. (2001) Chemical and isotopic studies of sediments from the Arabian Sea and Bay of Bengal. Unpublished Ph.D. Thesis, M.S. University, Udaipur, 131p.

Agnihotri, R., Dutta, K., Bhushan, R. and Somayajulu, B.L.K. (2002) Evidence for solar forcing on the Indian Monsoon during the last millennium. Earth Planet Sci. Lett., v.198, pp.521-527.

Agnihotri, R. and Dutta, K. (2003) Centennial scale variations in monsoonal rainfall (Indian, east equatorial and Chinese monsoons): Manifestations of solar variability. Curr. Sci., v.85, pp.459–463.

Agnihotri, R., Bhattacharya, S.K., Sarin, M.M. and Somayajulu, B.L.K. (2003a) Changes in surface productivity, sub-surface denitrification and SW monsoon during the Holocene: a multi proxy record from the eastern Arabian Sea. Holocene, v.13, pp.701-713.

Agnihotri, R., Sarin, M.M., Somayajulu, B.L.K., Jull, A.J.T., Burr, G.S. and Sarkar A. (2003b) Late- Quaternary Paleo-productivity and Organic carbon deposition record from the Eastern Arabian Sea. Paleogeo. Paleoclim. Paleoeco., v.197, pp.43-60.

Agnihotri, R., Kurian, S., Fernandes, M., Reshma, K., Souza, W. D. and Naqvi, S.W.A. (2008a) Variability of subsurface denitrification and surface productivity in the coastal eastern Arabian Sea over the past seven centuries. Holocene, v.18, pp.755-764.

Agnihotri, R., Altabet, M. A., Herbert, T. D. and Tierney, J. E. (2008b) Subdecadally resolved paleoceanography of the Peru margin during the last two millennia. Geochem. Geophys. Geosyst., v.9, Q05013, doi:10.1029/2007GC001744.

Altabet, M.A., Higginson, M.J. and Murray, D. W (2002) The effect of millennial-scale changes in Arabian Sea denitrification on atmospheric CO_2. Nature, v.415, pp.159-162.

Anderson, D.M., Overpeck, J.T. and Gupta, A.K. (2002) Increase in the Asian SW Monsoon during the past four centuries. Science, v.297, pp.596–599.

Berger, W. H., Killingley, J. S., Metzler C. V. and Vincent, E. (1985) ^{14}C dated high resolution $\delta^{18}O$ records from the tropical Atlantic. Quat. Res., v.23, pp.258-271.

Bhattacharyya, S. and Narasimha, R. (2005) Possible association between Indian monsoon rainfall and solar activity. Geophys. Res. Lett. V. 32, L05813, doi:10.1029/2004GL021044

Bhushan, R., Dutta, K. and Somayajulu, B.L.K. (2001) Concentrations and burial fluxes of organic and inorganic carbon on the eastern margins of the Arabian Sea. Mar. Geol., v.178, pp.95-113.

Bonatti, E., Fisher, D. E., Joensuu O. and Rydell, H. S. (1971) Postdepositional mobility of some transition elements, phosphorous, uranum and throrium in deep sea sediments. Geochim. Cosmochim. Acta, v.35, pp. 189-201.

Bond, G., Showers, W., Cheseby, M., Lotti, R., Kajda I. and Bonnani, G. (1997) A pervasive millennial-scale cycle in the north Atlantic Holocene and glacial climate. Science, v.278, pp.1257-1266.

Bond, G. Kromer, B., Beer, J., Muscheler,R., Evans, M.N., Showers, W., Hoffmann, S., Bond, R.L., Hajdas, I. and Bonani, G. (2001) Persistent solar influence on North Atlantic climate during the Holocene. Science, v.294, pp.2130–2136.

Boon, J.J, Rijpstra, W.I.C., de Lange, F., de Leeuw, J.W., Yoshioka, M. and Shimizu, Y. (1979) Black Sea sterol - a molecular fossil for dinoflagellate blooms. Nature, v.277, pp.125–127.

Brassell, S. C., Eglinton, G., Marlowe, I. T., Pflaumann, U. and Sarnthein, M. (1986) Molecular stratigraphy: A new tool for climatic assessment. Nature, v.320, pp.129–133.

Calvert, S.E., Pedersen, T.F., Naidu, P.D. and von Stackelberg, U. (1995) On the organic carbon maximum on the continental slope of the eastern Arabian Sea. J. Mar. Res., v.53, pp. 269–296.

Dean, W. E., Gardner J. V. and Piper, D. Z. (1997) Inorganic geochemical indicators of glacial-interglacial changes in productivity and anoxia on the California continental margin. Geochim. Cosmochim. Acta, v.61, pp.4507-4518.

Duplessy, J.-C. (1982) Glacial to interglacial contrasts in the northern Indian Ocean. Nature, v.295, pp.494-498.

Duplessy, J. -C., Bé, A. W. H. and Blanc, P. L. (1981) Oxygen and Carbon isotopic composition and biogeographic distribution of the planktonic foraminifera in the Indian ocean. Paleogeo. Paleoclim. Paleoecol., v.33, pp.9-46.

Dymond, J., Suess, E. and Lyle, M. (1992) Barium in deep-sea sediment: A geochemical proxy for paleoproductivity. Paleoceanogr., v.7, pp.163-181.

Fleitmann, D., Burns, S.J., Mudelsee, M., Neff, U., Kramers, J., Mangini, A. and Matter, A. (2003) Holocene forcing of the Indian Monsoon recorded in a stalagmite from southern Oman. Science, v.300, pp.1737–1739.

Fleitmann, D., Burns, S.J., Mangini, A., Mudelsee, M., Kramers, J.,Villa, I., Neff, U., Al-Subbary, A.A., Buettner, A., Hippler, D. and Matter, A. (2007) Holocene ITCZ and Indian monsoon dynamics recorded in stalagmites from Oman and Yemen (Socotra). Quat. Sci. Rev., v.26, pp.170–188.

Fontugne, M.R. and Duplessy, J.C. (1986) Variations of the monsoon regime during the upper Quaternary: evidence from carbon isotopic record of organic matter in north Indian Ocean sediment cores. Palaeogeogr. Paleoclim. Palaeoecol., v.56, pp.69-88.

Ganeshram, R.S., Pedersen, T.F., Calvert, S.E. and McNeil, G.W. (2000) Glacial-interglacial variability in denitrification in the world's oceans: Causes and consequences. Paleoceanogr., v.15, pp.361-376 .

Goes, J.I., Thoppil, P.G., Gomes, H.R. and Fasullo, J.T. (2005) Warming of the Eurasian landmass is making the Arabian Sea more productive. Science, v.308, pp.545–547.

Goswami, B.N., Venugopal, V., Sengupta, D., Madhusoodanan, M.S. and Xavier, P.K. (2006) Increasing trend of Extreme Rain Events over India in a Warming Environment. Science, v.314, pp.1442-1445.

Gupta, A. K., Anderson, D. M. and Overpeck, J. T. (2003) Abrupt changes in the Asian southwest monsoon during the Holocene and their links to the North Atlantic Ocean. Nature, v.421, pp.354-357.

Gupta, A.K., Das, M. and Anderson, D.M. (2005) Solar Influence on the Indian summer monsoon during the Holocene. Geophys. Res. Lett., v.32, L17703.

Haigh, J. D. (2001) Climate variability and the influence of the Sun. Science, v.294, pp.2109 - 2111.

Hu, F.S., Kaufman, D., Yoneji, S., Nelson, D., Shemesh, A., Huang, Y., Tian, J., Bond, G. and Brown, B.T. (2003) Cyclic Variation and Solar Forcing of Holocene Climate in the Alaskan Subarctic. Science, v.301, pp.1890-1893.

Jaffé, R., Rushdi, A.I., Medeiros, P.M., and Simoneit, B.R.T. (2006) Natural product biomarkers as indicators of sources and transport of sedimentary organic matter in a subtropical river. Chemosphere, v.64, pp.1870–1884.

Kodera, K. (2004) Solar influence on the Indian Ocean Monsoon through dynamical processes, Geophys. Res. Lett., v.31, L24209, doi:10.1029/2004GL020928.

Kumar, M.D., Naqvi, S.W.A., George, M.D. and Jayakumar, D.A. (1996) A sink for atmospheric carbon dioxide in the northeast Indian Ocean. J. Geophys. Res. (C: Oceans): v.101, pp. 18, 121-18,125.

Kurian, S., Agnihotri, R., Borole, D. V., Naqvi, S.W.A., Ferreira, A.M. and Vale, C. (2008) Solar control of primary production along the Indian west coast on decadal to centennial time scales. J. Quat. Sci. (press) DOI: 10.1002/jqs.1193.

Lockyer, N., (1900) Solar changes of temperature and variations in rainfall in the region surrounding the Indian Ocean. Science, v.12, pp.915–918.

Lückge, A., Rolinski, H. D., Khan, A. A., Schulz, H. and von Rad, U. (2001) Monsoonal variability in the northeastern Arabian Sea during the past 5000 years: geochemical evidence from laminated sediments. Palaeogeo. Palaeoclim. Palaeoecol., v.167, pp.273-286.

Madhupratap, M., Kumar, S. P., Bhattathiri, P. M. A., Kumar, M. D., Raghukumar, S., Nair, K. K. C. and Ramaiah, N. (1996) Mechanism of the biological response to the winter cooling in the northeastern Arabian Sea. Nature, v.384, pp.549–552.

Mangini, A., Spotl, C. and Verdes, P. (2005) Reconstruction of temperature in the Central Alps during the past 2000 yr from a d18O stalagmite record. Earth Planet Sci. Lett., v.235, pp.741-751.

Mann, M.E. (2007) Climate over the past two Millennia. Annu. Rev. Earth Planet. Sci., v.35, pp.111–136.

Meldrum, C. (1873) On a periodicity of rainfall in connection with the Sun-spot periodicity. Proc. R. Soc. Lond., v.21, pp.297–308

Mudge, S.M. and Norris, C.E. (1997) Lipid biomarkers in the Conwy Estuary (North Wales, U.K.): a comparison between fatty alcohols and sterols. Mar. Chem., v.57, pp.61– 84.

Naqvi, S. W. A. (1987) Some aspects of the oxygen deficient conditions and denitrification in the Arabian Sea. J. Mar. Res., v.29, pp.459-469.

Naqvi, S.W.A., Jayakumar, D.A., Narvekar, P.V., Naik, H., Sarma, V.V.S.S., D'Souza, W., Joseph, S. and George, M.D. (2000) Increased marine production of N_2O due to intensifying anoxia on the Indian continental shelf. Nature, v.408, pp.346-349.

Naqvi, S.W.A, Naik, H., Jayakumar, D.A., Shailaja, M.S., and Narvekar, P.V. (2006a) Seasonal oxygen deficiency over the western continental shelf of India. In: *Neretin, l. (ed.),* Past and Present Water Column Anoxia, NATO Science Series, IV. Earth and Environmental Sciences – v.64, Springer, pp. 195-224.

Naqvi, S.W.A., Naik, H., Pratihary, A., D'Souza, W., Narvekar, P.V., Jayakumar, D.A., Devol, A.H., Yoshinari, T. and Saino, T. (2006b) Coastal versus open-ocean denitrification in the Arabian Sea. Biogeosci., v.3, pp.621-633.

Naqvi, S.W.A., Narvekar P.V. and Desa, E. (2006c) Coastal biogeochemical processes in the North Indian Ocean (14, S-W). In: *Robinson, A.R. and Brink, K.H. (eds.)* The Sea, v.14B, Harvard University Press, pp. 723-781.

Parthasarathy, B., Munot, A.A. and Kothawale, D.R. (1995) Monthly and seasonal rainfall series for All-India homogeneous regions and meteorological subdivisions: 1871-1994.Contributions from Indian Institute of Tropical Meteorology, Research Report RR-065, Pune, INDIA.

Prahl F. G. and Wakeham, S. G. (1987) Calibration of unsaturation patterns in long-chain ketone compositions for palaeotemperature assessment. Nature, v.330, pp.367–369.

Prahl, F.G, Dymond, J. and Sparrow, M.A (2000) Annual biomarker record for export production in the central Arabian Sea. Deep-Sea Res. II, v. 47, pp.1581-1604.

Prahl, F.G. and Muehlhausen, L.A. (1989) Lipid biomarkers as geochemical tools for paleoceanographic study. In: *Berger, W.H., Smetacek, V.S., and Wefer, G. (eds.),* Productivity of the Ocean: present and past. Wiley, New York, pp. 271-289.

Prell, W. L. (1984) Variations of monsoon upwelling: A response to changing solar radiation. Geophys. Monograph Ser., AGU, v.29, pp. 48-57.

Rao, V.P., Montaggioni, L.,Vora, K.H., Almeida, F, Rao, K.M. and Rajagopalan G. (2003) Significance of relict carbonate deposits along the central and southwestern Margin of India for late Quaternary environmental and sea level changes. Sediment. Geol., v. 159, pp. 95-111.

Reichart, G.J. (1997) Late Quaternary variability of the Arabian Sea monsoon and oxygen minimum zone, Ph. D. thesis, University of Utrecht, Netherlands.

Reichart, G.J., Lourens, L.J. and Zachariasse, W.J. (1998) Temporal variability in the northern Arabian Sea Oxygen Minimum Zone (OMZ) during the last 225,000 years. Paleoceanogr., v.13, pp.607-621.

Rostek, F., Bard, E., Beufort L., Sonzogni, C. and Ganssen, G. (1997) Sea surface temperature and productivity records for the past 240 kyr in the Arabian Sea. Deep Sea Res., v.44, pp.1461-1480.

Saliot, A., Laureillard, J., Scribe, P. and Sicre, M.A. (1991) Evolutionary trends in the lipid biomarker approach for investigating the biogeochemistry of organic matter in the marine environment. Mar. Chem., v.36, pp.233–248.

Sarkar, A., Ramesh, R., Bhattacharya, S. K. and Rajagopalan, G. (1990) Oxygen isotope evidence for a stronger winter monsoon current during the last glaciation. Nature, v.343, pp.549-551.

Sarkar, A., Ramesh, R., Somayajulu, B. L. K., Agnihotri, R., Jull, A. J. T. and Burr, G.S. (2000) High resolution monsoon record from the Eastern Arabian Sea. Earth. Planet. Sci. Lett., v.177, pp.209-218.

Schubert, C.J., Villanueva, J., Calvert, S.E., Cowie, G.L, von Rad, U., Schulz, H., Berner, U., and Erlenkeuser, H. (1998) Stable phytoplankton community structure in the Arabian Sea over the past 200,000 years. Nature, v.394, pp.563–566.

Schulte, S., Rostek, F., Bard, E., Rullkotte, J. and Marchal, O. (1999) Variations of oxygen-minimum and primary productivity recorded in sediments of the Arabian Sea. Earth Planet Sci. Lett., v.173, pp.205–221.

Schulz, H., von Rad, U. and Erlenkeuser, H. (1998) Correlation between Arabian Sea and Greenland Climate oscillations of the past 110,000 years. Nature, v.393, pp.54-57.

Shetye, S.R. and Gouveia, A.D. (1998) Coastal circulation in the North Indian Ocean. Coastal segment (14, S-W). In: *Robinson, A.R. and Brink, K.H.(eds.)* The Sea, v.11, John Wiley & Sons, New York, pp.523-556.

Shindell, D., Rind, D., Balachandran, N., Lean, J. and Lonergn, P. (1999) Solar cycle variability, ozone, and climate. Science, v.284, pp.305-308

Sinninghe Damstè J.S, Kuypers, M.M.M., Schouten, S., Schulte, S. and Rullkötter, J. (2003) The lycopane/C31 n-alkane ratio as a proxy to assess palaeoxicity during sediment deposition. Earth Planet. Sci. Lett., v.209, pp.215-226.

Sinninghe Damste´, J.S., Rijpstra, W.I.C. and Reichart, G.J. (2002) The influence of oxic degradation on the sedimentary biomarker record II. Evidence from Arabian Sea sediments. Geochim. Cosmochim. Acta, v.66, pp.2737-2754.

Sirocko, F., Sarnthein, M., Erlenkreuser, H., Lange, H., Arnold, M. and Duplessy, J. -C. (1993) Century scale events in monsoon climate over the past 24, 000 years. Nature, v.364, pp.322-324.

Sirocko, F., Sconberg, D. G., Mclntyre, A. and Molfino, B. (1996) Teleconnections Between the Subtropical Monsoons and High -Latitude Climates During the Last Deglaciation. Science, v.272, pp.526-529.

Sirocko, F., Schonberg D. G. and Devey C. (2000) Processes controlling trace element geochemistry of Arabian Sea sediments during the last 25,000 years. Global and Planet. Change, v.26, pp.217-303.

Solanki, S. K., Usoskin, I. G., Kromer, B., Schu¨ssler M. and Beer, J. (2004) Unusual activity of the Sun during recent decades compared to the previous 11,000 years. Nature, v.431, pp.1084-1087.

Somayajulu, B. L. K., Yadav, D. N. and Sarin, M. M. (1994) Recent Sedimentary records from the Arabian Sea. Proc. Indian. Acad. Sci., 103, pp. 315-327.

Suthhof, A., Ittekkot, V. and Haake, B.G. (2001) Millennial scale oscillation of denitrification intensity in the Arabian Sea during the late Quaternary and its potential influence on atmospheric N_2O and global climate. Glob. Biogeochem. Cyc., v.15, pp.637-649.

Thamban, M., Rao, V. P., Schneider, R. R. and Grootes, P. M. (2001) Glacial to Holocene fluctuations in hydrography and productivity along the southwestern continental margin of India. Palaeogeogr. Palaeoclimatol. Palaeoecol., v.165, pp.113-127.

Thamban, M., Rao, V.P. and Schneider, R.R. (2002) Reconstruction of late Quaternary monsoon oscillations based on clay mineral proxies using sediment cores from the western margin of India. Mar. Geol., v.186, pp.527-539.

Thamban, M., Kawahata, H. and Rao, V.P, (2007) Indian summer monsoon variability during the Holocene as recorded in sediments of the Arabian Sea: timing and implications. J. of Oceanogr., v.63, pp.1009 - 1020.

Thamban, M. and Rao,V.P. (2001) New stable isotope records of sediment cores from the SE Arabian Sea – Inferences on the variations in monsoon regime during the late Quaternary. Curr. Sci., v.80, pp.1432-1436.

Tiwari, M., Ramesh, R., Somayajulu, B. L. K., Jull, A. J. T. and Burr, G.S. (2005a) Early deglacial (19–17 ka) strengthening of the northeast monsoon. Geophys. Res. Lett., v.32, L19712, doi:10.1029/2005GL024070.

Tiwari, M., Ramesh, R., Somayajulu, B. L. K., Jull, A. J. T. and Burr, G. S. (2005b) Solar control of southwest monsoon on centennial timescales. Curr. Sci., v.89, pp.1583-1588.

Tiwari, M., Ramesh, R., Yadava, M. G., Somayajulu, B. L. K., Jull, A. J. T. and Burr, G. S. (2006) Is there a persistent control of monsoon winds by precipitation during the late Holocene? Geochem. Geophys. Geosyst., v.7, Q03001, doi:10.1029/2005GC001095.

Tiwari, M. and Ramesh, R. (2007) Solar variability in the past and palaeoclimate data pertaining to the southwest monsoon. Curr. Sci., v.93, pp.477-487.

Volkman, J.K., Eglinton,G., Corner, E.D.S. and Sargent, J.R. (1980) Novel unsaturated straight-chain C37-C39 methyl and ethyl ketones in marine sediments and a coccolithophore *Emiliania huxley*. In: *A. G. Douglas and J. R. Maxwell (eds.),* Advances in Organic Geochemistry, pp. 219-227. Pergamon Press.

Volkman, J.K. (1986) A review of sterol markers for marine and terrigenous organic matter. Org. Geochem., v.9, pp.83–99.

Volkman, J.K., Barrett, S.M., Blackburn, S.I., Mansour, M,P, Sikes EL, and Francois G. (1998) Microalgal biomarkers: A review of recent research developments. Org. Geochem., v.29, pp.1163–1179.

Von Rad, U., Schaaf, M., Michels, K.H., Schulz, H., Berger, W.H. and Sirocko, F. (1999) A 5000–yr record of climate change in varved sediments from the oxygen minimum zone off Pakistan, northeastern Arabian Sea. Quat. Res., v.51, pp. 39–53.

Wagle, B.G. (1990) Beach rocks of the Central West Coast of India. Geomar. Lett., v.10, pp.111-115.

Wang, L., Sarnthein, M., Erlenkeuser, H., Grimalt, J., Grootes, P., Heiling, S., Ivanova, E., Kienast M., Pelejero, C. and Pflaumann U. (1999) East Asian monsoon climate during the late Pleistocene: high resolution sediment records from the south China Sea. Mar. Geol., v.156, pp. 245-284

Wang, Y., Cheng H., Edwards, R.L., He, Y., Kong X., An, J., Kelly, M.J., Dykoski, C.A. and Li, X. (2005) The Holocene Asian Monsoon: Links to solar changes and north Atlantic climate. Science, v.308, pp. 854-857.

Yadav, D.N. (1996) Manganese mobilization from the Western Continental margin of India. Curr. Sci. v.71, pp.900-905.

❑❑❑

13

Marine ^{14}C Reservoir Age and Suess Effect in the Indian Ocean

KOUSHIK DUTTA
AMS Radiocarbon Laboratory, Institute of Physics
Sachivalaya Marg, P.O. Sainik School, Bhubaneswar – 751 005, India
Email: koushikc14@gmail.com

Abstract

Apparent radiocarbon (^{14}C) ages of marine biogenic samples that derive their carbon from surface seawater dissolved inorganic carbon are on the average about 400 years older than contemporary terrestrial woods or global atmospheric CO_2. This age offset is due to mixing of old carbon from the deep ocean and is referred to marine ^{14}C reservoir effect. Both regional and temporal variations of ocean circulation pattern causes significant spatial and temporal variations in marine ^{14}C reservoir ages and hence of biogenic surface marine samples. Knowledge of reservoir age is very important to accurately calibrate ^{14}C-ages of biogenic carbonates and sediment organic matter from marine sediments that are frequently used in paleoceanographic studies. Here the concept of marine ^{14}C reservoir ages and their quantification are discussed and data for the Indian Ocean region are reviewed. Regional variations of marine ^{14}C reservoir ages and fossil fuel Suess effect for the Indian Ocean and the South China Sea are analyzed.

Introduction

The only naturally occurring radioactive isotope of carbon is ^{14}C, commonly known as radiocarbon. ^{14}C is introduced in the environment through its production in the upper atmosphere by the reaction of atmospheric nitrogen with thermal neutrons that are produced from cosmic ray spallation reaction on other atmospheric components.

$$^{14}N_7 + {}^1n_0 = {}^{14}C_6 + {}^1H_1 + \textit{(energy)}$$

The energetic ^{14}C atoms freshly produced in the atmosphere are soon oxidized in presence of atmospheric oxygen to form carbon monoxide which eventually oxidises to carbon dioxide.

$$^{14}C_6 + O = {}^{14}CO$$
$$^{14}CO + O = {}^{14}CO_2$$

Radioactive carbon dioxide $^{14}CO_2$ being indistinguishable from other forms of CO_2 ($^{12}CO_2$ or $^{13}CO_2$), it eventually enters the biosphere through photosynthesis by terrestrial plants, and then to the entire food chain through herbivorous animals. ^{14}C activities of most terrestrial living organisms are therefore in equilibrium with that in the atmosphere, through continuous exchange of ^{14}C by photosynthesis or food intake and respiration. When an organism dies the ^{14}C exchange halts, and the ^{14}C in the dead tissues start to decrease exponentially through radioactive decay. ^{14}C forms stable nitrogen through beta decay and half-life of 5730±40 years.

$$^{14}C_6 = {}^{14}N_7 + \beta^-$$
$$(T\tfrac{1}{2} = 5730 \pm 40 \text{ years})$$

There are various conventions for expressing ^{14}C concentrations in natural samples (Stuiver and Polach, 1977; Mook and van der Plicht, 1999). ^{14}C concentration in a sample can be expressed either as age in years before present or BP, where 'present' is actually the year AD 1950. For geochemical applications ^{14}C is commonly expressed as $\Delta^{14}C$ which is permil (‰) deviation of $^{14}C/^{12}C$ ratio of the sample with respect to a modern ^{14}C standard (whose ^{14}C activity equals to that of pre-industrial wood of AD 1890), the ratios being normalised for isotopic fractionation of the ^{13}C isotope and corrected for decay of ^{14}C. Old carbon containing samples (such as foraminifers from deep sea cores) have less ^{14}C than that present in a modern sample and therefore have positive ^{14}C-ages in BP and negative $\Delta^{14}C$ ‰ values; whereas most present day terrestrial wood samples with higher ^{14}C than the modern ^{14}C standard show negative ^{14}C-ages in BP and positive Δ14C ‰ values. Geologically old carbon samples (such as marble or old limestone, coal and other fossil fuels) are '^{14}C-dead' from which nearly all ^{14}C got decayed away. Such samples have infinite ^{14}C-age and $\Delta^{14}C$ close to –1000‰.

Oceans are important reservoir of exchangeable carbon (40,000 Gt), most of which are present as dissolved inorganic carbon (DIC). The natural steady-state concentration of ^{14}C in the DIC of surface ocean waters is in quasi-equilibrium with atmospheric ^{14}C production, air-sea exchange of atmospheric CO_2, and mixing with deeper ocean water (Fig.1). The mixing time-scale of the global oceans through large-scale thermohaline circulation is about 1000 years. This causes aging of a given water mass during its residence in the ocean since its last equilibration with the atmosphere when it was closer to the surface, and results in ^{14}C-ages of deep ocean water close to 1000 BP ($\Delta^{14}C$ about –200‰) (Bien, *et al.*, 1965; Stuiver and Quay, 1983). Diffusive and advective mixing with ^{14}C-depleted deep ocean waters causes the ^{14}C concentration for the surface oceans to be 5% lower on the average than in the contemporary atmosphere. Hence, the ^{14}C-ages of surface ocean waters are apparently older than contemporary terrestrial plants by about 400 years (Taylor and Berger, 1967; Bard, 1988; Hughen *et al.*, 2004). This phenomenon is known as marine ^{14}C reservoir effect. While the long-term temporal variations of surface ocean ^{14}C-ages are a function of the contemporary atmospheric ^{14}C levels and vertical mixing rates of the global ocean, their spatial variations also depend on the regional circulation patterns within the thermocline and regional air-sea CO_2 exchange rates. Local variations of marine ^{14}C-ages can take place due to input of hard water whose DIC is depleted in ^{14}C. Sources of hard water include rivers draining a geologically old carbonate terrain or ground water flowing through old carbonate aquifer. All these factors cause considerable spatial variations in the steady-sate (or natural) distribution of ^{14}C in the surface oceans and regional offsets of marine reservoir ^{14}C-ages from the global mean value, and make it difficult to use a common ^{14}C calibration curve for ^{14}C-dating of all marine samples. This is unlike ^{14}C-dating of terrestrial samples, which depend on past ^{14}C levels of rapidly mixing global atmosphere with a small difference between the two hemispheres.

The introduction of anthropogenic ^{14}C produced from atmospheric nuclear tests of the late 1950's and early 1960's had nearly doubled the ^{14}C levels of the atmosphere CO_2, which eventually transferred to other exchangeable carbon reservoirs as the biosphere and the oceans (Rafter and Fergusson, 1957). Mixing of bomb-^{14}C contaminated atmospheric CO_2 resulted in steady increase of the level of ^{14}C in the surface oceans. This injection of bomb ^{14}C in surface waters though obliterated the natural (pre-nuclear) ^{14}C signatures and its inventory in the water column provided a means of determining the air-sea exchange rates of CO_2. The results of state-of-the-art ocean general circulation models (GCMs), which predict the future rise of greenhouse gases in the atmosphere, must be verified through observations before they can be used for planning of policies to control anthropogenic CO_2 emissions. Such models can be also used to simulate steady state distributions of pre-nuclear $\Delta^{14}C$ for the global oceans (Maier-Reimer, 1993; Toggweiler *et al.*, 1989). Comparison of these model results with direct measurements of ^{14}C in surface dwelling marine calcareous organisms from the pre-nuclear era offer a way to check the validity of these models and testing their predictive capabilities.

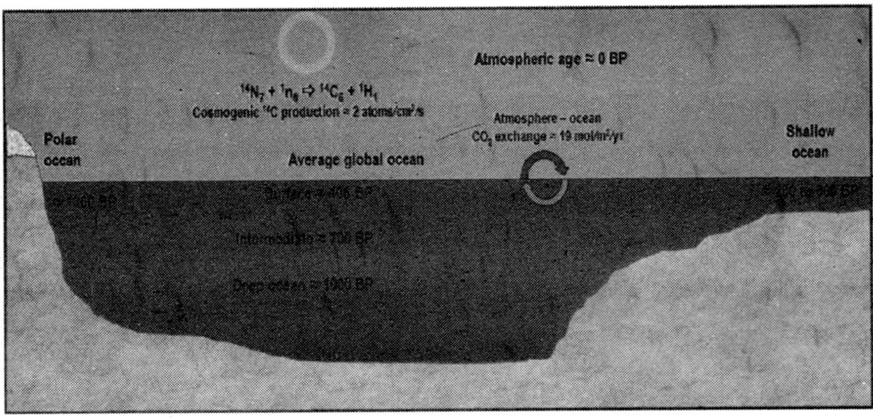

Fig. 1: Natural radiocarbon age distribution in the marine environment.

Knowledge about the natural (or pre-nuclear) level of ^{14}C in the surface ocean waters is necessary to determine:

(i) Offsets of regional reservoir ages from the modelled global mean—to calibrate the ^{14}C-ages of marine calcareous fossils, which grew during the pre-nuclear period and used for dating of marine sediments

(ii) Temporal variations of bomb-^{14}C in water column and its inventory—to assess the air-sea CO_2 exchange rates and decadal circulation in the thermocline

First oceanic measurements of ^{14}C started during the mid 1950's—about the same time when bomb-^{14}C started penetrating the world oceans. Soon, it was realized that bomb-^{14}C can be used as a powerful tracer to study air-sea exchange rates of CO_2 and mixing in the thermocline region on decadal time scales, utilizing the temporal variations of ^{14}C concentrations (Rafter and O'Brien, 1973; Broecker *et al.*, 1978; Broecker and Peng, 1982; Broecker, *et al.*, 1995). The need for information about the pre-nuclear surface ocean ^{14}C levels was felt, while attempting to determine the increase of the surface ^{14}C levels from the pre-nuclear values. To determine the penetration depth and inventory of bomb-^{14}C in oceans, it is essential to have knowledge about the surface pre-nuclear ^{14}C (Broecker, *et al.*, 1985). Several approaches have been made to determine these pre-nuclear surface ocean ^{14}C values from the analyses of suitable marine samples from the pre-nuclear era. Broecker *et al.*, (1985) estimated the pre-nuclear $\Delta^{14}C$ values for different oceanic regions of the world within ±10‰. These values range from a minimum of –140‰ in the polar oceans to –50‰ in the mid latitudes. However, the surface ocean circulations for the global oceans are far from simple to adopt a uniform scheme for meridional pre-nuclear $\Delta^{14}C$ variations for different oceans.

Chronological Applications of ^{14}C

Since carbon is ubiquitous in all living beings, the radioactive decay of ^{14}C can be conveniently used for age determination of organic or biogenic samples. The science of ^{14}C-dating was established by Prof. Willard Frank Libby of the University of Chicago, who for the first time demonstrated the potential of ^{14}C in dating of archaeological artefacts (Libby et al., *1949;* Libby, 1955), and was awarded the Nobel Prize in Chemistry in 1960 for this pioneering discovery. The half-life of ^{14}C (5730±40 years) is suitable for dating of samples that are within 50,000 years old, thus making it the most widely used dating method in Quaternary geochronology. To determine accumulation rates of marine sediments, the most common method in use is ^{14}C-dating of planktonic foraminifers. Adult species (250 to 500 ì) of surface dwelling foraminifera, such as *Globigerinoides ruber, Globigerinoides sacculifer, Neogloboquadrina dutertrie* and *Orbulina universa* are best suited for ^{14}C-dating, which grow within the top 100 m of the seawater column. In special cases such as in near shore areas where planktonic species are scarce, calcareous benthic foraminifera species may also be used. Marine organic matter may also be used for ^{14}C-dating of sediments when calcareous fossils are totally absent. However, as will be discussed later the last two methods may need special attention for their ^{14}C-age calibration. All samples chosen for analysis are carefully scrutinized and are thoroughly cleaned from all forms of contamination which may potentially alter their ^{14}C-age. Since the amount of datable carbon from sediments are usually small—typically few tens of milligrams from a gram of sediment, accelerator mass spectrometry (AMS) method is essential for ^{14}C analysis of these samples. For accurate calibration of marine ^{14}C-dates their initial ^{14}C-ages must be well known, that depend on regional marine reservoir effect.

Concept of marine ^{14}C reservoir age and ΔR correction values

For a given oceanic region at any given time, the ^{14}C reservoir age is defined as the difference between the measured conventional ^{14}C-age of the reservoir (seawater) and that of the contemporary atmospheric CO_2 (derived from tree-rings). ^{14}C reservoir ages can be determined from ^{14}C analysis of pre-nuclear (pre-1950s) marine calcareous samples of known calendar age. Considerable spatial variability is seen in the regional marine reservoir ^{14}C-ages from the global mean value, mainly due to variations in local ocean circulation patterns and regional exchange rates of CO_2 with the atmosphere. For a given region at any given time, the difference between the regional marine ^{14}C-age and that of the modelled global surface ocean is expressed by the term ΔR, which accounts for regional deviations of reservoir ages from the global mean value (Stuiver and Braziunas, 1993). The regional ΔR correction values can be determined from the difference of regional ^{14}C reservoir age and that of the modelled global ocean.

Marine reservoir ^{14}C-ages exhibit considerable spatial variations due to change in ocean circulation regimes. In those oceanic regions where favourable conditions exist for mixing with deeper ocean water (with older ^{14}C-ages) through wind induced upwelling, surface marine reservoir ^{14}C-ages will be higher than the global average value. Therefore, ΔR for such regions will have a positive value. Examples of such region in the Indian Ocean are the eastern and the western Arabian Sea, where upwelling takes place through seasonal monsoon induced winds. DIC inputs from rivers that leach geologically old carbonate rocks may result in large regional reservoir ^{14}C-ages (Little, 1993). On the other hand, ^{14}C in seawater DIC are higher than the average global ocean in shallow oceanic regions or in enclosed lagoons, where mixing with ^{14}C depleted water from deeper ocean is limited.

Such regions have negative ΔR values, examples of which include the Gulf of Kutch in the northeastern Arabian Sea, and the Chilika Lake in the northern Bay of Bengal. Shallow oceanic regions of the western Pacific such as the southern part of the South China Sea also have negative ΔR values, where the water depth is less than 600 m. According to the latest compilation of Global Marine Reservoir Correction Database (Reimer and Reimer, 2001), ΔR values for the global oceanic regions excluding the Southern Ocean vary from -200 year to +800 year, and majority of them fall between -50 years to +250 years. As discussed later, rapid vertical mixing off the Antarctica coast causes unusually high ΔR values for the Southern Ocean.

The Marine 04 calibration curve (Hughen *et al.,* 2004), is shown in Fig.2(a) which has been determined using the decadal tree-ring ^{14}C data and the carbon cycle model after Oescheger *et al.,* (1975). Regional ΔR correction values must be added to this curve in Fig-2(a) to obtain the regional marine ^{14}C calibration curves shown in Fig.2(b). Reconstruction of regional marine ^{14}C-age calibration curves for three hypothetical oceanic regions are shown in Fig-2(b). The Marine04 calibration curve extends back to 26,000 years BP. For calibrating older ^{14}C-dates up to 50,000 years BP one may use the calibration curve after Fairbanks *et al.,* (2005), which uses high-precision ^{14}C data of very old unaltered corals absolute dated with U-Th series isotopic method.

Knowledge of regional ΔR values is necessary for accurate calibration of ^{14}C-ages of marine samples. It is implicit in the definition of ΔR, that temporal variations of regional reservoir ages will parallel those of the global ocean, thus ΔR is assumed to be time independent for any given region (Stuiver *et al.,* 1986, 1998). The reservoir ^{14}C-ages and ΔR correction values reported from various oceanic regions are valid for the modern state of ocean circulation, since these were usually derived from samples that grown within past few hundred years or so. Since climate induced changes in ocean circulation may effect both mixing with deep water and atmosphere-ocean CO_2 exchange, regional reservoir ^{14}C-ages may respond to major shifts in climate, thus temporal variations of ΔR is inevitable. Reservoir ^{14}C-ages change at glacial-interglacial time scales, due to major shifts in ocean circulation pattern. Model simulations of past reservoir ^{14}C-ages have indicated that 30% reduction in Atlantic meridional circulation may enhance reservoir ages in high latitudes by 500 years (Franke *et al.,* 2008). If the changes in regional reservoir ^{14}C-ages are significantly different from that of the global modelled ocean, the ΔR values would also change. Thus, the assumption that ΔR values are constant for a given region is not strictly valid. Staubwasser *et al.*, (2002) and Ascough *et al.*, (2004, 2006) reported such changes indeed took place during the Holocene in the Arabian Sea and in the north Atlantic Ocean.

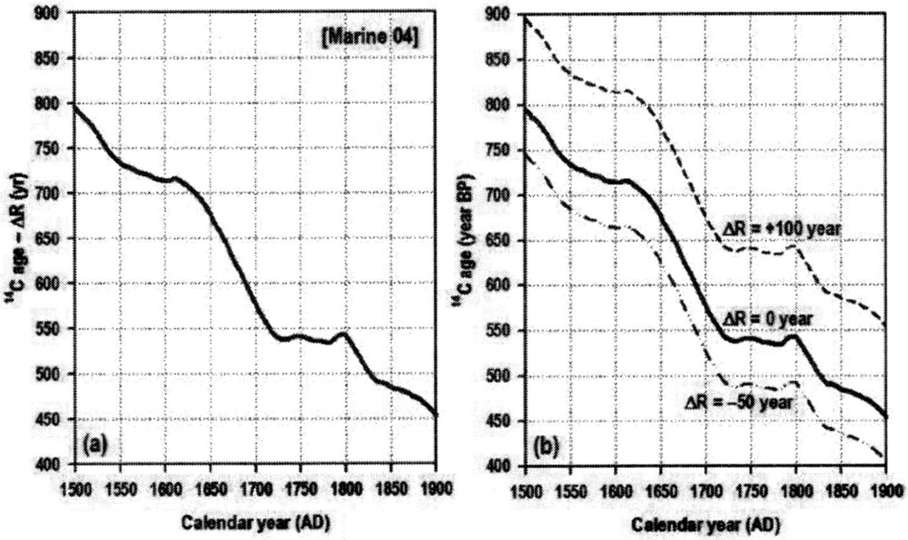

Fig 2 : Illustration of the concept of marine reservoir effect and ΔR correction values. (a) The Marine04 calibration curve (Hughen *et al.*, 2004). (b) Reconstructed marine ^{14}C-ages at three hypothetical oceanic regions. Top curve: in an upwelling region for which reservoir ^{14}C-age is older than the global mean (ΔR = +100 years). Middle curve: in an oceanic area that resembles the average world ocean (ΔR = 0 year). Bottom curve: an oceanic area with lower vertical mixing with deeper waters where reservoir ^{14}C-age is younger than the global mean (ΔR = –50 years).

Determination of marine ^{14}C reservoir effect and regional ΔR values

An ideal approach to determine the ^{14}C-age of seawater is to measure directly ^{14}C in its DIC (Fonselius and Östlund, 1959). However, direct measurements of ^{14}C in oceans began too late—only after the onset of nuclear weapons testing in the early 1950's (Rubin and Suess, 1955). Therefore, nearly all the seawater samples collected for ^{14}C determinations were contaminated with bomb ^{14}C. To determine pre-nuclear surface ocean Δ^{14}C, reservoir ^{14}C-ages and ΔR correction values, one must rely on the analysis ^{14}C in marine samples of known age, which were grown or collected before 1950. Some of the commonly used methods for determination of marine ^{14}C reservoir ages and ΔR correction values are described here.

(a) Marine shells

Measurement of ^{14}C in archived marine calcareous shells (*e.g.*, bivalves or gastropods) collected prior to the nuclear-testing era is the most common approach to determine marine ^{14}C reservoir ages or pre-nuclear levels of surface water $\Delta^{14}C$ (Stuiver *et al.*, 1986; Siani *et al.*, 2000; Dutta *et al.*, 2001). Short-lived and epifaunal (non-burrowing) mollusc species are preferred with well-documented collection dates. One can also measure ^{14}C in long-lived giant calms (e.g., *Tridacna gigas* or *Arctica islandica*), which form well-developed annual growth bands (Weidman and Jones, 1993). Shell derived marine ^{14}C reservoir ages often yield anomalous old ages due to incorporation of geologically old detrital carbonates in the shells (Dye, 1994). Several studies have indicated that ^{14}C-ages of various mollusc species differ due to difference in their habitat and diet. However, Ascough *et al.*, (2005a) concluded that any mollusc species may be suitable for the purpose if no geologically old carbonate rocks or sources of hard water are present.

(b) Corals

Hermatypic or reef building corals form annual growth bands made of aragonite ($CaCO_3$) with $^{14}C/^{12}C$ in equilibrium with the ambient seawater DIC. Once deposited these aragonite bands do not exchange their carbonate with external source unless they are very old causing recrystallisation. In a live collected coral these bands can be precisely dated back in time by counting them just as in tree-rings. Past changes in $^{14}C/^{12}C$ ratio of the DIC of surface-oceans can be determined from ^{14}C measurements in coral growth bands. Some of the earliest ^{14}C measurements in coral skeletons were reported by Moore *et al.*, (1973), Moore and Krishnaswami (1974) and Buddemeier *et al.*, (1974). Continuous records of ^{14}C variations in surface oceans can be deciphered from long corals, extending back to several hundred years in the pre-nuclear era (Nozaki *et al.*, 1978; Druffel and Linick, 1978; Druffel, 1981; Druffel and Suess, 1983; Druffel and Griffin, 1993; Grumet *et al.*, 2002; Hua *et al.*, 2004). The major disadvantage of using corals is their restricted geographical distribution. Their habitat is confined mainly within the tropical or limited sub-tropical oceanic regions.

(c) Charcoal and Marine Shell Pairs

Direct determination of the atmosphere-ocean ^{14}C-age difference can be obtained from ^{14}C measurements in co-existing terrestrial charcoal and marine shell pairs collected from same stratigraphic horizon, thus capturing the oceanic and terrestrial events simultaneously (Southon *et al.*, 1990, 1992; Talma, 1990; Little, 1993; Ingram, 1998). This method has potential to determine past variations of regional reservoir ages in response to the ocean circulation changes for longer timescales. To measure past reservoir ages, Bard (1988) had suggested comparing the ^{14}C-age of a short and well-defined marker or event both on the land and in the deep-sea sediments (*e.g.* deposition of a volcanic ash layer).

(d) Otoliths

Otoliths are hard calcareous deposits that grow within the inner ears of teleost or bony fish, which primarily function as gravity and auditory receptors. Similar to corals, otoliths are made of aragonite and form annual growth layers. AMS ^{14}C measurements in otoliths of known age marine fishes can be used for reconstructing the ^{14}C evolution in surface water (Higham and Hogg, 1995; Kalish, 1993; Kalish *et al.*, 2000). Advantage of using fish otoliths over other methods is their large geographical range. Fish being nektonic (or free swimming), their otoliths can integrate the past ^{14}C signals over much wider oceanic region, unlike other archives as in shells or corals that record oceanic ^{14}C variations only at their growth location.

Most of the methods described above have their own merits and demerits, and are often fraught with problems. Ascough *et al.* (2005b) reviewed on various intricacies involved in determination of marine reservoir ^{14}C-ages. Reservoir ages and ΔR correction values are usually applicable for surface marine samples, but may not be valid for samples such as benthic foraminifers which live at the sediment surface. However, the same ΔR correction values may be applicable for benthic samples too, if the water depth of their occurrence is within 50 m or so as often the case for near shore samples. Reservoir ages derived from surface marine biogenic carbonates are also applicable to marine organic matter, since they derive their carbon from the same reservoir (seawater DIC). However, complications may arise if the organic matter is a mixture of terrestrial and marine organic matters. Organic molecules of purely marine origin will register the marine reservoir effects in their ^{14}C-ages, while terrestrial organic matters may show a range of ages that can be either too old or too young. In such cases, it may be beneficial to assess the contribution of each organic matter types from geochemical (C/N wt.) or isotopic ($\delta^{13}C$) ratios, or even better to isolate individual organic compounds of purely marine origin. All reservoir ^{14}C-age estimates are long-term average, ignoring any seasonal DIC ^{14}C changes that may occur due to seasonal change in ocean circulation pattern. Most known age marine biogenic carbonate samples used for regional reservoir age determinations are usually collected from coastal locations or small islands, which are then extrapolated for use with open ocean samples. Reservoir age corrections as derived using samples from small islands found to match fairly well to that derived from nearby coastal locations, since the surface waters of the coastal ocean are normally get replenished by the open ocean water in a short time scale.

Inspite of the availability of several established methods, pre-nuclear surface water ^{14}C data for many key oceanic regions are still sparse mainly due to scarcity of suitable known-age marine samples. The Indian Ocean region is an area of considerable interest from the perspective of paleoclimate and paleoceanographic studies. Unfortunately, spatial data of ^{14}C reservoir ages for this region is too meagre, essential for reporting reliable ^{14}C chronologies crucial for such studies.

Studies of Reservoir ^{14}C-ages and ΔR Correction Values for the Indian Ocean

The earliest seawater DIC ^{14}C measurements in the tropical Indian Ocean region were done during the expeditions named Monsoon (10°S, 99°E; November, 1960), Lusiad (8°N, 71°E; October, 1962) and Dodo (6°S, 55°E; September, 1964) by the Scripps Institute of Oceanography, USA (Linick, 1975). The surface ocean $\Delta^{14}C$ values measured during these periods ranged from –38 to –24‰, for the latitude band of 24°S to 11°S during the Monsoon expedition. These results, however, could not provide pre-nuclear $\Delta^{14}C$ values for the region. These values are significantly higher than those obtained from pre-nuclear $\Delta^{14}C$ in biological archives – indicating the penetration of bomb ^{14}C even during early 1960s. More extensive studies of surface water $\Delta^{14}C$ measurements were made between 1977 and 1978 as a part of the international GEOSECS Indian Ocean expedition (Stuiver and Östlund, 1983). To calculate the bomb ^{14}C inventories from the GEOSECS ^{14}C measurements, Broecker *et al.*, (1985) assumed pre-nuclear surface $\Delta^{14}C$

values of -60 to -65‰ for the northern Indian Ocean. In a later report, Broecker *et al.,* (1995) assumed surface natural $\Delta^{14}C$ -59‰ in the equatorial Indian Ocean (GEOSECS 448), and a steadily decreasing $\Delta^{14}C$ trend was followed towards the north, assuming minimum $\Delta^{14}C$ value of -68‰ for the northern Arabian Sea (GEOSECS 416). Moore and Krishnaswami (1974) first reported measurement of ^{14}C in the growth bands of corals from the Gulf of Kutch. $\Delta^{14}C$ of coral growth bands grown during or before 1950s have been reported for corals from the Red Sea and Gulf of Aden (Cember, 1989), the Gulf of Kutch (Chakraborty, 1993) and the western Indian Ocean (Grumet *et al.,* 2002).

First systematic studies of marine ^{14}C-reservoir ages and ΔR correction values for the northern Indian Ocean were reported by Dutta *et al.,* (2001), from ^{14}C measurements in archived marine mollusc shells collected between 1930 and 1954. The ΔR correction values were calculated from the ^{14}C data of annual growth bands of corals from the Gulf of Kutch (Chakraborty, 1993; Bhushan *et al.,* 1994), and AMS ^{14}C dates of annually laminated sediments in the northeastern Arabian Sea (von Rad *et al.,* 1999). Southon *et al.,* (2002) reported ΔR correction values for several locations in the Indian Ocean and the South China Sea mainly using museum archived marine shells. Hua *et al.,* (2004) reported ΔR correction values for the Cocos (Keeling) Islands in the eastern Indian Ocean using annual coral bands.

Spatial and Temporal Variations of Marine ^{14}C Reservoir Ages in the Indian Ocean

The data of marine ^{14}C reservoir ages and ΔR correction values for the Indian Ocean and South China Sea are compiled here and shown in Fig-3. The ^{14}C-reservoir ages for the open eastern Arabian Sea range from 622 to 390 years, which are in general higher when compared to the open Bay of Bengal values which range from 315 to 474 years. Reservoir ages in the shallow marine areas of the Gulf of Kutch and the Chilika Lake lagoon are 241 years and 229 years, respectively (Dutta *et al.,* 2001). One sample collected near Pondicherry in the western Bay of Bengal yielded anomalous high reservoir age of 831 years, possibly due to incorporation of geological carbonate (Southon *et al.,* 2002). These two shallow marine samples and the anomalous old sample from Pondicherry were excluded for determining regional ΔR correction values of the Indian Ocean. The reservoir ages are younger in the shallow marine areas of the South

China Sea. The Red Sea, the Persian Gulf, the western Arabian Sea and the western Indian Ocean near Madagascar have rather narrow range of reservoir ^{14}C-ages, mostly lying between 400 and 500 years. The eastern Indian Ocean south of Indonesia and west of Australia has reservoir ^{14}C-ages between 323 and 377 years, similar to the Bay of Bengal values.

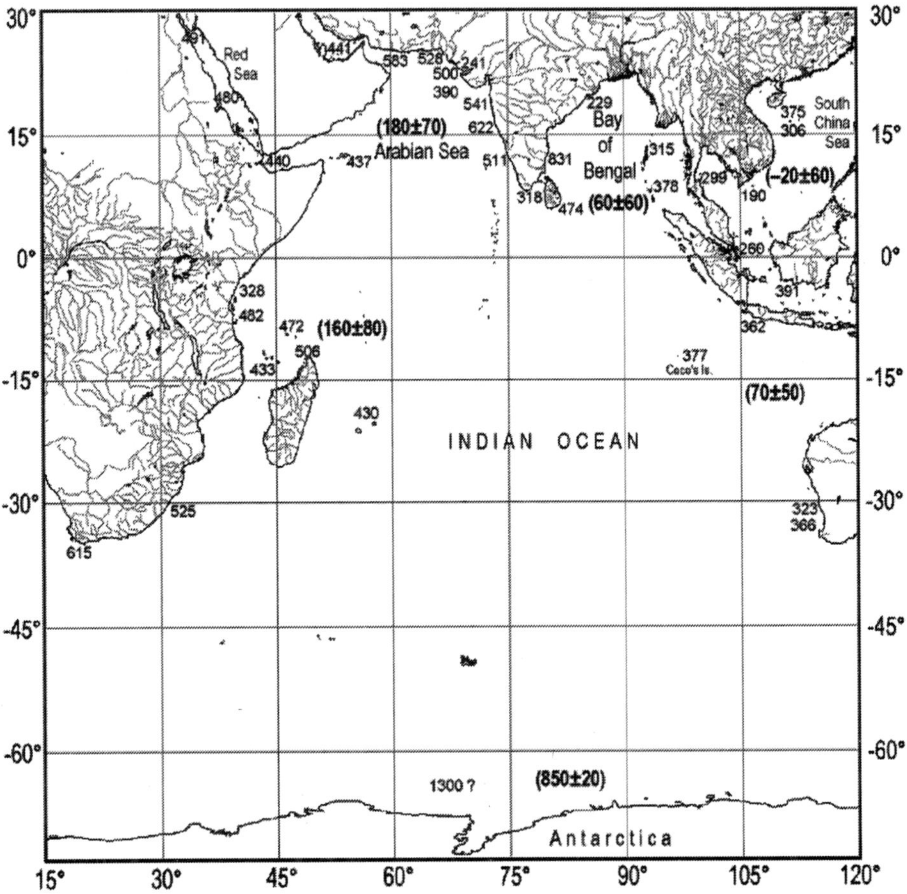

Fig. 3 : Spatial variation of marine ^{14}C reservoir ages for the entire Indian Ocean obtained from Global Marine Reservoir Correction Database (Reimer and Reimer, 2001). Reservoir ^{14}C-ages (in years) reported from selected locations are shown, while regional average ΔR values (in years) are in given bold with 1-s standard deviation. Indian Ocean reservoir ^{14}C-age data compiled from Bowman (1985a, 1985b), Cember (1989), Bhushan *et al.*, (1994), von Rad *et al.*, (1999), Dutta *et al.*, (2001), Southon *et al.*, (2002), Grumet *et al.*, (2002), and Hua *et al.*, (2004). Values for the southern Indian Ocean south of 60°S (in parentheses) have been estimated from Berkman and Forman (1996).

The observed pattern of ^{14}C reservoir ages in the Indian Ocean can be explained in terms of regional variation of circulation patterns within the thermocline. In the northeastern Arabian Sea vertical mixing is favoured by (i) seasonal upwelling during southwest monsoon and (ii) convective processes associated with winter cooling (Madhupratap *et al.*, 1996). The average reservoir ^{14}C-ages for the Arabian Sea is older than that of the modelled world ocean (mean $\Delta R \sim 180$ years), due to upwelling induced mixing with deeper ^{14}C-depleted water derived from Antarctic Bottom Water from the south. The Bay of Bengal, which receives large amount of fresh water from several major rivers (Milliman and Mead, 1983) has steep gradients of the isopycnal (equal density) surfaces within the top 200 m. Steep isopycnal gradient in the Bay of Bengal may greatly impede the vertical mixing rate, retarding the advection of deeper ^{14}C depleted water. This resulted in relatively younger reservoir ^{14}C-ages ($\Delta R \sim 60$ years) for the Bay of Bengal compared to the Arabian Sea. Input of riverine DIC depleted in ^{14}C may tend to counteract the effect of reduced mixing (Little, 1993), due to the hard water effect. Higher reservoir ages in the Arabian Sea compared to the Bay of Bengal shows that greater vertical mixing is the dominant process, which determines the observed pattern of ^{14}C reservoir ages in the northern Indian Ocean. Interestingly, the modelled pre-nuclear surface ocean ^{14}C distribution by Toggweiler *et al.*, (1989) and Guilderson *et al.*, (2000) predicts younger ^{14}C-ages in the Arabian Sea than in the Bay of Bengal. This is in contrary to the pattern observed from the ^{14}C data of pre-nuclear marine carbonates.

These ΔR values would be useful in calibrating the ^{14}C-ages of marine microfossils for dating marine sediments and also for dating coastal calcareous archeological samples from the Indian Ocean region. The regional average ΔR correction values given in Fig.3 have taken into account all valid reservoir ^{14}C-age data, and are only indicative of the large scale pattern. In order to calibrate a given marine ^{14}C date, ΔR data from the nearest locations must be chosen. Pooled mean (error weighted) ΔR correction values from these locations must be used for the calibration.

Significant climate induced changes in the northern Indian Ocean circulation took place during the glacial and interglacial periods (Duplessy, 1982). Since the marine ^{14}C reservoir age of the northern Indian Ocean, particularly in the Arabian Sea is controlled by summer monsoon upwelling, winter monsoon convection and the reservoir age of the Arabian Sea thermocline water which is derived from

Antarctic Bottom Water (Staubwasser *et al.*, 2002), variations in any of these factors due to global climate shifts can potentially alter the regional reservoir ^{14}C-ages. Staubwasser *et al.*, (2002) reported that the surface Arabian Sea reservoir ages were more than 1000 years during the deglaciation, and they varied between 780 and 1120 years during the early Holocene.

Marine ^{14}C Reservoir Ages of the Southern Indian Ocean off Antarctica

The southern Indian Ocean is a region for which knowledge of pre-nuclear surface water ^{14}C is too meagre. The earliest seawater ^{14}C measurements in the southern Indian Ocean region were compiled by Delibrias *et al.*, (1980). Reports of marine reservoir ^{14}C-ages close to Antarctica range from 932 BP to 1632 BP, corresponding to mean ΔR of 849±13 year (n=14). Till date, no systematic study on marine reservoir ^{14}C-ages had been done specifically for the Southern Indian Ocean near Antarctica. This oceanic region is an area of considerable interest from the perspective of paleoclimate studies from multi-proxy geochemical analyses of marine sediments. Such studies must rely on firm ^{14}C-chronology, for which accurate knowledge on reservoir ^{14}C-age is essential. Berkman and Forman (1996) recommended use of reservoir ^{14}C-ages of 1300±100 year for the Southern Ocean region. However, it is unlikely that reservoir age of the Southern Indian Ocean off Antarctica coast are significantly different than in other areas of the Southern Ocean, due to fast zonal mixing by the circum-polar currents.

Marine ^{14}C Suess Effect in the Northern Indian Ocean and the South China Sea

Following the beginning of industrial era during the late 19th century and before the onset of large scale atmospheric tests in the late 1950's, ^{14}C in the atmospheric CO_2 steadily decreased by about -25‰ through dilution by ^{14}C depleted fossil fuel CO_2. Invasion of fossil fuel CO_2 to the oceans caused significant depletion of surface ocean $\Delta^{14}C$ values between 1900 and early 1950s, and phenomena known as marine Suess effect (Druffel, 1981; Druffel and Suess, 1983; Toggweiler *et al.*, 1989). The observed depletion was typically few permil per decade, but variable among different oceanic areas (Druffel and Griffin, 1993). The $\Delta^{14}C$ values of the samples reported from the northern Indian Ocean and the South China Sea are analyzed in this study, which also show a decreasing trend between AD 1900 and 1955 (Fig.4). The results show the magnitude of marine ^{14}C Suess effect of -3.0±1.8‰ per decade for the Arabian Sea, -2.1±0.5‰ per decade for the Bay of Bengal, and -2.0±1.3‰ per decade for the South China Sea.

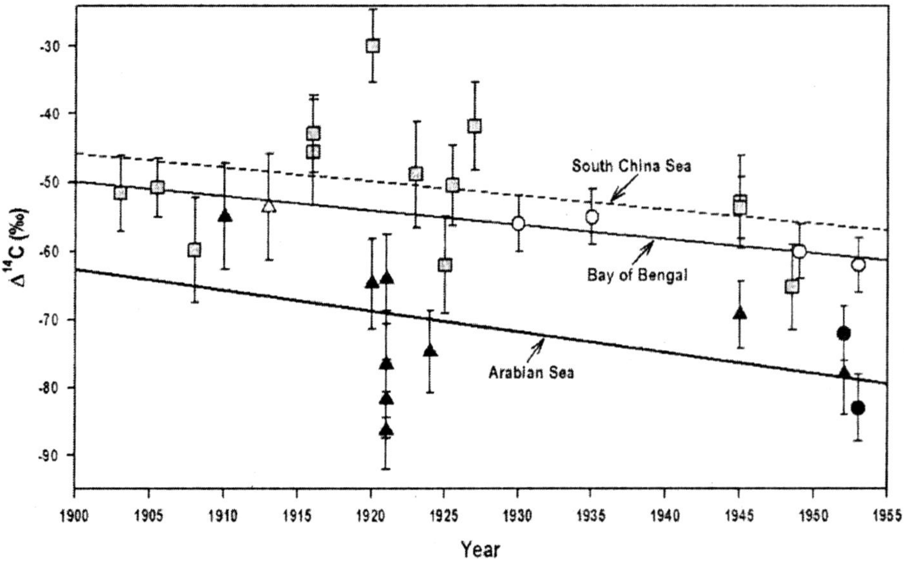

Fig. 4 : Suess effect in the Northern Indian Ocean and the South China Sea. Data from Dutta *et al.*, (2001) plotted as circles and from Southon *et al.*, (2002) plotted as triangles and squares. Symbol fills: white – Bay of Bengal, black – Arabian Sea; and grey – South China Sea. Straight lines indicate linear fit to the $\Delta^{14}C$ data between 1900 and 1955.

Even though the uncertainties in individual reservoir age estimates are of the order of ±5‰, several samples were considered to evaluate the Suess effect for a large oceanic area. Thus, the small values of the Suess effect (2 to 3±1‰) reported in this study are statistically significant. Magnitude of Suess effect for various ocean basins depends mainly on the penetration of anthropogenic CO_2 to the ocean. The present analysis shows similar magnitude of Suess effect for the Arabian Sea and the Bay of Bengal, as indicated by the parallel trends of the linear fits, though the vertical scales of these lines are offset depending on their reservoir ages.

Further measurements from the Indian Ocean region are needed to assess and reduce the uncertainty in marine ^{14}C Suess effect. This can be better understood from ^{14}C analysis in long coral cores dating back to the pre-industrial era. Druffel (1981) reported post-industrial and pre-nuclear $\Delta^{14}C$ recorded in corals from the Urvina Bay, Galapagos Island, in the eastern equatorial Pacific Ocean, which also show similar magnitude of Suess effect of about - 6‰ between 1930 and 1953.

Conclusions

^{14}C is commonly used for dating of marine sediments that are up to 50,000 years old. Initial ^{14}C-ages depend on regional marine ^{14}C-ages of surface oceans and any variations thereof must be known to accurately determine ^{14}C-dates of marine samples. Information on natural or pre-bomb ^{14}C in surface oceans is also important for studies of oceanic carbon cycle that use anthropogenic or bomb-^{14}C as a tracer. Reservoir ^{14}C-ages has been reported from many places in the Indian Ocean analyzing known age marine biogenic carbonates, but data on its spatial and temporal variability is still not adequate. The $\Delta^{14}C$ values between 1900 and 1955 from the Indian Ocean and the South China Sea show a decreasing trend, indicating marine ^{14}C Suess effect in the Bay of Bengal. The average magnitude of this Suess effect is -3±2‰ per decade for the Arabian Sea and -2±1‰ per decade for the Bay of Bengal and the South China Sea.

Acknowledgements

This research work was carried out at Physical Research Laboratory, Ahmedabad with a research fellowship from Dept. of Space, Govt. of India, and at Institute of Physics, Bhubaneswar, supported by Dept. of Atomic Energy, Govt. of India. Insightful comments by an anonymous reviewer greatly helped in improving the quality of this article.

References

Ascough, P.L., Cook, G.T., Dugmore, A.J., Barber, J., Higney, E. and Scott, E.M. (2004) Holocene variations in the Scottish marine radiocarbon reservoir effect. Radiocarbon, v.46, pp.611-620.

Ascough, P.L., Cook, G.T., Dugmore, A.J., Scott, E.M. and Freeman, S.P.H.T. (2005a) Influence of mollusk species on marine ÄR determinations. Radiocarbon, v.46, pp.611-620.

Ascough, P.L., Cook, G. and Dugmore, A. (2005b) Methodological approaches to determining the marine radiocarbon reservoir effect. Prog. Phys. Geogr., v.29, 532, doi: 10.1191/0309133305pp461ra.

Ascough, P.L., Cook, G.T., Church, M.J., Dugmore, A.J., Arge, S.V. and McGovern, T.H. (2006) Variability in North Atlantic marine radiocarbon reservoir effects at c. AD 1000. The Holocene, v.16, 131, doi: 10.1191/0959683606hl913rr.

Bard, E. (1988) Correction of accelerator mass spectrometry ^{14}C-ages measured in planktonic foraminifera: Paleoceanographic implications. Paleoceanogr., v.3, pp.635–645.

Berkman, P.A. and Forman, S.L. (1996) Pre-bomb radiocarbon and the reservoir correction for calcareous marine species in the southern ocean. Geophys. Res. Lett., v.23, pp.363-366.

Bien, G.S., Rakestraw, N.W. and Suess, H. (1965) Radiocarbon concentration in the Pacific and Indian Oceans and its relation to deep water movements. Limnology and Oceanogr., v.10, pp.25–37.

Bhushan, R., Chakraborty, S. and Krishnaswami, S. (1994) Physical Research Laboratory (Chemistry) radiocarbon date list I. Radiocarbon, v.36, pp.251–256.

Bowman, G.M. (1985a) Revised radiocarbon oceanic reservoir correction for Southern Australia. Search,' v.16, pp.164–165.

Bowman, G.M. (1985b) Oceanic reservoir correction for marine radiocarbon dates from northwestern Australia. Aus. Arch., v.20, pp.58–67.

Broecker, W.S., Peng T.–H. and Stuiver, M. (1978) An estimate of the upwelling rate in the equatorial Atlantic based on the distribution of bomb radiocarbon. Jnltribution of bomb radiocarbon. J. Geophys. Res., v.83, pp.6,179–6,186.

Broecker, W.S. and Peng, T.–H. (1982) Tracers in the Sea. Lamont Doherty Geological Observatory Press, Palisades, New York.

Broecker, W.S., Peng, T.–H., Östlund, H.G. and Stuiver, M. (1985) The distribution of bomb radiocarbon in the ocean. J. Geophys. Res., v.90, pp.6,953–6,970.

Broecker, W.S., Sutherland, S. Smethie, W., Peng, T.–H. and Östlund, H.G. (1995) Oceanic radiocarbon: Separation of the natural and bomb components. Glob. Biogeochem. Cyc., v.9, pp.263–288.

Buddemeier, R.W., Maragos, J.E. and Knutson, D.W. (1974) Radiographic studies of reef coral exoskeletons: rates and patterns of coral growth. J. Exp. Mar. Bio. and Ecol., v.14, pp.179–200.

Cember, R. (1989) Bomb radiocarbon in the Red Sea: A medium-scale gas exchange experiment. J. Geophys. Res., v.94, pp.2,111–2,123.

Chakraborty, S. (1993) Environmental significance of isotopic and trace elemental variations in banded corals. Ph.D.–Thesis, MS University of Baroda, Vadodara, India.

Delibrias, G. (1980) Carbon-14 in the Southern Indian Ocean. Radiocarbon, v.22, pp.684–692.

Druffel, E.R.M. and Linick, T.W. (1978) Radiocarbon in annual coral rings of Florida. Geophys. Res. Lett., v.5, pp.913–916.

Druffel, E.R.M. (1981) Radiocarbon in annual coral rings from the eastern tropical Pacific Ocean. Geophys. Res. Lett., v.8, pp.59–62.

Druffel, E.R.M. and Suess, H. (1983) On the radiocarbon record in the banded corals: Exchange parameters and net transport of $^{14}CO_2$ between atmosphere and surface ocean. J. Geophys. Res., v.88, 1,271-1,280.

Druffel, E.R.M. and Griffin, S. (1993) Large variations of surface ocean radiocarbon: Evidence of circulation changes in the southwestern Pacific. J . Geophys. Res., v.98, pp.249–259.

Dupplessy, J.-C. (1982) Glacial to interglacial contrasts in the northern Indian Ocean. Nature, v.295, pp.494–498.

Dutta, K., Bhushan, R. and Somayajulu, B.L.K. (2001) ΔR correction values for the Northern Indian Ocean. Radiocarbon, v.43, pp.483–485.

Dye, T. (1994) Apparent ages of marine shells: implications for archeological dating in Hawai'i. Radiocarbon, v.36, pp.51–57.

Fairbanks, R.G., Mortlock, R.A., Chiu, T.-C., Cao, L., Kaplan, A., Guilderson, T.P., Fairbanks, T.W. and Bloom, A.L. (2005) Marine Radiocarbon Calibration Curve Spanning 10,000 to 50,000 Years B.P. Based on Paired $^{230}Th/^{234}U/^{238}U$ and ^{14}C Dates on Pristine Corals. Quat. Sci. Rev., v.24, pp.781-796.

Fonselius, S. and Östlund, G. (1959) Natural radiocarbon measurements on surface water from the North Atlantic and the Arctic Sea. Tellus, v.11, pp.77–82.

Franke, J., Paul, A. and Schulz, M. (2008) Modelling variations of marine reservoir ages during the last 45000 years. Clim. Past, v.4, pp.125–136.

Grumet, N., Guilderson, T.P. and Dunbar, R.D. (2002) Meridional transport in the Indian Ocean traced by coral radiocarbon. J. Mar. Res., v.103, pp.24, 641–24,650.

Guilderson, T.P., Caldeira, K. and Duffy, P.B. (2000) Radiocarbon as diagnostic tracer in ocean and carbon cycle modeling. Glob. Biogeochem. Cyc., v.14, pp.887–902.

Higham, T.F.G. and Hogg, A.G. (1995) Radiocarbon dating of prehistoric shell from New Zealand and calculation of the ΔR value using fish otoliths. Radiocarbon, v.37, pp.409–416.

Hua, Q., Woodroffe, C.D., Barbetti, M., Smithers, S.G., Zoppi, U. and Fink, D. (2004) Marine reservoir correction for the Cocos (Keeling) Islands, Indian Ocean. Radiocarbon, v.46, pp.603–610.

Hughen, K.A., Baillie, M.G.L., Bard, E., et al. (2004) Marine04 marine radiocarbon age calibration, 0–26 cal kyr BP. Radiocarbon, v.46, pp.1,059–1,086.

Ingram, B.L. (1998) Differences in radiocarbon age between shell and charcoal from a Holocene shell mound in northern California. Quat. Res., v.49, pp.102–110.

Kalish, J.M. (1993) Pre- and post-bomb radiocarbon in fish otoliths. Earth Planet. Sci. Lett., v.114, pp.549–554.

Kalish, J.M., Nydal, R., Neadraas, K., Burr, G.S. and Eine, G.L. (2000) A 70 years long time series of radiocarbon from cod otoliths in the Barents Sea. Abstract, 17th International Radiocarbon Conference, June 18–23, 2000, Judean Hills, Israel, pp.94.

Libby, W.F., Anderson, E.C. and Arnold, J.R. (1949) Age determination by radiocarbon content: World-wide assay of natural radiocarbon. Science, v.109, pp.227–228.

Libby, W.F. (1955) Radiocarbon dating, 2nd edition. University of Chicago press, Chicago, U.S.A.

Linick, T.W. (1975) Uptake of Bomb-produced Carbon-14 by the Pacific Ocean. Ph.D.-Thesis, University of California, San Diego, U.S.A.

Little, E.A. (1993) Radiocarbon age calibration at archeological sites of coastal Massachusetts and vicinity. J. Archeol. Sci., v.20, pp.457–471.

Madhupratap, M., Kumar, S.P., Bhattathiri, P.M.A., Kumar, M.D., Raghukumar, S., Nair, K.K.C. and Ramaiah, N. (1996) Mechanism of the biological response to the winter cooling in the northeastern Arabian Sea. Nature, v.384, pp.549–552.

Maier-Reimer, E. (1993) Geochemical cycles in an ocean general circulation model: Preindustrial tracer distribution. Glob. Biogeochem. Cyc., v.7, pp.645–678.

Milliman, J.D. and Mead, R.H. (1983) World delivery of riverine sediments to the oceans. J. Geol., v.21, pp.1–21.

Mook, W.G. and van der Plicht, J. (1999) Reporting ^{14}C activities. Radiocarbon, v.41, pp.227-239.

Moore, W.S., Krishnaswami, S. and Bhat, S.G. (1973) Radiometric determination of coral growth rates. Bull. Mar. Sci., v.23, pp.157–176.

Moore, W.S. and Krishnaswami, S. (1974) Correlation of X-radiography revealed banding in corals with radiometric growth rates. In: Proceedings, 2nd International Coral Reef Symposium, Brisbane, Australia, pp.269–276.

Nozaki, Y., Rye, D.M., Turekian, K.K. and Dodge, R.E. (1978) A 200 year record of Carbon-13 and Carbon-14 variations in a Bermuda coral. Geophys. Res. Lett., v.5, pp.825–828.

Oescheger, H., Siegenthaler, U., Schotterer, U. and Gugelmann, A. (1975) A box diffusion model to study the carbon dioxide exchange in nature. Tellus, v.27, pp.168-192.

Rafter, T.A. and Fergusson, G.J. (1957) "Atomic bomb effect" – recent increase of carbon-14 content of the atmosphere and biosphere. Science, v.126, pp.557-558.

Rafter, T.A. and O'Brien, B.J. (1973) ^{14}C measurements in the atmosphere and in the South Pacific Ocean – A recalculation of the exchange rates between the atmosphere and the ocean. In: *Rafter, T.A., and Grant-Taylor, T. (eds.),* Proceedings of the 8th International Conference on radiocarbon dating, The Royal Society of New Zealand, v.1, pp.241-266.

Reimer, P.J. and Reimer, R.W. (2001) A marine reservoir correction database and on-line interface. Radiocarbon, v.43, pp.461-463.

Rubin, M. and Suess, H. (1955) US Geological Survey radiocarbon dates II. Science, v.121, 481-488.

Siani, G., Paterne, M., Arnold, M., Bard, E., Bernard, M., Tisnerat, N. and Bassinot, F. (2000) Radiocarbon reservoir ages in the Mediterranean Sea and the Black Sea. Radiocarbon, v.42, pp.271-280.

Southon, J.R., Nelson, D.E. and Vogel, J.S. (1990) A record of past ocean-atmosphere radiocarbon differences from the northeast Pacific. Paleoceanogr., v.5, pp.197-206.

Southon, J.R., Nelson, D.E. and Vogel, J.S. (1992) The determination of past ocean-atmosphere radiocarbon differences. In: *Bard, E., and Broecker, W.S. (eds.),* The Last Deglaciation: Absolute and Radiocarbon Chronologies. Springer-Verlag, Berlin, Germany, pp.219-227.

Southon, J.R., Kashgarian, M., Fontugne, M., Metivier, B. and Yim, W.W-S. (2002) Marine reservoir corrections for the Indian Ocean and southeast Asia. Radiocarbon, v.44, pp.167-180.

Staubwasser, M., Sirocko, F., Grootes, P.M. and Erlenkeuser, H. (2002) South Asian monsoon climate change and radiocarbon in the Arabian Sea during early and mid Holocene. Paleoceanogr., v.17, 1063, doi:10.1029/2000PA000608.

Stuiver, M. and Polach, H.A. (1977) Reporting of ^{14}C data. Radiocarbon, v.19, pp.355-363.

Stuiver, M. and Östlund H.G. (1983) GEOSECS Indian Ocean and Mediterranean radiocarbon. Radiocarbon, v.25, pp.1-29.

Stuiver, M. and Quay, P.D. (1983) Abyssal water Carbon-14 distribution and the age of the world oceans. Science, v.219, pp.849-851.

Stuiver, M., Pearson, G.W., and Braziunas, T.F. (1986) Radiocarbon age calibration of marine samples back to 9000 cal BP. Radiocarbon, v.28, pp.980-1,021.

Stuiver, M., and Braziunas, T.F. (1993) Modeling atmospheric ^{14}C influences and ^{14}C-ages of marine samples to 10,000 BC. Radiocarbon, v.35, pp.137-189.

Stuiver, M., Reimer, P.J., and Braziunas, T.F. (1998) High precision radiocarbon age calibration for terrestrial and marine samples. Radiocarbon, v.40, pp.1,127-1,083.

Talma, A.S. (1990) Radiocarbon age calibration of marine shells. In: Quarterly Report, Quaternary Dating Research Unit, CSIR, Pretoria, South Africa. pp.10.

Taylor, R.E., and Berger, R. (1967) Radiocarbon content of marine shells from the Pacific coasts of Central and South America. Science, v.158, pp.1,180-1,182.

Toggweiler, J.R., Dixon, K., and Bryan K. (1989) Simulations of radiocarbon in a coarse-resolution world ocean model, 1. Steady state prebomb distributions. J. Geophys. Res., v.94, pp.8,217-8,242.

von Rad, U., Schael, M., Michels K.H., Schulz, H., Berger, W.H., and Sirocko, F. (1999) A 5000-yr record of climate change in varved sediments from the oxygen minimum zone off Pakistan, northern Arabian Sea. Quat. Res., v.51, pp.39-53.

Weidman, C.R., and Jones, G.A. (1993) A shell-derived time history of bomb ^{14}C on Georges Bank and its Labrador Sea implications. J. Geophys. Res., v.98, pp.14,557-14,588.

□□□

14

Variation in Erythemal UV Dose for Indian Cities as Observed from Global Ozone Monitoring Experiment Data

NANDITA D. GANGULY
Department of Physics, St. Xavier's College, Ahmedabad-380009, India
Email: ganguly.nandita@gmail.com

Abstract

Variation in Erythemal UV Dose (EUVD) for five Indian cities and their expected critical limits has been studied using the data obtained from Global Ozone Monitoring Experiment (GOME). The EUVD values are found to be higher and peak earlier at places located at lower latitudes than at higher latitudes. They are higher during summer / monsoon compared to winter. Seasonal variation in EUVD is more pronounced at higher latitudes compared to lower latitudes. EUVD variations show a negative correlation with solar activity and are observed to be well within the statistically calculated critical limits for these five Indian cities.

Introduction

Every living organism has limits to the amount of solar ultraviolet irradiance, which it can endure, and this amount must be within appropriate levels for life to persist. According to ecologist Shelford (1977), each environmental factor has both minimum and maximum limits called critical limits beyond which a particular species finds it difficult to survive (Eugene, 1983). The solar UV radiation is classified as UV-A (320 - 400 nm), UV-B (280 - 320 nm) and UV-C (200 - 280 nm), based on the wavelength of radiation. While UV-C radiation is completely absorbed by atmospheric ozone and other gases, most of the UV-A radiation reaches the earth's surface, which is, however, not harmful to biological life on earth. The UV-B radiation, which is partially absorbed by ozone and reaches the earth's surface, is harmful to plants and animals. Of the global UV radiation at the ground, 94% is UV-A and 6% is UV-B. Due to the higher energy levels of UV-B, even this small fraction of UV dose is responsible for inducing sunburn and tanning, influencing the immune system, being the main cause of snow blindness, is an important factor in the induction of cataracts, contributes significantly to the aging of the skin and eyes, and is effective in causing skin cancer. However, decrease in UV - B beyond a certain limit would result in reducing the formation of vitamin D in the skin, which is essential for life on earth (Glerup, 2000). The solar UV-B irradiance at any location depends on the solar zenith angle, column ozone content, column aerosol content, cloud cover and the altitude of the observation site (Green, 1974). Singh *et al.,* (2006), have studied the effect of atmospheric aerosols on Erythemal UV Dose (EUVD) at a polluted urban location Delhi and a relatively cleaner site Port Blair (Andaman Islands) during March 2002. The observations at both sites were taken under clear sky conditions and the column ozone values were almost similar. They have observed that UV-B radiation intensities were 65% less at Delhi than at Port Blair, thus indicating that UV-B radiation forcing is influenced by Aerosol Optical Depth (AOD). Extensive ground based and satellite measurements and model calculations have been done to analyze the UV irradiance, particularly in the UV-B and erythemal region both in India and abroad (Blumthaler, *et al.,* 1994; Zerefos, *et al.,* 1995; Herman, *et al.,* 1996; Bodhaine, *et al.,* 1997; Herman, *et al.,* 1999). In view of these considerations, the variations in EUVD for five Indian cities: Chennai (13.08° N, 80.30° E, 60 meters above sea level), Hyderabad (17.37° N, 78.43° E, 536 meters above sea level), Mumbai (18.93° N, 72.85° E, 10 meters above sea level), Ahmedabad (23.05° N, 72.67° E, 55 meters above sea level) and New Delhi (28.62° N, 77.22° E, 239 meters above sea level) have been discussed in this paper using the data obtained from Global Ozone Monitoring Experiment (GOME), for the period 1996 - 2001. The expected upper and lower critical limits (UCL and LCL) for EUVD have also been calculated statistically.

Data and Analysis

GOME-based EUVD data for the period 1996 - 2001 have been developed by Ronald Van der A, Michiel van Weele and Jos van Geffen of TEMIS UV team and has been obtained from the website : www.temis.nl. Daily EUVD (kJ/m^2) has been estimated from the integration of erythemal UV index, as derived from satellite observations, from sunrise to sunset, with a time step of 10 minutes after taking the cloud cover information based on the International Satellite Cloud Climatology Project (ISCCP) database into account. The validation of the EUVD is, however, preliminary. The EUVD is computed at latitude/longitude grid with cells measuring 0.5 by 0.5 degrees, which amounts to about 50 x 50 km at the equator. The 'monthly climatology' is an average over all individual days of a given month for all available years. The 'monthly average' is an average over all individual days of the given month of the given year. Total cloud fraction for the period 1/1/2000 - 31/12/2000 has been obtained from the Moderate Resolution Imaging Spectroradiometer (MODIS). MODIS is one of the first passive satellite radiometers designed to observe aerosols from an altitude of about 700 km. It measures reflected solar radiance and terrestrial emissions in 36 wavelength bands between 0.405 and 14.385 μm, with resolutions between 250 m to 1 km. It was launched aboard Terra satellite in late 1999. The Goddard Chemistry Aerosol Radiation and Transport (GOCART) model contains gridded monthly averaged data of total cloud fraction with a spatial resolution of 2.5° x 2.0°.

Results and discussions

Variation of EUVD with Latitude and Season

The variations in EUVD for five Indian cities: Chennai , Hyderabad, Mumbai, Ahmedabad and New Delhi have been studied using the data obtained from GOME, for the period 1996 - 2001. The EUVD values are found to be higher for places located at lower latitudes than at higher latitudes for all months of the year (Fig. 1).

This is because when UV radiation passes through the atmosphere, it is partially absorbed by ozone and scattered by air molecules, aerosol particles and clouds. The total columnar ozone is minimum at the tropics where it is produced and increases at higher latitudes where it is transported. Apart from this, as the solar zenith angle is lower at the tropics and increases at higher latitudes, the UV - B radiation has to pass through a shorter path through the atmosphere in the tropics and is therefore less absorbed and scattered compared to higher latitudes, where it passes through a comparatively longer path through the atmosphere and is therefore absorbed and scattered to a greater extent.

Although the EUVD at all the five Indian cities is minimum in December (winter), it increases to reach a maximum in March at Chennai, in April at Mumbai and Hyderabad, in May at Ahmedabad and in June at New Delhi after which it decreases (Fig.1). Thus, the EUVD values are found to peak earlier at places located at lower latitudes than at higher latitudes. They are found to be higher during summer / monsoon months when the sun is directly overhead, compared to winter months when the sun is lower in the sky. Standard deviation for EUVD at Chennai, Hyderabad and Mumbai is maximum in February and at Ahmedabad and Delhi, it is maximum in March. Seasonal variation in EUVD are more pronounced at higher latitudes compared to lower latitudes, as observed from Fig.1 (Table-1). The difference between maximum (summer) and minimum (winter) values of EUVD is 3.7 kJ/m^2 at New Delhi, 3.07 kJ/m^2 at Ahmedabad, 2.65 kJ/m^2 at Mumbai, 2.53 kJ/m^2 at Hyderabad and 2.29 kJ/m^2 at Chennai. This is because the Solar Zenith Angle (SZA) at the lower latitudes is low and varies very little throughout the year resulting in high EUVD and lesser difference between summer and winter values compared to places at higher latitudes where the SZA is high and varies significantly from summer to winter.

Variation of EUVD with Solar Activity

The output radiation of the sun varies over a range of timescales of which the largest variation is due to the 11-year solar cycle that can be estimated by the average number of sunspots. The variation in solar irradiance is dependent on wavelength, with greater variability at shorter wavelengths (Solanki, *et al.*, 1998). Models and data estimates from the Upper Atmosphere Research Satellite Solar Stellar Irradiance Comparison Experiment instrument shows that although the total solar irradiance varies by only about 0.1% over the 11-year solar cycle, the amplitude of variation is as high as 8.3% for wavelengths in the 200 nm range and 0.85% for wavelengths in the 300 nm range (Lean, 2000). As the solar output increases with increase in solar activity, EUVD is also expected to increase with increase in solar activity. However, it is observed that the monthly mean EUVD (except for Delhi) exhibits a decreasing trend (Fig. 2; Table- 1) with increase in solar activity from 1996 - 2000. The variation in EUVD for the declining phase of solar activity from 2001 to 2006 could not be studied, as the monthly mean EUVD data from GOME was not available for this period.

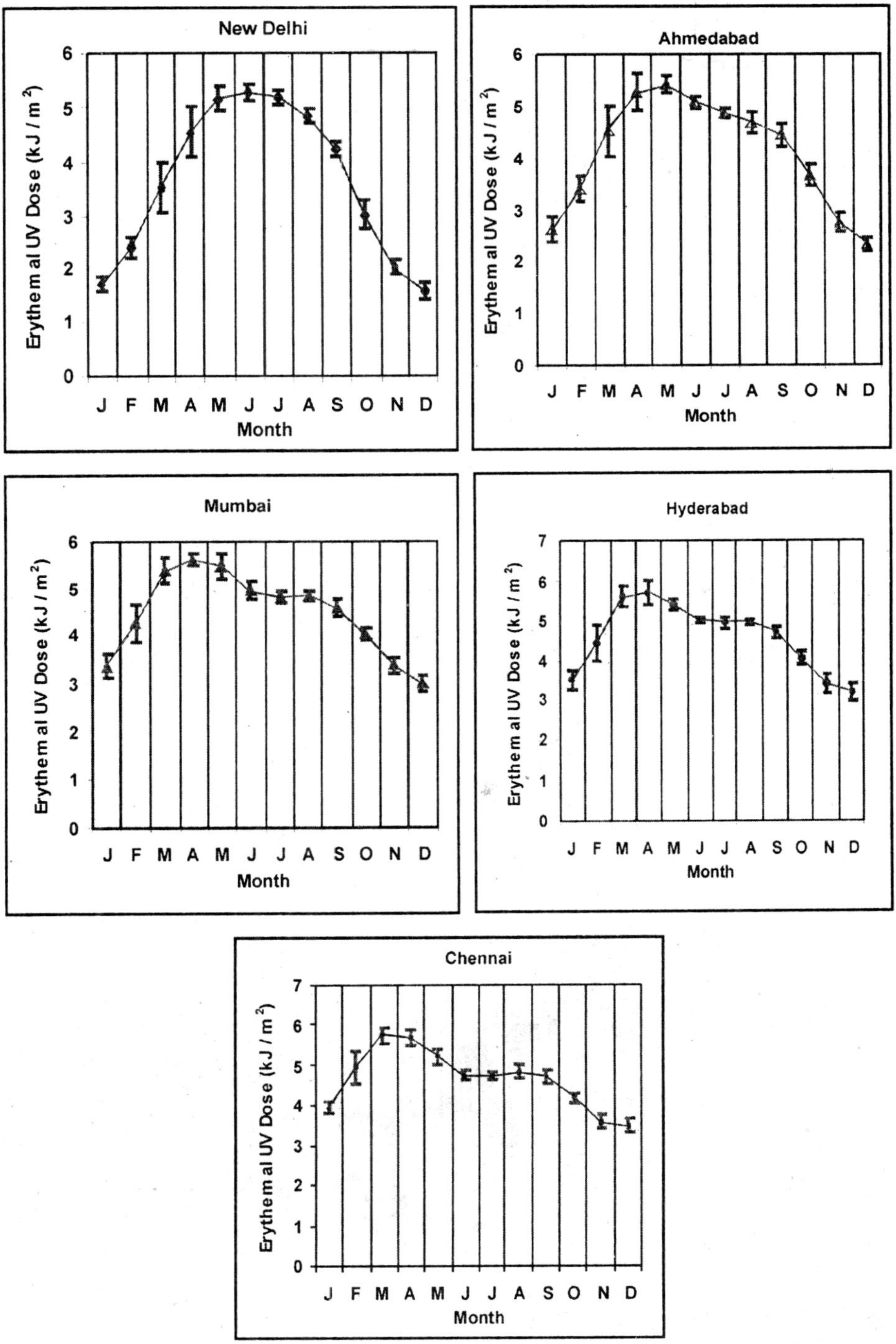

Fig. 1 : Monthly climatology for Erythemal UV Dose (EUVD) for different Indian cities (Source: http://www.temis.nl/).

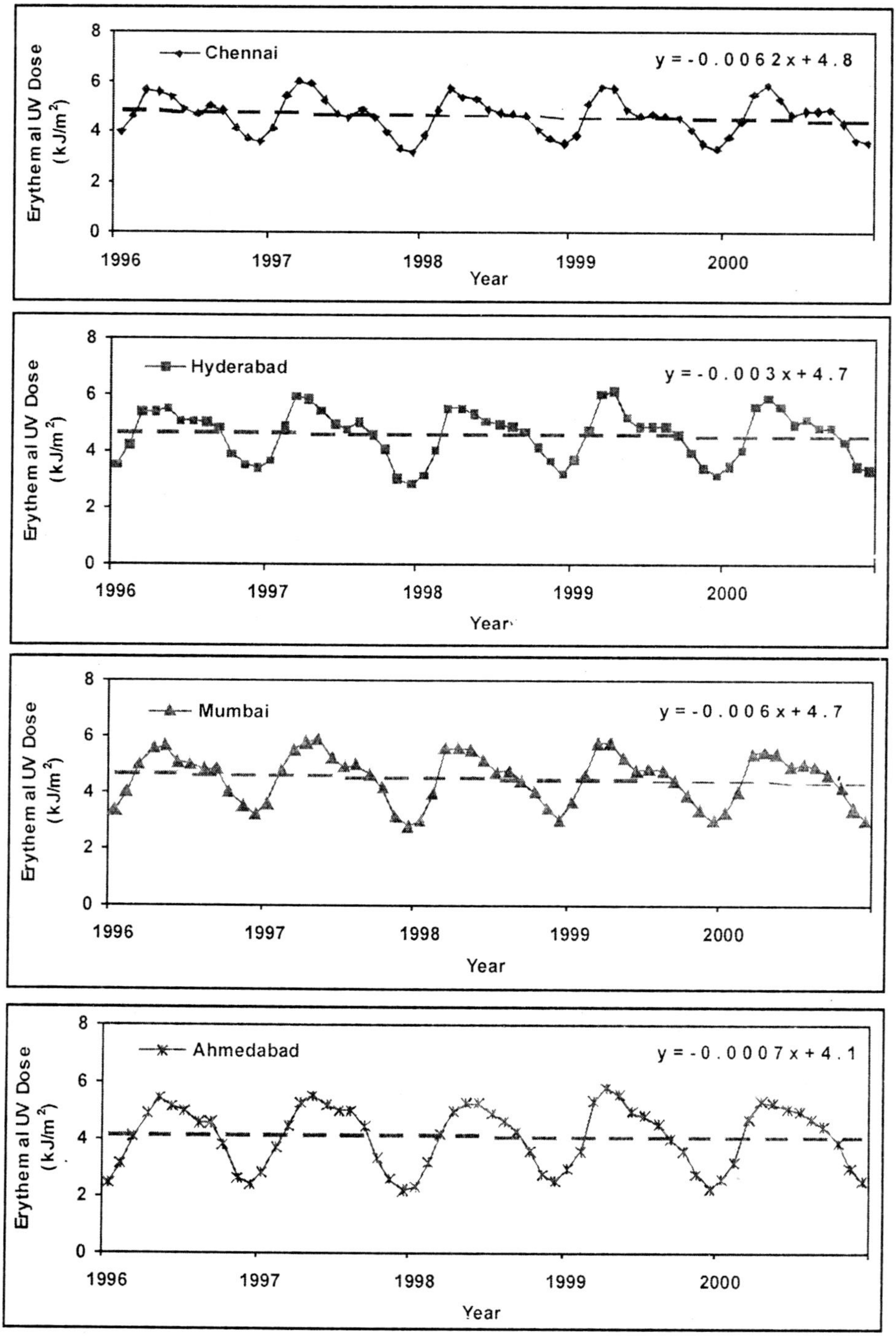

Fig. 2 : Trend of monthly average Erythemal UV Dose (EUVD) for different Indian cities from 1996 - 2000 (source: http://www.temis.nl/).

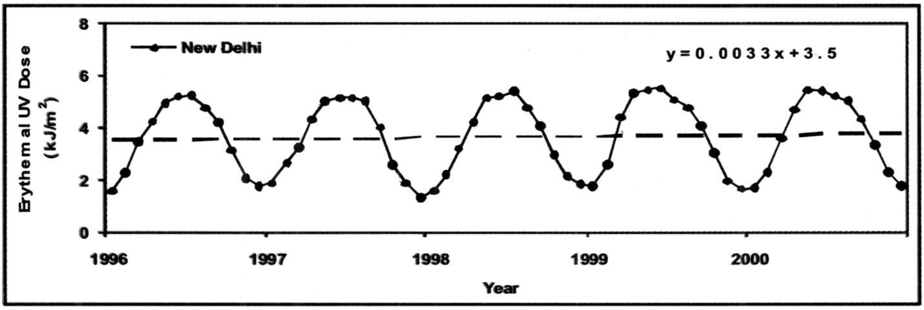

Fig. 2 : Contd... Trend of monthly average Erythemal UV Dose (EUVD) for different Indian cities from 1996 - 2000 (source: http://www.temis.nl/).

The small decreasing trend in EUVD with increasing solar activity may be attributed to an increase in population growth and rapid development at these urban cities, resulting in a gradual increase in pollutants and soot aerosols (Norris, 2001; Ramanathan *et al.*, 2005), which absorb and scatter a substantial amount of solar radiation. More over cloud cover has also increased over southern India in the recent years (Norris, 2001). Although the effect of clouds on UV radiation is difficult to quantify because of their rapid temporal variation, height, thickness and amount, a uniform cloud layer generally leads to a decrease in irradiance at the surface of the earth. However, local surface UV irradiance can be increased if clouds are not obstructing the disk of the sun and additional radiation is reflected from the side of a broken cloud field toward the ground (Nack and Green, 1974; Mims and Frederick, 1994). Basu (2001) has observed that cloud cover is less over the northern and central parts of India.

The total cloud fraction obtained from MODIS Goddard Chemistry Aerosol Radiation and Transport model (GOCART) for the period 1/1/2000 - 31/12/2000 (Fig. 3) in India shows that cloud fraction is higher at Chennai (0.65), Hyderabad (0.40), Mumbai (0.40) and Ahmedabad (0.30) compared to New Delhi (0.25). Thus, in general, the decreasing trend in EUVD is found to be higher for places having higher cloud fraction (except for Hyderabad and Mumbai). The higher decreasing trend in EUVD at Mumbai compared to Hyderabad may be because, although the cloud fraction for Hyderabad and Mumbai are same, the altitude of Hyderabad above sea level (536 meter) is higher than that of Mumbai (10 meter). More over, Mumbai being a more densely populated and developing city compared to Hyderabad, the level of aerosols and pollutants is also higher. Thus, gradual increase in pollutants, soot aerosols and cloud cover might be responsible for the observed decreasing trend in EUVD with increasing solar activity. Also a small uncertainty in the integration of Erythemal UV index, derived from satellite observations at these places, affecting the estimation of EUVD cannot be ruled out.

Table 1: Summary of Erythemal UV Dose (EUVD) and total cloud fraction observed at different Indian cities (sources: http://www.temis.nl/ and MODIS GOCART model).

City (Latitude)	Average monthly maximum EUVD (kJ/m ^{2}) for the period from 1996 - 2001	Average monthly minimum EUVD (kJ/m ^{2}) for the period from 1996 - 2001	Ratio of maximum EUVD for the period from 1996 - 2001	Trend of EUVD from 1996 - 2000	Total cloud fraction
Chennai (13.08 °N)	5.76 in Mar	3.47 in Dec	Che/Che = 1.00	Decreasing trend of 0.67 % / year	0.65
Hyderabad (17.37 °N)	5.70 in Apr	3.17 in Dec	Hyd/Che = 0.990	Decreasing trend of 0.31 % / year	0.40
Mumbai (18.93 °N)	5.62 in Apr	2.97 in Dec	Mum/Che = 0.976	Decreasing trend of 0.60 % / year	0.40
Ahmedabad (23.05 °N)	5.40 in May	2.33 in Dec	Ahm/Che = 0.934	Decreasing trend of 0.09 % / year	0.30
New Delhi (28.62 °N)	5.29 in Jun	1.59 in Dec	Del/Che = 0.918	Increasing trend of 0.48 % / year	0.25

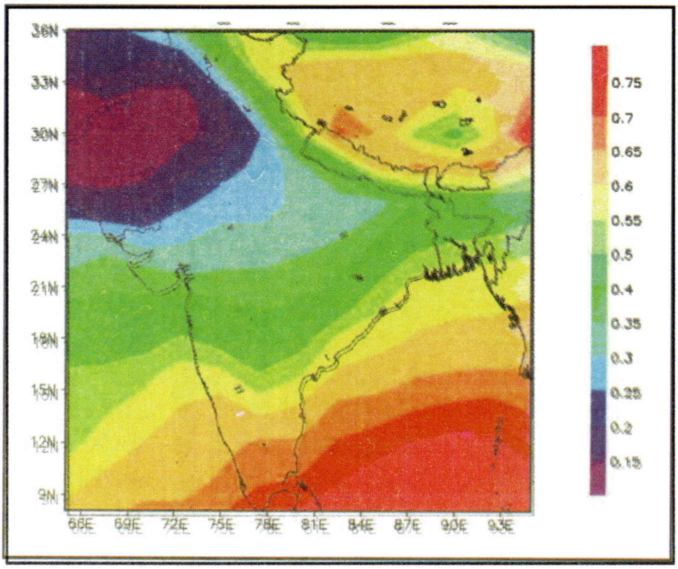

Fig. 3 : Total cloud fraction obtained from MODIS Goddard Chemistry Aerosol Radiation and Transport (GOCART) model for the period 1/1/2000 31/12/2000 in India.

Critical Limits for EUVD

Under equilibrium, living organisms and vegetation become gradually adapted to the prevalent natural dose of solar ultraviolet radiation. Living organisms and vegetation in the tropical region are able to tolerate a higher UVD compared to their counterparts at higher latitudes. Since increase in UV dose is responsible for inducing several negative effects on human beings and vegetation, but its decrease beyond a certain limit would result in reducing the formation of vitamin D in the skin, it is necessary to determine the upper and lower critical limits for EUVD. A statistical control chart, known as the 'Shewhart chart', separates the signal from noise. It is essentially a time plot of observations with critical limits added. If the variation in EUVD values are normal and in control, then all the points in the time plot should lie within the critical limits. The critical limits indicate that, when the variability of the process is large, some special cause (new and unanticipated sources of variation) is likely to be operating. When an observation exceeds the Upper Critical Limit (UCL), or lies below the Lower Critical Limit (LCL), a search for the special cause should be initiated. The LCL and UCL for EUVD are given by the relations:

$$\text{UCL} = (\text{Average EUVD value}) + (3 \cdot \text{Standard Deviation } \sigma)$$

$$\text{LCL} = (\text{Average EUVD value}) - (3 \cdot \text{Standard Deviation } \sigma)$$

Even when a process is in control (that is, no special causes are present in the system), there is approximately a 0.27% probability of a point exceeding the 3-sigma critical limits. For a Shewhart control chart using 3-sigma limits, this false alarm occurs on average once in every 370 observations. The upper and lower critical limits for the EUVD in different Indian cities have estimated statistically using the data from 1996 - 2001. The Shewhart charts for different Indian cities are plotted in figure 4.

According to the definition of "in control", the data points should meet the following four criteria:

(1) No sample point should lie outside the critical limits.

(2) Most points should lie near the average line.

(3) Nearly equal number of points should lie above and below the average line.

(4) The points should be randomly distributed.

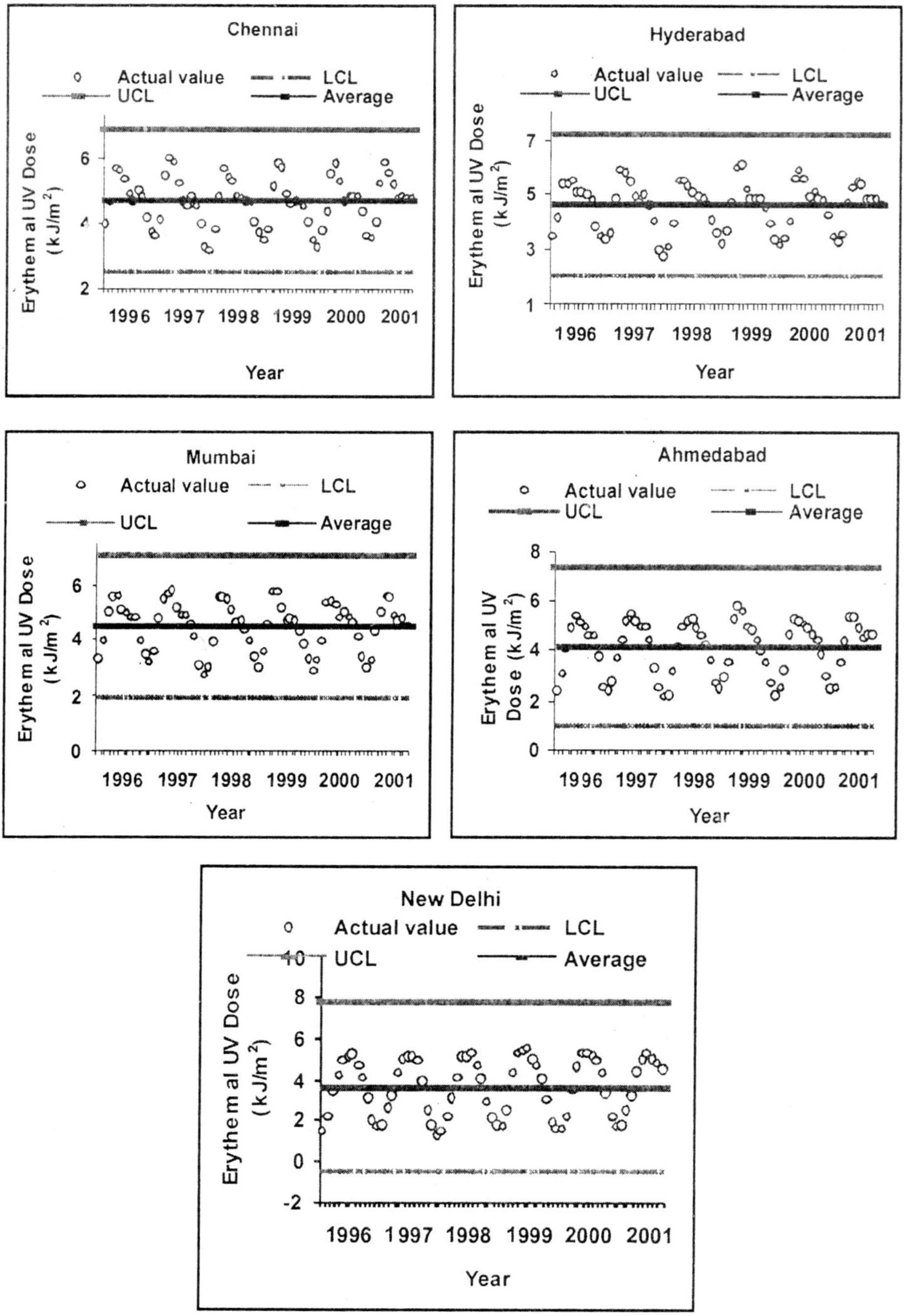

Fig. 4 : Shewhart chart (statistical control chart) representing the upper critical limit (UCL) and lower critical limit (LCL) for Erythemal UV dose in different Indian cities from 1996 – 2001. [http://www.temis.nl/].

Since the actual EUVD are observed to be well within the respective +/- 3σ (critical limits), and all the four criteria for control are met, fig. 4 suggests that the EUVD variations for these five Indian cities are under statistical control.

Conclusion

The variations in EUVD for five Indian cities have been studied using the data obtained from GOME, for the period 1996 - 2001. The EUVD values are found to be higher and peak earlier at places located in lower latitudes compared to higher latitudes and are found to be higher during summer / monsoon compared to winter. Seasonal variations in EUVD are more pronounced at higher latitudes compared to lower latitudes. EUVD variations (except for New Delhi) show a negative correlation with solar activity from 1996 - 2000. The EUVD variations for these five Indian cities are found to be under statistical control.

Acknowledgements

The author wishes to acknowledge her gratitude to the honorable reviewer for valuable suggestions, Ronald van der A, Michiel van Weele and Jos van Geffen of TEMIS UV team for providing the GOME-based Erythemal UVD data and Prof. J. N. Goswami, Director, Physical Research Laboratory, Ahmedabad for providing library facilities. The image for total cloud fraction from GOCART presented in the paper has been developed by NASA and has been obtained through GIOVANI.

References

Basu, B. K. (2001) Simulation of the Summer Monsoon over India in the Ensemble of Seasonal Simulations from the ECMWF Reanalyzed Data. J. of Climate, v.14 (7), pp. 1440- 1449.

Blumthaler, M., Webb, A.R., Seckmeyer, G., Bias, A.F., Huber, M. and Mayer, B. (1994) Simultaneous spectroradiometry: A study of solar UV irradiance at two altitudes. Geophysical Research Letters, v. 21, pp. 2805- 2808.

Bodhaine, B.A., Dutton, E.G., Hofmann, D.J., McKenzie, R.L. and Johnston, P.V. (1997) UV measurements at Mauna Loa. J. Geophysical Research, v. 102, pp.19265-19273.

Eugene O. (1983) Basic Ecology. Saunders College Publishing, Philadelphia.

Glerup, H., Mikkelsen, K., Poulsen, L., Hass, E., Overbeck, S., Thomsen, J., Charles, P. and Eriksen, E.F. (2000) Commonly recommended daily intake of vitamin

D is not sufficient if sunlight exposure is limited. J. Internal Medicine, v. 247, pp. 260-268.

Green, A. E. S., Mo, J. and Miller J. H. (1974) A study of solar erythema radiation doses. Photochemistry and Photobiology, v. 20, pp.473-482

Herman, J.R., Bhartia, P.K., Ziemke, J., Ahmad, Z. and Larko, D. (1996) UV-B increases (1979 - 1992) from decreases in total ozone. Geophysical Research Letters, v.23, pp.2117-2120.

Herman, J.R., Krotkov, N., Celarier, E., Larko, D. and Labow, G. (1999) The distribution of UV radiation at the earth's surface from TOMS measured UV backscattered radiances. J. Geophysical Research, v. 104, 12,059-12,076.

Lean, J. (2000) Evolution of the Sun's spectral irradiance since the Maunder minimum. J. Geophysical Research, v. 27(16), pp. 2425–2428.

McKinley, A. F. and Diffey, B. L. (1987) A reference action spectrum for ultraviolet induced erythema in human skin. CIE Journal, v. 6, pp. 17 - 22.

Mims, F. M. III and Frederick, J. E. (1994) Cumulus clouds and UV-B. Nature, v. 371, pp. 291.

Nack, M. L. and Green, A. E. S. (1974) Influence of clouds, haze, and smog on the middle ultraviolet reaching the ground. Applied Optics, v. 13, pp. 2405–2415.

Norris, J. R. (2001) Has Northern Indian Ocean cloud cover changed due to increasing anthropogenic aerosol. Geophysical Research Letters, v. 28, pp. 3271-3274.

Ramanathan, V., Chung, C., Kim, D., Bettge, T., Buja, L., Kiehl, J. T., Washington, W. M., Fu, Q., Sikka, D. R. and Wild, M. (2005) Atmospheric brown clouds: Impacts on South Asian climate and hydrological cycle special series of articles by members of the National Academy of Sciences, USA. PNAS, v. 102, pp. 5326-5333.

Shelford, V. E. (1977) Animal communities in temperate America: As illustrated in the Chicago region. Arno, New York.

Singh, R., Tanwar, R. S., Srivastava, M. K. and Singh, S. (2006) A comparative study of atmospheric aerosol effect on UV-B erythemal dosage at Delhi and Port Blair site, India.COSPAR2006-A-02780-1.

Solanki, S. and Unruh, Y. (1998) A model of the wavelength dependence of solar irradiance variations. Astronomy and Astrophysics, v. 329, pp. 747–753.

Zerefos, C. S., Bias, A. F., Meleti, C. and Ziomas, C. (1995) A note on the recent increase of solar UV-B radiation over northern middle latitudes, Geophysical Research Letters,v. 22, pp. 1245- 1247.

Website for Erythemal UVD: http://www.temis.nl/.

❑❑❑

15

Chemical Characteristics of Water Soluble Components of Fine Particulate Matter, $PM_{2.5}$ at Delhi, India

SURESH TIWARI[1], MANOJ K. SRIVASTAVA[2], and DEEWAN S. BISHT[1]

[1]Indian Institute of Tropical Meteorology, Pune, New Delhi Branch, Prof. Ram Nath Vij Marg, New Delhi-110060, India

[2]Department of Geophysics, Banaras Hindu University, Varanasi - 221005, India

E-mail : smbtiwari@yahoo.co.uk

Abstract

Aerosol samples in the size range up to 2.5μm were collected from January to December, 2005 at New Delhi, India, as a pilot experiment and analysed for organic (Oxalate and Formate) and inorganic (Sulfate, Ammonium, Nitrate, Potassium, Chloride, Sodium, Calcium and Magnesium) chemical components. Initial results show that the annual mean $PM_{2.5}$ concentration was 98.7μg/m^3, which varied between 38 and 285μg/m^3. The water soluble inorganic fraction constituted only 9% of $PM_{2.5}$ with SO_4^{2-}, NH_4^+ and NO_3^- being the dominant ions followed by potassium. Annual cycle shows higher concentration of $PM_{2.5}$ during winter season (October to January) and the lowest during monsoon. It is attributed to the enhanced production of aerosols and prevailing meteorological conditions. The $PM_{2.5}/PM_{10}$ ratio (0.86) coupled with the Hy SPLIT air-mass back trajectories indicated that $PM_{2.5}$ was dominated by fine particles, when the winds passed through the eastern azimuth, where many industries and major thermal power plants are located.

Introduction

Atmospheric aerosols have received much attention during the last two decades due to their potential influence on the global radiation budget and adverse impacts on human health and agriculture. Particulate matters with aerodynamic diameters less than 2.5μm ($PM_{2.5}$) have especially been found associated with increasing respiratory illness, carcinogens (Dockery and Pope, 1994), asthma (Anderson *et al.*, 1992) and ultimately in increasing number of premature deaths. Many epidemic studies have linked airborne concentrations of $PM_{2.5}$ and PM_{10} with a variety of health problems, including the morbidity as well as mortality (Wang *et al.*, 2003). The high levels of $PM_{2.5}$ have also been associated with amenity problems such as visibility degradation associated with haze (Milne *et al.*, 1982).

Aerosols, which are a complex mixture of elemental carbon, organic carbon, ammonium, nitrates, sulphates, mineral dust, trace elements, water vapour etc., have been examined for their composition and sources in urban and rural environments (*e.g.* Chow *et al.*, 1994). Water-soluble fraction of atmospheric aerosol of hygroscopic in nature, contains many important compounds, and can change size, composition, number-density and lifetime of aerosols (Novakov and Penner, 1993; IPCC, 1995; USEPA, 1996; Jacobson *et al.*, 2000). Also, water-soluble aerosols increase the solubility of toxic organic compounds by acting as surface active reagents and therefore increasing their toxicity to human health apart from altering their optical properties. Results show that, on an average, about 185 to 483×10^6 ton global aerosol load is emitted from anthropogenic activities per year, such as transportation, stationary combustion, industrial processes, solid waste disposal and other miscellaneous sources (Bridgman, 1990). Most of these sources are concentrated in the Northern Hemisphere, especially the areas downwind of the industrial sites, slash-and-burn agriculture regions, and overgrazed grasslands. Due to its characteristic economic, industrial and demographic nature, the atmospheric pollutants from the Indian region have different chemical compositions than the aerosols found in the developed parts of the globe. Some studies in Asian countries including India have even reported the organic, inorganic fractions, heavy metals and other chemical characterisation of the atmospheric aerosols (Khemani *et al.*, 1982, 1985; Smirnov *et al.*, 1998, Zhang and Friedlander, 2000, Kumar *et al.*, 2003; Sharma and Shaily, 2005; Patel *et al.*, 2006, Rengarajan and Sarin, 2006; and Rengarajan *et al.*, 2007).

In a city like Delhi, with over 15 million inhabitants contributing towards the anthropogenic aerosols, coupled with the desert dust aerosols from the north-western region (Singh *et al.*, 2006), a very high level of ambient particulate loadings is expected. An earlier study on air pollution and health in India from the World Bank (1995) shows that small particulate emission (PM_{10}) and gases (like SO_2) are responsible for 95% of health damage (Khillare *et al.*, 2004). A Delhi based non-government organization (NGO) has estimated an increase in premature deaths from 40,351 in 1991-92 to 51,779 in 1995 due to air related pollutions, out of which Delhi alone contributed to 7491 and 9859 during the same period suggesting one death per hour, on an average, due to the polluted air in the capital of India (Agarwal, 1992). A recent study conducted by Pande (1999) revealed that Delhi's air pollution is responsible for over 40% of the emergency hospital admissions with respiratory and heart problems. In the present study, we have studied the chemistry of $PM_{2.5}$ over Delhi using some discrete data samples. Particularly, the water–soluble chemical components have been studied in details.

Site Description and Aerosol Collection

Delhi (28^{0}35'N; 77^{0}12'E, 218 m above msl) experiences a severe weather swing between different seasons: from hot and humid weather in summer to cold and dry weather during winter. The prevailing wind throughout the year is easterly, northerly and northwesterly, and it is strongest during summer. Apart from such swings of weather during the annual cycle, the entire northern part of India, especially the Indo-Gangetic Plain, experiences a thick foggy weather during winter and show low boundary layer height. During such conditions, pollutants could not be dispersed or mix with free troposphere. The impact of such conditions is discernible as poor visibility and high levels of pollutants in this region. During the pre-summer and summer season, dust storm events also affect the climate of Delhi (Singh *et al.*, 2006).

The sampling of aerosols for this study was carried out at about 15m above the ground level, on the rooftop of a building situated in the thoroughly urbanized central part of Delhi. To be mentioned here that the sampling area is primarily a residential area, and no large pollutant source exists nearby which could have influenced the sampling site directly. Sampling location is given on the road map of Delhi in Fig.1.

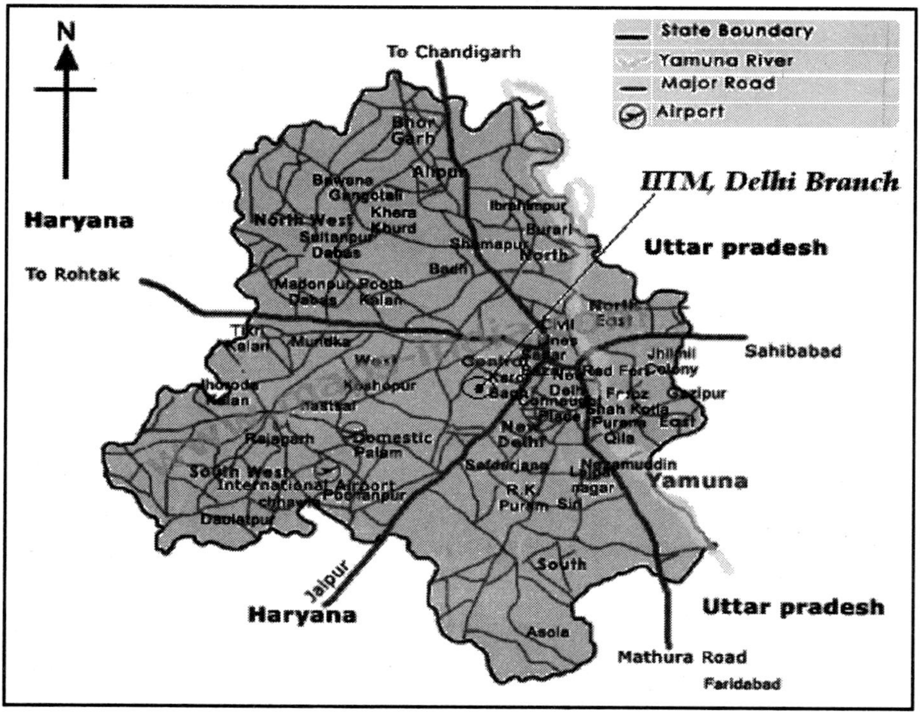

Fig.1: Sampling location in the road map of Delhi

Sampling of $PM_{2.5}$ was done by using Fine Particulate Sampler (APM 550; Envirotech Instruments Pvt. Ltd, design approved by USEPA). The initial flow rate was kept at about 1 m^3/ hr which subsequently reduced when the filters were loaded with fine particles. A 47mm quartz micro-filter (Whatman, UK) was used for sampling. The sampling duration for each sample was ~6 hrs [1100hrs to 1700hrs] and reflects the characteristics of the air pollution during the daytime. Total 14 samples of $PM_{2.5}$ were collected from January to December 2005. Table-1 shows the considered dates in this study, which are fairly distributed during various seasons.

For the chemical analysis, quartz micro-fibber filters were extracted with ultrapure water via ultrasonication and analyzed for water soluble ions. The ions, *viz.* organic (formate and oxalate), inorganic anions (chloride, nitrate and sulfate) and various cations (sodium, ammonium, potassium, calcium and magnesium) are obtained using Ion Chromatograph (IC). For quality assurance, calibrations of the IC were performed before and after the analysis. Field blanks were also obtained at regular intervals for improvement of the final results. These field blank filter papers are weighed and analyzed for the various chemical species. Values obtained by these field blanks were used for the correction of the final results.

Table 1: Dates of Observation

Sample Number	Dates
1	07.01.05
2	10.01.05
3	01.03.05
4	21.04.05
5	04.05.05
6	18.05.05
7	02.06.05
8	16.06.05
9	04.08.05
10	02.09.05
11	09.12.05
12	16.12.05
13	22.12.05
14	27.12.05

Results and Discussion

Variation in the Concentration of Ionic mass and Gravitational Mass of $PM_{2.5}$

Fig. 2 shows the daily variation in the mass concentration and measured ionic mass along with standards of $PM_{2.5}$. The average loading of $PM_{2.5}$ was found to be 98.71µg/m^3, which is approximately two and half times higher than the Indian National Ambient Air Quality Standard (NAAQS; 40µg/m^3) and almost six times higher than the United State Environmental Protection Agency (USEPA) standard (*i.e.* 15µg/m^3). The average loading of $PM_{2.5}$ showed highest deposition during winter (285µg/m^3) and lowest during monsoon (38µg/m^3) season. The high concentration of $PM_{2.5}$ during winter is expected to be caused by the combined effect of the elevated emissions from fossil fuel and coal burning and the prevailing meteorological conditions such as low temperature, wind speed, and mixing height. The lower inversion layer caused by low mixing height limits the dilution and dispersion of fine pollutants during winter season (Chatterjee *et al.,* 2007).

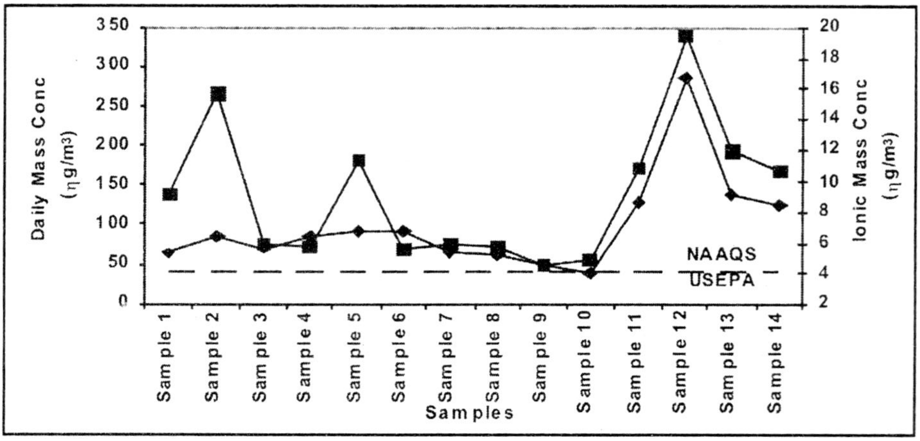

Fig. 2: Daily variation in the mass concentration and ionic mass along with the Indian and the US standard level of urban $PM_{2.5}$

Relation between gravimetrical $PM_{2.5}$ mass concentration and the summation of ionic mass of analyzed water soluble inorganic chemical constituents was studied and found that only 9% of $PM_{2.5}$ was made up by water soluble inorganic fraction. There are presumably the negative discrepancies due to the abundance of elemental carbon, organic matter, un-dissolved mineral dust, trace metals, and water etc. Similar negative discrepancy was also found by Hueglin *et al.*, (2005) for an urban side in Switzerland. The mass contributions for ionic mass (30%) were organic matter, black carbon (48%), mineral dust, trace metals (7%) and unknown and calculated water (15%).

The relative percentage contribution of each ionic constituent to the total ionic mass concentration was calculated and found that the contribution of secondary component, *viz.* SO_4^{2-}, was 51% (4.74μg/m^3) to the total ion mass, while the anthropogenic acidic components, which are in fine fraction, contributed approximately 75%. NH_4^+ and NO_3^- were the second and third highest water soluble components followed by potassium and oxalate. The contribution of alkaline inorganic components (Ca^{2+}, Mg^{2+}, Na^+, K^+, and NH_4^+) was approximately 25% to the total ionic mass. The calcium ion, which is mostly a soil generated components in the atmosphere, accounted for only ~2% of the total ion mass for $PM_{2.5}$. This is obvious in this size range as it is generally available in atmosphere in the coarse mode fraction. Similarly, the contribution of magnesium (1%) was also negligible in $PM_{2.5}$.

Ionic Variability

The ionic abundance in $PM_{2.5}$ ($\mu g/m^3$), on an average, showed the general trend for inorganic ions as $SO_4^{2-} > NH_4^+ > NO_3^- > K^+ > Cl^- > Na^+ > Ca^{2+} > Mg^{2+}$ whereas for the organic species as oxalate > formate. The variation of measured ionic species of $PM_{2.5}$ on different days during the study period is shown in Fig. 3. Variation in the mass concentration of chloride and sodium were observed to be significant. Being far from the ocean, non-sea salt ions are dominating the aerosol chemistry comprising about 91% of the total water soluble ions. Sulfate, ammonium, oxalate and potassium showed similarity in their variation. However, calcium showed the different variation and formate and nitrate showed the similar behavior. As obvious from Fig. 3, SO_4^{2-} and NH_4^+ formed the most abundant part of water soluble fraction of inorganic species in $PM_{2.5}$. Together, SO_4^{2-} and NH_4^+ accounted for ~6 % of overall $PM_{2.5}$ mass, but they were ~62% of the $PM_{2.5}$ water soluble inorganic ion mass. About 97 % of the total sulfate was from non-sea salt fraction. The SO_4^{2-} was significantly correlated with NH_4^+ ($r=0.91$, $p<0.0001$, Fig. 4). The inclination and the low intercept of the regression line strongly suggested that NH_4^+ and the most of SO_4^{2-} were probably present as $(NH_4)_2SO_4$ particles. It indicates often sufficient ammonia in the ambient air of Delhi, to neutralize the sulphuric acid formed from oxidation of sulphur dioxide. The mean NH_4^+ concentration during the study period was 5.76$\mu g/m^3$ and it further suggests that both these ions were originated from the local anthropogenic sources in Delhi. Whereas, some part of these ions might have also been incorporated in the ambiance of Delhi from the nearby industrial locations.

Trajectory Analysis

In order to identify the transport patterns of air parcels for higher $PM_{2.5}$ loading that may be arriving from different source regions to the sampling site, five-day HYSPLIT back trajectories were analyzed for all the considered days. Norman *et al.* (2001) have studied air mass trajectories during the monsoon period and found that the transport of chemical species was influenced by the prevailing winds. The wind associated with the low pressure system originated in the Bay of Bengal and moving towards inland or the easterly wind near the foothills of Himalayas, associated with the monsoon trough could carry the pollutants from the industrial belt located in the central part of India, eastern Uttar Pradesh and its neighborhood to the areas in downwind. Two days (on 16th and 22nd December 2005) were chosen to show the trajectory analysis since these days show higher concentration of $PM_{2.5}$. The obvious dates during winter seasons showed higher $PM_{2.5}$ mass, amongst the measured dates, due to prevailing meteorological concentrations, such as lower mixing layer height. Trajectories for the considered two days (Fig. 5a and 5b, respectively) at 500m, 1000m and 1500m, when $PM_{2.5}$ mass concentration found higher, were computed using the Hybrid Single Particle Lagrangian Integrated Trajectory (HYSPLIT, Version 4) model developed by National Oceanic and Air Administration/Air Resources Laboratory (NOAA/ ARL,USA) (Draxler and Hess, 1997). Air mass trajectory analysis showed that air parcels were impacted with the emissions from the surrounding industrial locations. As seen from the figure 5a, five days long air mass trajectories ending on December 16 were originated from west and north-west azimuth for all the height levels, which turned to be from the east azimuth in last phase. It was found that the concentration of $PM_{2.5}$ was the highest (285ìg/m^3) on December 16. This indicates towards the impact of sources from farther locations situated at cross-country boundary as well as adjoining industrialized places, such as Panipat (Haryana) and Ghaziabad (Uttar Pradesh) and other locations in the east of the sampling site. Similarly, air mass trajectories ending on December 22 at Delhi showed that winds were predominating easterlies with respect to Delhi, where the three thermal power plants (Rajghat, Indraprastha and Badarpur) are located. For this day, the $PM_{2.5}$ concentration was 137ìg/m^3. Chemical analysis showed that the concentrations of sulphate and ammonium were higher when wind blew from east azimuth as compared to those from the other directions and indicates the influence of emissions of precursor gases from sources located in the east.

Mass Ratio Between $PM_{2.5}$ and PM_{10}

The ratio of $PM_{2.5}$ to PM_{10} is used to determine the contribution of fine particulate to the respirable particulate matter with aerodynamic particle size less than 10µm. Fig.6 shows the ratio of $PM_{2.5}$ / PM_{10} for different stations in India as well as important places around the globe. The data for PM_{10} at Delhi was taken from the Central Pollution Control Board data bank. The mean $PM_{2.5}$ / PM_{10} ratio was found to be 0.86, which is remarkably high. Such higher values are expected to be due to the high concentration of $PM_{2.5}$ in the atmosphere at Delhi region attributed mainly to the combustion of fossil fuel during winter. As mentioned in Section 5.1, the low level inversion is one of the important causes responsible for higher concentration of $PM_{2.5}$ during winter.

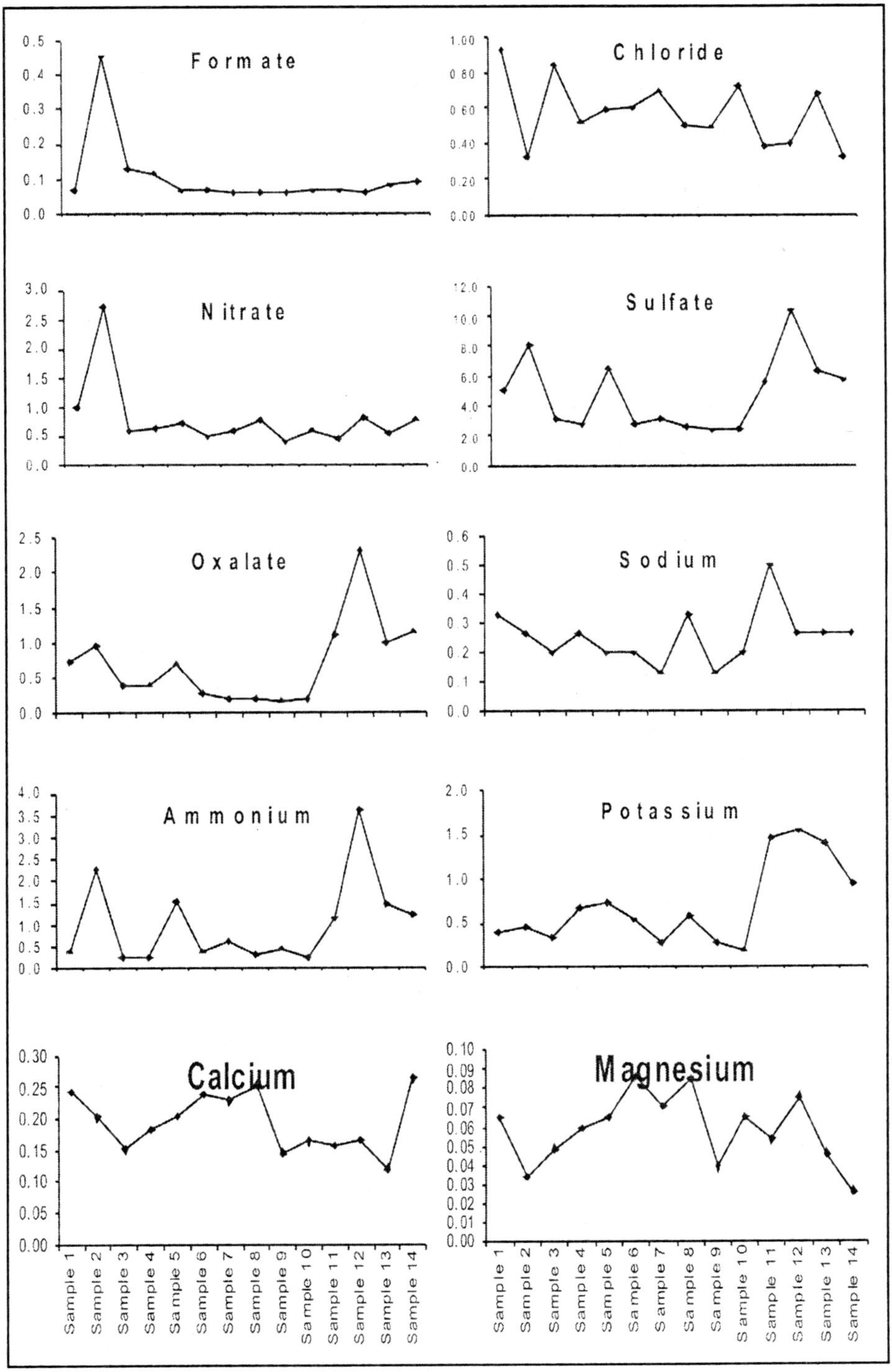

Fig.3: Daily variation of measured ionic species of $PM_{2.5}$

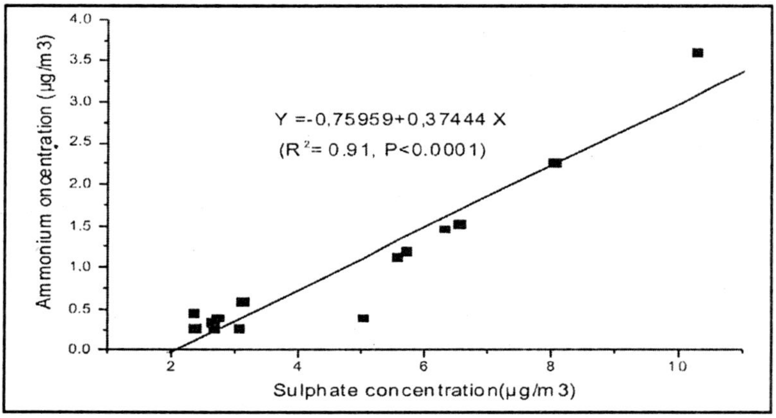

Fig. 4: Relation between Sulfate and Ammonium in $PM_{2.5}$

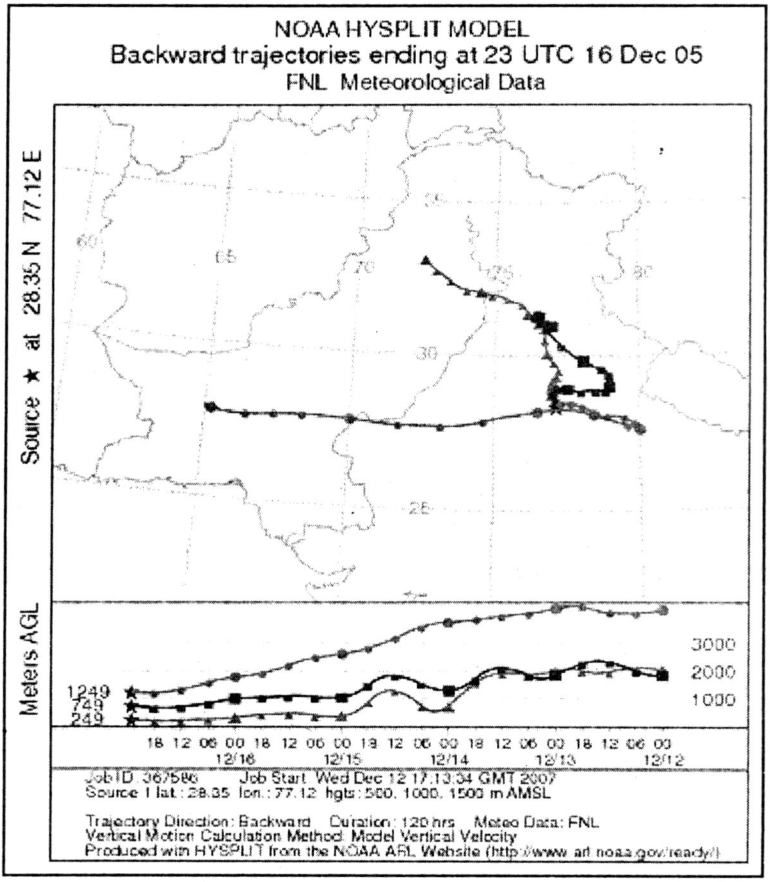

Fig. 5a : Air mass back trajectories at different heights for Delhi on December 16th, 2005.

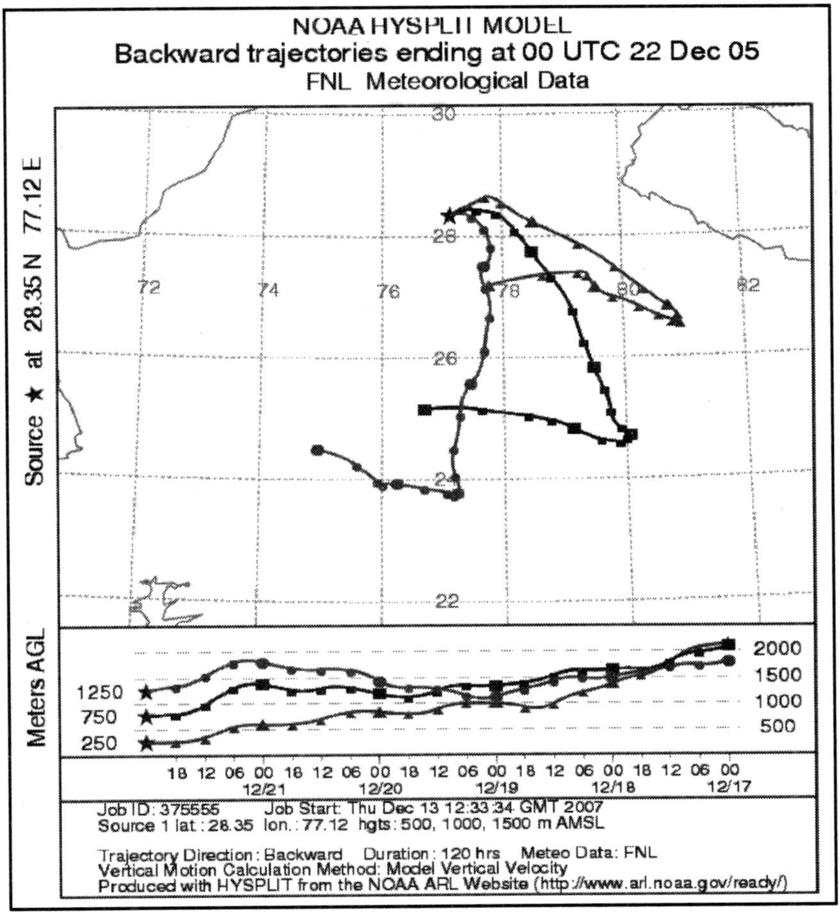

Fig. 5b : Air mass back trajectories at different heights for Delhi on December 22nd, 2005.

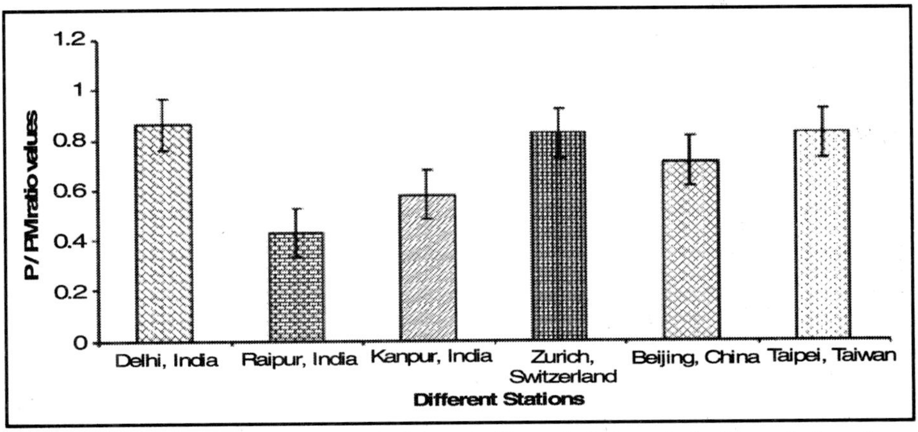

Fig. 6: The ratio of $PM_{2.5}$/ PM_{10} at Delhi and different stations in India and elsewhere

In the Indian region, however, the ratio values are low, in general, due to the dominance of PM_{10}, which are contributed by the wind-driven desert-dust. However, the $PM_{2.5}$ concentrations are mainly due to the combustion processes. The ratio values at Delhi are comparably higher than those reported at Raipur (Patel *et al.*, 2006), and Kanpur (Sharma and Shaily, 2005) where enormous burning of fossil fuels by various industries as well as thermal power plants and various industries like tanneries are located, respectively. This higher value also indicates towards very high contribution of $PM_{2.5}$ particle formation at Delhi. Ratio values at Zurich (Hueglin *et al.*, 2005), Beijing (Wang *et al.*, 2007), and Taipei (Mao *et al.*, 2007) were almost comparable with those observed at Delhi. This indicates more concentrations of fine size fraction in the ambiance of Delhi with respect to the total aerosol loadings and suggests more anthropogenic activities for the cause. At Beijing and Taipei, although the fine particle concentration was more as compared to Delhi, however, it is due to equally high concentration of coarse particle; that the ratio of $PM_{2.5}$ to PM_{10} at these two cities was comparable to that observed at Delhi.

Factor Analysis

In order to study the origin of these aerosols and the components dominating in it, factor analysis was undertaken. Factor loadings for each variable have been calculated by principal component analysis (Khare *et al.*, 2004). In general, factor loadings greater than 0.5 are considered significant in source apportionment studies. The factor loadings with the variance accounted by the various factors for Eigen value >1 are presented in Table-2. The data is synthesized into three major factors, explaining about 77.3% of the total variance. The division of information into different groups (common factors) can be interpreted as the variables listed within the same factor that might have the same origin.

The first factor comprises approximately half of the total variance (43%) and consists of potassium, ammonium, nitrate, oxalate and sulfate. The high loading of these ions in this factor is associated with the influence of secondary pollutants in the atmosphere, from the industries such as from thermal power plant as well as from automobile emissions. Ammonia (NH_3) is the most abundant gaseous alkaline component in the atmosphere. Particulate ammonium (NH_4^+) found in the atmosphere originates from NH_3 and its mean concentration in $PM_{2.5}$ was found to be 1.01ìg/m^3. The neutralization of acidic species like SO_4^{2-} and NO_3^- by NH_4^+ is very critical from acid deposition point of view. The major sources for ammonia gas include animal farming, use of fertilizers and organic decomposition. Ammonia is not transported to very long distances as it is rapidly converted into NH_4^+ aerosols at a rate of 30% h^{-1} (Asman and Van Jaarsveld, 1991). As per early study pointed out by Khemani *et al.* (1982) for this region, the cation accumulates in the fine particle size and correlates well with SO_4^{2-}, suggesting that NH_4^+ ion were present in the atmosphere in the form of salt such as $(NH_4)_2SO_4$.

Table 2 : Factor analysis of ionic mass concentration of water soluble components in $PM_{2.5}$

	Factor 1	Factor 2	Factor 3
Calcium	-	-	0. 86
Chloride	-	-	-
Formate		0. 85	-
Potassium	0.75	-	-
Magnesium	-	-	0. 59
Sodium	-	-	-
Ammonium	0. 92	-	-
Nitrate	0. 51	0. 81	-
Oxalate	0. 91	-	-
Sulphate	0. 94	-	-
Eigenvalue	4.3	2.2	1.3
% Variance	43.0	21.9	12.4
Cumulative	43.0	65.0	77.4

The second factor showing 21.9 % of the total variance also represents a 'strong man- made activities factor' and is dominated by NO_3^- and formate. The high loadings of NO_3^- in this factor suggest an influence of nitrogen dioxide emitted by incompletely combusted fossil fuels from automobiles/ diesel generators / tractors, and biomass burning which are abundant in this area, where the oxidation the gas is reported.

The third factor is comparatively minor and accounts for about 12.4% of the total variance with high loadings for natural sources such as soil containing calcium and magnesium. These ions are frequently found in soil dust. The Thar Desert, located in the west with respect to the observation location, is the major source of these alkaline soil derived components from where they are transported by wind. Recently, Rengarajan *et al.*, (2007) have found the soil contribution of nearly one fourth by weight in the total suspended particle with the dominant ions for another north Indian urban station, Hissar, being SO_4^{2-}, NO_3^- and NH_4^+.

Fig.7 shows the contribution of ionic sources in measured mass to the daily average divided on the basis of factor analysis. The contribution of anthropogenic species was dominant (81%) in the $PM_{2.5}$ ionic mass concentration. It is produced by human activities such as burning of coal, wood, kerosene oil, and plants leaves during winter period to protect from cold (Ali *et al.,* 2004). Apart from this, coal is also major fuel which is used in industrial activities around National Capital Region of Delhi. The results from factor analysis indicate towards the role of man-made activities in the composition of fine size aerosols over Delhi whereas contribution of mineral dust and soil aerosols, which are generally in coarse size, is very meager in the formation of fine size aerosols.

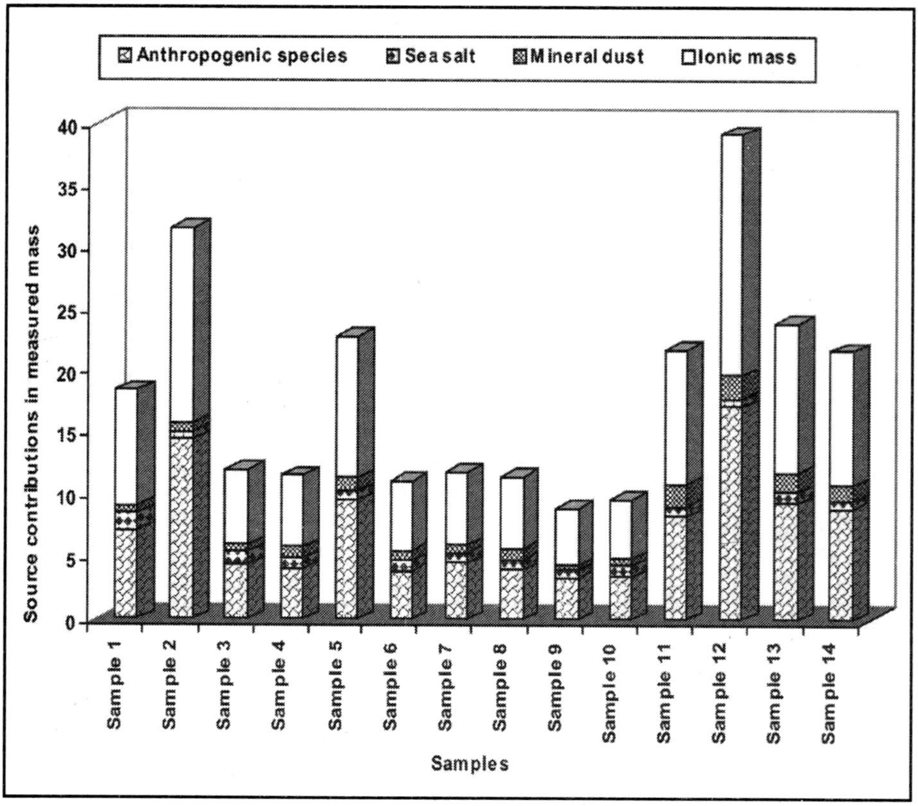

Fig. 7: Contribution of ionic sources in measured mass of $PM_{2.5}$

Conclusions

In summary, the study related to the investigation of the characteristics of the major chemical species ($PM_{2.5}$) at Delhi indicated that SO_4^{2-}, which is produced from man-made sources, contributed approximately 75%.The annual average mass concentrations of $PM_{2.5}$ was 98.71μg/m^3, which is approximately two and half times and six times higher than the Indian NAAQS and USEPA standards, respectively, at urbanized region. It suggests that particle pollution in Delhi is rather severe, especially due to fine particles. Variations of these particles were regulated mainly by the strengths of sources and the prevailing meteorological conditions. The ratio between $PM_{2.5}$ to PM_{10} was found to be high (0.86) in comparison to various other locations in India and was comparable to those reported from other locations in the world, indicating the dominance of $PM_{2.5,}$ especially during the winter period. The concentrations of SO_4^{2-}, NO_3^- and NH_4^+ were significantly high in the total water soluble inorganic fraction during winter. Being far from the ocean, non-sea salt ions dominated the aerosol chemistry (about 91% of the total water soluble inorganic ions). Among the detected ions, SO_4^{2-} was the major chemical component of $PM_{2.5}$ (4.74μg/m^3, 51% to the total ions), while NH_4^+ and NO_3^- were the second and third highest water soluble components followed by potassium. The mean $(NH_4)_2SO_4$ concentration during the study period was 5.76μg/m^3 and suggests that secondary aerosols, such as SO_4^{2-}, NO_3^- and NH_4^+ were locally generated in Delhi and only a minor contribution is made by transported pollutants from the nearby industrial locations. Using the air mass trajectory analysis, the study confirmed that air parcels during winter passed through the major industrial areas located in the east of sampling site in Delhi; and sometimes, from west/north west regions. It is speculated that more attention should be paid to the regional sources of secondary pollutants while formulating pollution mitigation programmes at Delhi because these species may play a critical role in the occurrence of heath hazards like increase of mortality and different types of respiratory and pulmonary diseases.

Acknowledgements

The authors are extremely thankful to Prof. B. N. Goswami, Director, IITM and Dr. P. C. S. Devara, Scientist F and Head, P. M. & A. Division for their guidance and unstinted support throughout the study. We are also thankful to Dr. R. Bala, University of Singapore, Singapore, for helping in chemical analysis of $PM_{2.5}$ aerosol samples at their laboratory.

References

Agarwal, A. (1992) Analysis: When will India be able to control Pollution? Down to Earth. April 15, pp. 22–26.

Ali, K., Momin, G. A., Tiwari, S., Safai P. D., Chate D. M. and Rao, P. S. P. (2004) Fog and precipitation chemistry at Delhi, North India. Atmospheric Environment, v. 38, pp. 4215-5422.

Anderson, K. R., Avol, E. L., Edwards, S. A., Shamoo, D. A., Peng, R. C., Linn, W. S. and Hackney, J. D. (1992) Controlled exposures of volunteers to respirable carbon and sulphuric acid aerosols. J. Air Waste Management Association, v. 42, pp. 771.

Asman, W. and Van Jaarsveld, H. (1991) A variable-resolution transport model applied for NH_x in Europe. Atmospheric Environment, 24A, 2, pp. 445–464.

Bridgman, H. A. (1990). Global Air Pollution: Problems for the 1990s. Belhaven Press, London.

Chatterjee, A., Dutta, C., Sen, S., Ghosh, K., Biswas, N., Ganguly, D. and Jana, T. K. (2007) Formation, transformation, and removal of aerosol over a tropical mangrove forest. J. Geophys. Res., v.111, No. D24, D24302, 10.1029/2006JD007144.

Chow, J. C., Watson, J. G., Fujita, E. M., Lu, Z. and Lawson, D. R. (1994) Temporal and spatial variations of $PM_{2.5}$ and PM_{10} aerosol in the Southern California Air Quality Study. Atmospheric Environment v. 28, pp. 2061–2080.

Dockery, D. W. and Pope, C.A. (1994) Acute respiratory effects of particulate air pollution. Annual Reviews of Public Health, v. 15, pp. 107–132.

Draxler, R. R. and Hess, G. D. (1997) Description of the HYSPLIT- 4 modeling system. NOAA Technical Memorandum, ERL ARL-224.

Hueglin, C., Robert G., Urs B., Martin G., Christian M. and Heinz V. (2005) Chemical characteristics of PM2.5 and PM10 and coarse particles at urban, near city and rural sites in Switzerland. Atmospheric Environment, v. 39, pp. 637-651.

Intergovernmental Panel on Climate Change (IPCC) (1995) Climate Change. Cambridge University Press, New York.

Jacobson, M. C., Hansson, H. C., Noone, K. J. and Charlson, R. J. (2000) Organic atmospheric aerosols: review and state of the science. Reviews of Geophysics, v. 38, pp. 267–294.

Khare, P., Goel, A., Patel, D. and Behari J. (2004) Chemical characteristics of rain water at a developing habitat of Northern India. Atmospheric. Environment, v. 69, pp. 135-145.

Khemani, L. T. and Mehta, S. Naik, Momin G. A, Kumar, R., Chatterjee R. N., Singh G. and Ramamurty, V. (1985) Trace element in the atmospheric Aerosols at Delhi, North India. J. Atmospheric Chemistry, v. 2, pp. 273-285.

Khemani, L. T., Momin, G. A., Naik, M. S., Vijayakumar, R. and Ramana Murty, Bh. V. (1982) Chemical composition and size distribution of atmospheric aerosols over the Deccan Plateau, India. Tellus, v. 34, pp. 151–158

Khillare, P. S., Pandey, Rajni and Balachandran, S. (2004) Characterization of Indoor PM_{10} in Residential Areas of Delhi. Indoor Built Environ, v. 13, pp. 139–147.

Kumar, R., Rani, A., Kumari, K. M. and Srivastava, S. S. (2003) Direct measurement of atmospheric dry deposition to natural surfaces in a semiarid region of north central India. J. Geophys. Res., v. 108 (D20), 4625, doi:10.1029/2002JD003194, 2003.

Mao, I-F., Chien-Nan, C., Yi-Chang, L. and Mei-Lien, C. (2007) Airborne particles $PM_{2.5}$ / PM_{10} mass distribution and particle-bound PAH concentrations near a medical waste incinerator. Atmospheric Environment, v. 41, pp. 2467-2475.

Milne, J. W., Roberts, D. B., Walk, S. J. and William, D. J. (1982) Sources of Sydney brown haze. In: The Urban Atmosphere — Sydney, A case Study, CSIRO, Australia.

Norman, M., Das, S. N., Pillai, A. G., Granat, L. and Rodhe, H. (2001) Influence of mass trajectories on the chemical composition of precipitation in India. Atmospheric Environment, v. 35, pp. 4223–4235.

Novakov, T. and Penner, J. E. (1993) Large contribution of organic aerosols to cloud-condensation-nuclei concentrations. Nature, v. 365, pp. 823–826.

Pande, J. N. (1999). Analysis: Deadly particles, relevance to India. Down to Earth; Nov. 15, pp. 42–45.

Patel, K. S., Jaiswal, N. K., Saathoff, H., Schurath, U. and Leisner, T. (2006) Particulate Pollution in Raipur, Chattisgarh. Proceedings of brain storming workshop on aerosols and its impact on climate with special reference to Indo- Gangetic Plains, Nov. 10-11, IIT, Kanpur, pp. 36-39.

Rengarajan, R. and Sarin, M. M. (2006) Chemistry of aerosols over a semi-arid region: Evidence for acid neutralization by mineral dust. Geophys. Res. Lett., v.33, L23815, doi: 10.1029/2006GL027708.

Rengarajan, R., Sarin, M. M. and Sudheer, A. K. (2007) Carbonaceous and inorganic species in atmospheric aerosols during wintertime over urban and high-altitude sites in North India. J. Geophys. Res., v. 112, D21307, doi:10.1029/2006JD008150.

Sharma, M. and Shaily M. (2005) Assessment of ambient air $PM_{2.5}$ and PM_{10} and characterization of PM_{10} in the city of Kanpur, India. Atmospheric Environment, v. 39, pp. 6015–6026.

Singh, S., Nath, S., Kohli, R. and Singh, R. (2006) Aerosols over Delhi during pre-monsoon months: Characteristics and effects on surface radiation forcing. Geophys. Res. Lett., v. 32, L13808, doi:10.1029/2005GL023062.

Smirnov, V. V., Radinov, V. F. and Savchenko, A. V. (1998) Variability in aerosol and air ion composition in the Arctic spring atmosphere. Atmospheric Research, v. 49, pp. 163–176.

U. S. Environmental Protection Agency, Criteria document on fine particles (1996) EPA/600/P-95/001CF, Washington, DC.

Wang, Y., Guoshun Z. C., Zhisheng A. and Aihua Z. Z. (2007) Characteristics and sources of formic, acetic and oxalic acids in PM2.5 and PM10 aerosols in Beijing, China. Atmospheric Research, v. 84, pp. 169 -181.

Wang, G., Wang, H., Yu, Y., Gao, S., Feng, J., Gao, S. and Wang, L. (2003) Chemical characterization of water-soluble components of PM_{10} and $PM_{2.5}$ atmospheric aerosols in five locations of Nanjing, China. Atmospheric Environment, v. 37, pp. 2893-2902.

Zhang, Z. Q. and Friedlander, S. K. (2000) A comparative study of chemical databases for fine particle Chinese aerosol. Environmental Science and Technology, v. 34, pp. 4687–4694.

□□□

16

Cosmic Rays and Earth's Atmospheric Processes : A Review

DEVENDRAA SIINGH
Indian Institute of Tropical Meteorology, Pune-411 008, India
Email: devendraasiingh@tropmet.res.in

Abstract

The modulation of cosmic ray flux incident on the Earth's upper atmosphere by solar activity results in changing the global electrical properties of the atmosphere, which is in turn believed to affect weather and climate. In this review paper an attempt has been made to summarize the present understanding of Earth's atmospheric processes that are affected by Cosmic rays.

Introduction

The idea that the cosmic rays could affect the weather received due importance when a positive correlation was reported between the monthly galactic cosmic rays and the satellite-retrieved low cloud amount from 1983 to 1994 (Marsh and Svensmark, 2000). Close correlation has also been reported between cosmogenic isotopes and paleo- climate records such as monsoon activity (Neff *et al.*, 2001) and ocean temperature (Bond *et al.*, 2001). Cosmic rays help the formation of dense clouds in the lower atmosphere while having a small negative effect on cloud cover in the upper atmosphere. The low clouds retain more surface energy, keeping the surrounding air hot, while the high clouds reflect more sunlight into space keeping the upper atmosphere cooler (Yu, 2002). The solar magnetic field is the major parameter needed to reconstruct the secular variation of the cosmic rays flux impinging on the terrestrial atmosphere; this is because a stronger solar magnetic field more efficiently shields the earth from cosmic rays. Further, it is believed that the cosmic rays affect the total cloud cover of the earth and thus provide a driver for the terrestrial climate, although the physical mechanism underlying the link is still poorly understood (Svensmark and Friis-Christensen, 1997; Siingh et *al.*, 2005, 2007).

A conventional mechanism involves ion induced formation of aerosol particles followed by new particle growth to a size of cloud condensation nuclei (Carslaw *et. al.*, 2002; Arnold, 2006) and thereby influence the climate both in direct and indirect ways. Therefore, any variation in the cosmic ray flux can be expected to affect the distribution of lightning around the world and also the climate.

Modelling work (Yu and Turco, 2001) suggested that the small ions can produce a source of atmospheric cloud condensation nuclei, which indicates a potential effect on clouds and ultimately climate (Carslaw *et al.*, 2002). In this area many studies have been carried out which suggest that variations in cosmic rays on the scales of days and years influences global cloudiness (Svensmark and Friss-Christensen, 1997), cloud cover (Todd and Kniveton, 2001), mid-latitude cyclones (Tinsley and Deen, 1991) and high level cloud (Pudovkin and Veretenenko, 1995).

In India, the cosmic ray research has recently grown as one of the largest activities covering all aspects of the radiation. In 1937, Prof. Homi Bhabha has published a paper entitled "On the penetrating component of cosmic radiation" published in 'Proceeding of Royal Society". In this he concluded that a breakdown of the higher energies, as proposed by some theorists, would not explain the experimental results on the latitude effect of cosmic rays and the shape of the transition curve of large cosmic rays bursts. During the period, a new particle called 'meson' was discovered. Prof. Homi Bhabha predicted that the 'meson' would be unstable and would probably decay into an electron and neutrino. After that, the Indian Institute of Science at Bangalore constructed a 12" diameter cloud chamber that was used by Prof. M.S. Sinha to study the scattering characteristics of 'meson'.

Time distribution of cosmic rays has been tested and investigated by using different Geiger counter (Vikram Sarabhai, 1942). He has found that the arriving of cosmic rays follow a law of complete time randomness and their behaviour is similar to that shown by radiation from radioactive sources. Prof. Vikram Sarabhai also discussed the possibility of detecting deviation from time randomness in the case of cosmic rays. Prof. Piara Singh Gill and his students studied time variations of cosmic rays and extensive air showers at high latitude station Gulmarg by using the Geiger Muller counters.

Ahluwalia (1962) studied the semidiurnal variation of the cosmic rays intensity for geomagnetically disturbed days based on the data at the equatorial stations of Ahmedabad (India) and Huancayo (Peru) during 1957 and 1958. He showed that the low average amplitude of the semidiurnal variation observed on geomagnetically disturbed days, compared with geomagnetically quiet days, is primarily due to the large variability of the time of maximum of the semidiurnal variation on disturbed days. He also suggested that any theory which seeks to explain the origin of daily solar variation of cosmic rays must take account of semidiurnal as well as diurnal variations.

Cosmic Rays and Electrical System of the Earth Atmosphere

The variables of electrical system of the earth's atmosphere are highly dependent on a wide range of atmospheric processes, ranging from scales of a few meters on the edges of clouds up to global scale. A possible connection between electrical environment and climate of the earth atmosphere including the modulation of electrical conductivity and cloud nucleation rates by cosmic radiation have been discussed by Carslaw *et al.,* (2002) and Singh *et at.,* (2004). CTR Wilson in early days has discussed the importance of cosmic rays in the electrical properties of the atmosphere and cloud formation. Variation in ionization was suggested as an explanation for changes in weather and climate (Ney, 1959; Bering 1995).

The profiles of the ion production rate for cosmic rays for various geomagnetic latitudes during the solar cycle minimum using the data from Nehar (1967) is presented in Fig. 1. The ionization rate increases with geomagnetic latitude, and both the height of the peak and the slope of the ionization rate after the peak also increase. Near the ground there is about 20% variation between the equatorial region and the higher latitudes. The additional charge pairs produced by cosmic rays enhance conductivity of the lower atmosphere and hence enhance the thunderstorm charging current by facilitating enhanced charge transfer. Ionospheric potential is also enhanced in the process.

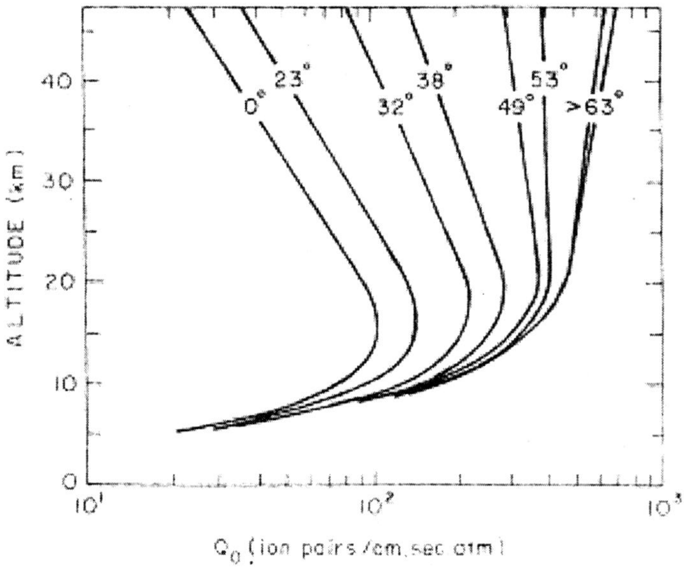

Fig. 1 : Vertical profiles of the cosmic ray ion production rate at various geomagnetic latitudes during solar minima (Neher, 1967).

Ion produced by cosmic rays in lower atmosphere had shown the largest modulation through a solar cycle leading to similar modulation in aerosols and water vapor condensation (Rogers and Yau, 1989). The aerosol particles may initiate the nucleation process in a slightly super-saturated water vapor. When the aerosol is dissolved in a tiny haze particle, the vapor pressure of the drop is lowered, increasing the probability of drop growth. The size of the aerosol is dependent on the statistical distribution of charges on the aerosols (Hoppel *et al.*, 1986; Gringel *et al.*, 1986).

Serrano *et al.* (2006) using the electric field data recorded at the Portela meteorological station (Lisbon) have shown a negative correlation with cosmic radiation flux and also analysis of the seasonal behavior of this correlation indicated that it is strong in wintertime and mild in summertime, while it follows the annual mean in autumn and in spring. They have given a probable explanation by considering that cosmic rays, instead of acting directly through air ionization may have indirectly acted by enhancing droplet and cloud formation by ion capture and formation of a negative layer in the lower atmosphere, which would reduce the electric field. This is similar to the reduced field hypothesis (GCR-CN-CCN-cloud hypothesis) proposed by Yu (2002).

Bazilevskaya (2000) reviewed the cosmic rays variability and emphasized that the difference between GCR and solar cosmic rays could be used to investigate possible atmospheric responses. He also discussed how ions could affect ice nucleation, atmospheric chemical reactions, droplet nucleation and global atmospheric electric circuit. Harrison (2000) had evaluated the effect of image charge on aerosol collection by droplets. The microphysics of the aerosols on cloud boundaries as a link between high-energy particles and the global circuit was summarized by Tinsley (2000) and Siingh *et al.* (2005).

Cosmic rays ionize the atmospheric gas constituents and hence modify the atmosphere's columnar resistance and ionospheric potential (Markson, 1981). The effect of cosmic rays on ionospheric potential was originally identified from solar modulation of low energy cosmic rays in the form of eleven-year cycle. The correlation between galactic cosmic rays and cloud cover for an 11-year solar cycle basis (Carslaw *et al.*, 2002) has enhanced the interest in the study of relationships between solar variations and weather changes. The cosmic ray flux penetrating down to the earth's lower atmosphere during active period of the Sun decreases due to its interaction with solar wind. The steady increase in solar activity during the twentieth century has led to a secular decline in cosmic rays (Carslaw *et al.*, 2002) and the expected global circuit response has been identified in surface measurements of potential gradient in the United Kingdom (Marcz and Harrison, 2003), although aerosol changes have also been suggested to be responsible agent (Williams, 2005). Harrison and Ingram (2005) have also reported a decrease in the air-earth current at Kew and in the potential gradient in mountain air in the 1970s.

The cosmic rays may also affect the climate / weather involving cloud processes such as condensation of nucleus abundances (Wilcox *et al.*, 1974), thunderstorms electrification and thermodynamics (Markson and Muir, 1980), ice formation in cyclones (Tinsley, 1996), *etc.* Svensmark and Friis-Christensen (1997) have analyzed data and showed a correlation between cosmic rays and the earth's cloud cover over a cycle. Turco *et al.* (1998) and Marsh and Sevensmark (2000) suggested that galactic cosmic rays could generate aerosol particles that can act as cloud condensation nuclei and affect, particularly over ocean, the formation and thickness of cloud. They also found a strong association between low clouds, at around 3 km altitude, and cosmic rays flux. Thus, it is likely that the cosmic ray influences the GEC as well as climate/weather. This early suggestion of Ney (1959) still remains to be explored in depth. In fact the observation of a correlation between cosmic ray intensity and cloudiness provides an opportunity to investigate ion-aerosol-cloud interaction, because variation in ion production rate due to cosmic rays may impact aerosol distribution and cloud formation. Rainfall is an important controlling factor of average cloudiness of a region via cloud lifetime. Even ice-particles growth induces rainfall because liquid clouds are highly supersaturated with respect to ice. Both these aspects can be studied in detail through ion-aerosol-cloud interactions in which ion production is governed by galactic cosmic rays.

Cosmic rays provide the sole source of ions away from terrestrial sources of radioisotopes such as radon, and cosmic ray variations directly influence the global atmospheric electric circuit on cloud formation (Tinsley, 2000). Variations in atmospheric electric fields maintained within the clouds due to the solar activity are proposed as a means of modifying the rate of droplet formation, which would in turn affect the latent heat release rates and transfer properties. Global circulation affects the temperature and thunderstorm activity as part of this process and thus causes change in the local thermodynamics. The global circuit links the upper atmosphere to the surface changes in the electrical potential of the ionosphere, which are communicated directly by the air-earth current flowing through cloud layers to the surface (Gray *et al.*, 2005; Singh *et al.*, 2007 and references there in). The physical consequence of this coupling for climate is not clear till now. Electrically induced aerosol microphysics on the boundaries of clouds may be important, such as coagulation and scavenging. This involves complex processes, which are not yet fully developed as compared to the particle nucleation process. Gray *et al.*, (2005) summarized the range of processes that can result from the solar modulation of cosmic rays and link global atmospheric electrical circuit with global climate. It is clearly seen that the cosmic rays and shower clouds control atmospheric electrical system which affect cloud microphysics leading to change in ice cloud and stratiform cloud and hence change in climate. The atmospheric electrical system also affects radiative transfer mechanism of the atmosphere leading to global temperature changes which control thunderstorm activity, and hence electrical system of the atmosphere, thus forming a closed system. The global temperature variation also affects cloud microphysics, ice cloud and stratiform cloud formation.

Cosmic Rays and the Earth's Climate

The ionization variation influences the optical transparency of the atmosphere by changing the aerosol formation and also by influencing the transition between different phases of water (Ney, 1959; Tinsley, 1996). Cloud reflects more energy than they stop and lead to a cooling in the range 17-35 Wm^{-1} (Ramanathan *et al.*, 1989). A close correlation has been observed between cloud coverage and cosmic ray activity (Svensmark and Friis-Christiansen, 1997). The cloud coverage may control atmospheric heating profile because cloud could reflect incoming short wave radiation leading to cooling, where as it may reflect the outgoing long wave radiation and produces heating.

Yu (2002) pointed out that the global warming in 20th century coincides with decreased cosmic ray intensity. Since cosmic ray intensity reaching the earth depends on solar winds, which is controlled by solar conditions. This suggests that the sun has also some role to play in global warming. Under suitable conditions, cosmic ray-induced cloud changes may warm the earth's surface and cool the lower troposphere and thus provide an explanation of the earth's varying temperature trends. The light ions produced during cosmic ray ionization also act as a source of cloud condensation nuclei (Arnold, 2006) leading to the fact that cosmic rays induces cloud formation.

Based on the premise that aerosols ionized by cosmic rays are more effective as ice nuclei and cause freezing of super cooled water in cloud, Brian Tinsley (1996, 2000) has developed a more detailed mechanism for the link between cosmic rays and cloudiness. It is not easy to estimate the net change in radiative forcing from a solar modulation of the cloud cover. The main problem is that it is not known which part of the cloud volume is effective in heat energy transport / reflection. However, net effect of clouds on planet is either to cool or to warm depending upon the thickness, density and altitude location of the cloud above the earth's surface, atmosphere, and therefore, one could imagine that an increase in cosmic rays flux could lead to a complex phenomena involving heat transfer and under some situation may lead to warming.

Cloud Variation

The cosmic rays and low altitude cloud (<3 km altitude) observed by global satellite are well correlated around the cosmic rays minimum of 1990 (Svensmark and Friis-Christensen, 1997; Marsh and Svensmark, 2000). The limitations of the data analysis and acquisitions have been the subject of considerable discussions (Kernthaler *et al.*, 1999). Kuang *et al.*, (1998) repeated Svensmark and Friis-Christensen's analysis of International Satellite Cloud Climatology Project (ISCCP) data and showed high correlations with an El Nino-Southern Oscillation (ENSO) index difficult to distinguish from the GCR flux. The pattern of change of cloudiness over the period, particularly in the Pacific Ocean, corresponds to what would be expected for the atmospheric circulation changes characteristic of E1 Nino (Farrar, 2000). Kernthaler *et al.* (1999) have also studied the ISCCP data set, using both geostationary and polar orbiter data and suggested that the correlation with cosmic rays flux is reduced if high latitude data are included. Jorgensen and Hansen (2000) also noted that a mechanism whereby cosmic rays resulted in greater cloud cover would be most likely to affect high cloud as ionization is greatest at these altitudes. Kristjansson and Kristiansen (2000) have analyzed additional data set of the ISCCP D2 satellite recorded during 1989 to 1993, and found little statistical evidence of a relationship between GCRs and cloud cover with the possible exception of low marine clouds in mid latitudes. They also could not find correlation between outgoing long-wave radiation and cloud cover as represented in ERBE data and GCRs. Harrison and Stephenson (2006) analyzed data recorded at different meteorological sites in UK and GCR data and confirmed the existence of cosmic ray effects on clouds on long time scales with less variability than the considerable variability of daily cloudiness. Yu (2002) has explained differential cosmic rays effects on low and high clouds in terms of ion-induced particle production and its recombination rates which are proportional to their concentrations.

The ion life time is greater at lower altitude where the cosmic ray produced ionization rate is smaller. In the lower atmosphere aerosol particle density is larger and hence cloud condensation nuclei would be formed easily. A negative correlation between cosmic rays and global temperature has been reported by Shaviv and Veizer (2003) for cosmic ray changes generated by the solar system and passing through the spiral arms of the galaxy. An attempt was made to explain it using a mechanism similar to ion-induced CCN formation (Marsh and

Svensmark, 2000), in which the CCN lead to more cloudiness, reduced insolation and lower temperatures. Thus, the evidence for a cosmic rays impact on cloudiness could not be accepted with certainty. If there is a systematic variation in low cloud properties caused by solar variability, it could have important implications for the valuation of Earth climate. It is essential to exactly pin-point the control of solar variability on cloud cover under different conditions and then to obtain the role of cosmic rays on the cloud cover and control of global temperature.

Carslaw *et al.* (2002) proposed two mechanisms linking cosmic ray produced ions to clouds named as clear air mechanism and near-cloud mechanism. In the clear air mechanism, ions produce ultra fine particles (Harrison, 2002), which is based on the examination of aircraft exhaust (Yu *et al.*, 1998). Extending the same work, Yu and Turco (2000) showed that charged molecular clusters can grow significantly faster than neutral clusters through the process of condensational growth and coagulation from molecular cluster scale to micron size, large enough to become stable particles. Charge may also act to stabilize small clusters, reducing evaporative loss. The primary particle in the above process is ultra-fine aerosol whose density is controlled by a competition between new particle formation and scavenging by the large pre-existing background aerosol. Based on the above physical process, Yu (2002) argued that increases in cosmic ray would lead to differential particle production with increases in the lower troposphere but decreases in the upper troposphere because of the altitude dependence of the sensitivity of particle production to the ionization rate. Using this fact one can explain the observed different correlations between cosmic ray variations and low, middle and high cloud anomalies. This hypothesis can be extended also to explain the observed global surface and troposphere temperature trends. The above physical processes could explain the measurements reported by many authors (Aplin and Harrison, 2001; Harrison and Aplin, 2001; Laasko *et al.*, 2004; Wilhelm *et al.*, 2004).

The near-cloud mechanism involves changes in electric field adjacent to a horizontal layer of cloud/aerosol, which may arise from the vertical conduction current flowing in the global electric circuit. This change in electric field causes the upper part of a thin stratiform cloud to become more positively charged than the clear air above it because of decrease in conductivity due to removal of ions in the layer. In fact the vertical electric field increases in the low conductivity region to maintain a constant conduction current. Tinsley and Heelis (1993) suggested that the electric field could enhance the effectiveness of aerosol as ice-forming nuclei. Tripathi and Harrison (2001, 2002) have confirmed it and shown that this electro-scavenging effect increases as the aerosol radius decrease below one micron. This effect is independent of sign of the aerosol charge, because electrical image force dominated at the near by distances from the droplet (Tinsley *et al.*, 2000). The observed peak in precipitation at cosmic ray maxima is the only argument to support cosmic rays related-scavenging (Kniveton and Todd, 2001), where as the physical mechanism should involve irradiance changes, which has not yet been studied. Gray *et al.* (2005) have asserted that the electrical effects are very much smaller in stratiform clouds than in thunder storms and the local electric fields and charge densities are modulated by cosmic rays and global circuit changes. Further, in the near-cloud mechanism, the aerosol charging is coupled with global electric circuit changes and hence the aerosol-cloud microphysics at long distances away from the source disturbance could be affected. This point has not been explored systematically. Through GEC short term solar changes may be communicated to cloud formation processes and hence to the climate variability. Further study in this direction is required.

Temperature Influence

Recent developments showed that global warming may result in enhanced convective activity of thunderstorm and in turn increased thunderstorm production on a global scale (Markson and Price, 1999). During the last century the temperature on the Earth's surface increased by ~ 0.6° C (Hansen *et al.*, 1999), where as its effect on lightning is not very well known. Even the causes which could explain these phenomena have not been found till now. Stozhkov (2002) had suggested two mechanisms for heating namely solar influence on the weather and climate, and the influence of human activities on atmospheric processes (such as green house effect). The physical mechanism of the green house gases has been understood where as the mechanism of solar influence on weather and climate requires further investigation.

Lean *et al.* (1995) have calculated the correlation between Northern hemisphere surface temperature and solar irradiance (reconstructed from solar indices) from 1610 to 1994 and showed the coefficient to be 0.86 in the pre-industrial period from 1610 to 1800.

Extending this correlation, they have suggested that solar forcing may have contributed about half of the observed 0.6C° surface warming since 1860 and one third of the warming since 1970. The remaining change in global temperature during the pre-industrial period may be due to cosmic rays forcing. It is possible to compute the variation in cosmic ray flux and the temperature change based on simple assumptions. From numerical modeling and satellite observations it is found that a 1% change in the total composition of the Earth's cloud cover corresponds to 0.5 W/m^2 change in net radiative forcing (Rossow and Cairns, 1995). Global cloudiness changed approximately 3.0%, which can estimate to 1.5 W/m^2 from the year 1987-1990 (Marsh and Svensmark, 2000). In the same period cosmic rays, as measured from ion chamber, has changed by ~ 3.5% (Marsh and Svensmark, 2000). He also calculated the approximate radiative forcing by taking in to account the running mean 11-years average increase of cosmic rays from 1975-1989 in between 0.6-1.2%, which comes out to be 0.3-0.5 W/m^2 change in cloud forcing. The direct influence of changes in solar irradiance is estimated to be only 0.1°C (Lean *et al.*, 1995). The cloud forcing however, gives for the above sensitivity, 0.2-0.5°C. In this calculation it is assumed that the whole cloud volume is affected by solar activity. From the above discussion it appears that an increase in cloud cover results in lower temperatures.

Maunder Minimum

The annual variation of sunspot (darker cooler regions of the Sun's surface associated with high magnetic flux) number shows a remarkable feature in Sun's history during the period from 1645 to 1715, when the sun spot numbers almost became zero. This is called as Maunder Minimum (MM). Another minimum in sun spot activity is observed between 1800 and 1825 called Dalton Minimum as is evident from Fig. 2 which summarized the 400 years of regular sunspot number observations.

The presence of Maunder minimum and Dalton minimum shows that the sun was less active. The variation in activity suggests that the sun may have been more active some time before than we have seen in the last 400 years. In future also Sun's activity may considerably change (increase or decrease). These upheavals in solar behavior may have been accompanied by significant long-term changes in radiative output. During Maunder minima, the absence of strong magnetic field region on the surface of the Sun affects the solar wind flow and hence modifies the characteristic features of cosmic rays incident on the Earth's atmosphere. The long term changes in solar radiative output and cosmic ray features may result in to long term climate changes on the Earth's surface.

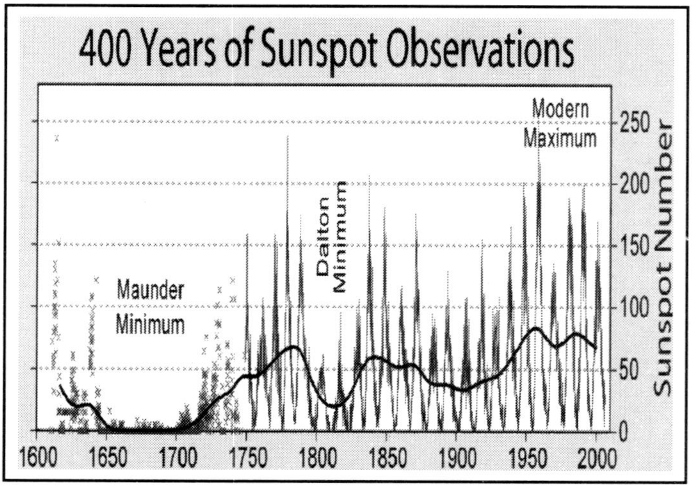

Fig. 2 : Summarizes the 400 years of regular sunspot number observations. The source of the figure is (http://en.wikipedia.org/wiki/Maunder_minimum).

The cause of these variations are not well understood, but because sunspots and associated faculae affect the brightness of the sun, solar luminosity is lower during the period of low sunspot activity. During Maunder minima period, a striking similarity between ^{10}Be curve and the temperature curve has been pointed out (Marsh and Svensmark, 2000). There seem to be a cyclic behaviour in both the magnetic and temperature data. In fact the decade 1690-1700 is the coldest during the last 1000 years, at the same time ^{10}Be concentration has the largest peak. A good agreement between temperature and ^{10}Be concentration suggests that cosmic rays are important in sum / climate link. It is general perception that the low activity during the MM and earlier periods may be among the principal cause of the Little Ice Age (Lean *et al.,* 1995 and references there in). Extending the same hypothesis, some scientists have proposed that the modern maximum may be partly responsible for global warming especially the reported temperature increase between 1900 and 1950 (Eddy, 1976). The above discussion shows that the sun may play significant changes in the climate in the time to come and even the destruction of civilization may be possible.

Fig. 3 (http://en.wikipedia.org/wiki/Maunder_minimum) shows two different proxies of solar activity during the last several hundred years. The lower curve in the figure shows sunspot number (Rg) as reconstructed from historical observations by Hoyt and Schatten (1998) and the upper curve shows the Beryllium -10 concentration (10^4 atoms/(gram ice-core) which is a cosmogenic isotope created in the atmosphere by galactic cosmic rays as measured from Dye-3 Greenland (Beer *et al.*, 1998). Both of the curve match well except some microscopic changes.

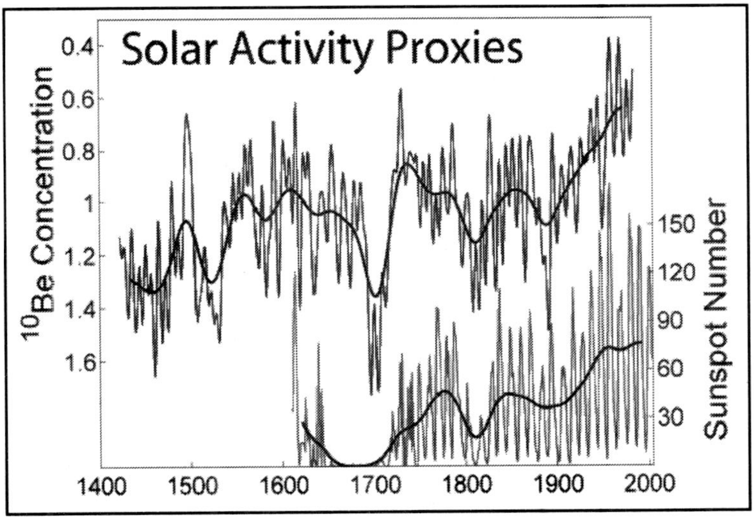

Fig. 3 : The lower panel of the figure shown sunspot Number (Rg) as reconstructed from historical observations (Hoyt and Schatten,1998) and upper panel of the figure shown the Beryllium-10 concentration (10^4 atmos/(gram ice-core) (Beer *et al.*, 1994). Both of the proxy is related to solar magnetic activity. The source of the figure is (http://en.wikipedia.org/wiki/Maunder_minimum).

Cosmic Rays and Aerosol

The nature and size of aerosol particles are height dependent and also have a significant effect on the earth atmosphere heat balance. The observation of spectra in the region of low cloud formation indicates that participating aerosols are produced locally. In the troposphere ionization, is the main contributor to the gas particle formation of ultra-fine particles (<20nm) aerosol, and the subsequent growth into the matured aerosol distributions which act as CCN (Yu and Turco, 2000). Ionization density height profile of the atmosphere and the major ion species produced is shown in Fig. 4 (Viggiano and Arnold,

1995, Marsh and Svensmark, 2000). It is clearly seen from this figure that the complexity of ions increases with decreasing altitude and ions produced within the troposphere by cosmic rays are potentially important for aerosol production in the atmosphere (Marsh and Svensmark, 2000). However, it is not simple and straight forward to show that the variability in atmospheric ionization due to the GCR flux could have a significant effect on either aerosol production or droplet growth. Simulation models have revealed that nucleation through ion-ion recombination is capable of maintaining a background aerosol distribution with realistic concentrations in the troposphere (Turco *et al.*, 1998). In the absence of other potential sources, we may expect that a systematic change in cosmic ray ionization could lead to a change in aerosol population acting as CCN and thus influence cloud formation process which may be different at different altitudes. For example, in the upper troposphere the short lived organic precursor gases may be less important because they may be destroyed during transport from the surface to the upper troposphere. Further, the atmospheric trace gas ammonia, which promotes sulphuric acid nucleation in continental ground level air may not be abundantly available in the upper atmosphere due to its large solubility in liquid water and hence wet removal.

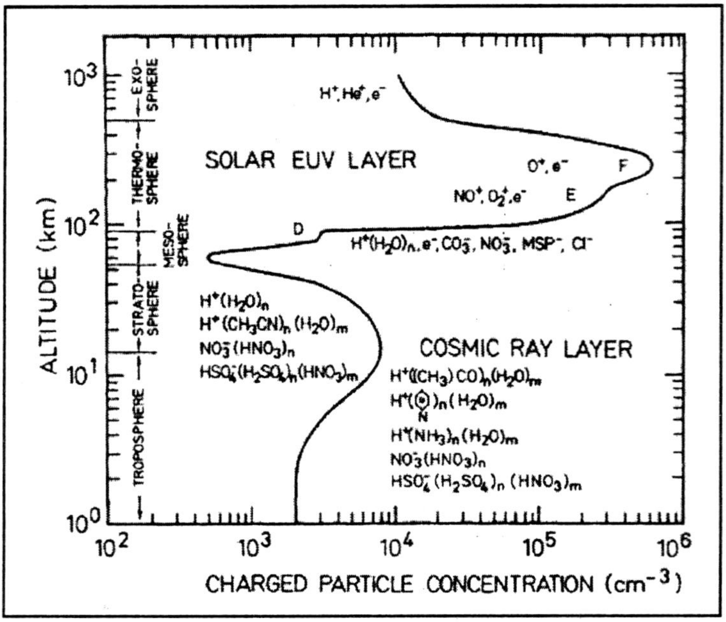

Fig. 4 : Ionization density height profile of the Earth atmosphere and the major ion species at various altitudes (Viggiano and Arnold, 1995).

Cosmic Rays and Ozone

The solar energetic particle events including GCR affect the atmospheric ozone. During these events a shower of high energy charged particles penetrate the terrestrial atmosphere and perturb the stability of its chemistry. Specially, the balance of nitrogen (NOx) and hydrogen (HOx) components are changed and the O_3 destruction begins via catalytic process (Jackman *et al.*, 1995). The high energy charged particle's effect can be synthesized as follows:

i. ozone response to solar proton event is generally observed at high latitudes where the shielding action of the geomagnetic field is reduced

ii. solar cosmic rays (SCRs) with energies greater than 10 MeV affect the stratosphere chemistry

iii. ozone depletion starts within few hours of the arrival of charged particles

iv. solar particle induced effects in the atmosphere could last days or weeks, but no relevant long-lived effects were claimed. However, by the downward transport of chemical components, NOx is primarily found in the troposphere as HNO_3, which dissolves in water vapor is finally removed with snow (Kudela *et al.*, 2000 with references therein)

Kozin *et al.* (1995) showed that during these events the total ozone content, registered by 29 stations situated in the latitudinal range 35° - 60 °, decreased practically synchronously with the galactic CR intensity. However, after the depletion, a positive phase of the disturbance occurs with a maximum about 9 – 11 days later. Shumilov *et al.* (1997) have shown an increase of the total ozone up to 10% at high latitudes during Forbush decreases (Fds) while insignificant effect are suggested for middle latitudes. However, Watanabe (1999) reported very small change in the stratosphere and troposphere conditions during Fds and geomagnetic activity at 69° S, 40° E (Sywa). The total ozone response to major geomagnetic storms (Lastovicka and Milch, 1999) were evaluated which was found to be significant, is probably caused by charges in the atmosphere dynamics. Marcucci *et al.*, (1999) studied the possible relationship between auroral activity and total ozone depletion over Antarctica during the period of the 1979- 1987 and 1990 August, and showed a delayed positive correlation between enhanced auroral activity and total ozone depletion over the Austral region. This delay time ranges between 8 to 13 days. These results seem to be in contrast with the ones for Fds and ozone variability. Moreover, high relativistic electron precipitation events, which occur during magnetic storms, leave their imprints in the mesospheric and upper stratospheric ozone content (Callis *et al.*, 1996). These results clearly show the control of cosmic rays over ozone distribution and its chemistry in the terrestrial atmosphere.

Cosmic Rays and Space Weather

Space weather is a new field of science dedicated to the understanding of the Solar Interplanetary-Terrestrial system. According to Lundstedt (1998) "Space weather refers to conditions on the sun and in the solar wind, magnetosphere, ionosphere, and thermosphere that can influence the performance and reliability of space-borne and ground-based technology system and can endanger human life or health". There are two ways by which the cosmic rays affect our space weather systems: (i) intensity increase caused by the acceleration of particles in solar flares and/ or in the interplanetary space lead to direct changes in materials exposed to the particle radiation both in space and on the surface of the Earth, (ii) CRs detected by the network of neutron monitor (NMs) reveal the anisotropies relevant for space weather research.

Cosmic rays and energetic charged particles in space affect the phenomena occurring in solar atmosphere, heliosphere and geosphere. These particles are relevant for (i) studying changes in the physical state of the magnetosphere and the near Earth inter-planetary medium, (ii) studying effects on space-craft and air-craft electronics, (iii) understanding of the electromagnetic propagation in the Earth environment. In fact high energy particles can deposit enough charge in the electronic components to affect the memory state (SEU: single event upset) or to induce false signals, where as less energetic events affect components by causing permanent damage (Hilgers and Daly, 1998; Johnson and Dyryklev, 1998; Kudela *et al.*, 2000). High energy electrons accumulate charges in poorly conductive material of the space-craft and may lead to dielectric breakdown and material damage. Dorman *et al.* (1993) have discussed the use of CR monitoring in complex studies of phenomena which are considered to be dangerous for the Earth's civilization. Space weather events leave their imprints on the total ozone content of the terrestrial atmosphere, which is also affected by cosmic rays. Thus, space weather events, ozone content and CR are interred linked and a detail study is required to understand the physical processes under laying the coupled system.

Cosmic Rays and Human Health

It is known that any type of radiation is dangerous to human body. But there is always a level below which the amount of radiation received is within the safe zone. However, it is difficult to quantify this level in the case of cosmic rays. Further, the amount of the cosmic radiation received is a function of altitude above the surface of the Earth, geomagnetic latitude and the solar activity cycle. Because of this it is difficult to model accurately the dose level for the human health. As the cosmic ray intensity increases with height, it becomes a cause of concern for those who are frequently flying and also for air craft crew members. The dosage units are μ Sv (Sievert Symbol, 1Sv = 1J/Kg = $1m^{-2}s^{-2}$) per hour and at sea level the dose rate is approximately 0.05 μ Sv per hour where as at commercial aircraft altitudes (~35000 ft), the doze rate between of the order of 4 μ Sv per hour. The accumulated radiation exposure from a flight schedule of 100 hours per month for six months comes out to be 2·4 mSv which is higher than the recommended limit of 2.0m Sv for a pregnant women (Shea and Smart, 1998). In general the recommended limit for human beings is about 20 mSv per year. US Federal Aviation Administration has developed a computer code to quantify the radiation exposure on any specific air route. Using this code one can easily compute the amount of expected radiation dose in a flight.

Acknowledgement

The author is thankful to anonymous reviewer for valuable suggestions. Part of the work for the article was done when DS was at Center for Sun-Climate Research (CSCR), Danish National Space Institute, and Copenhagen, Denmark. DS thanks to Dr Jens Olaf P. Pedersen invite at SCSR, Denmark for short period. DS acknowledges the financial support from Ministry of Earth Sciences (MoES), Government of India, New Delhi.

References

Ahluwalia, H. S. (1962) Semidiurnal variation of cosmic rays on geomagnetic disturbed days. Proc. Phys. Soc., v.80, pp.472-478.

Aplin, K.L. and Harrison, R.G. (2001) A self-calibrating programmable mobility spectrometer for atmospheric ion measurements. Rev. Sci. Instrum., v.72, pp.3467-3469.

Arnold, F. (2006) Atmospheric aerosol and cloud condensation nuclei formation: A possible influence of Cosmic rays? Space Sci. Rev., v.125, pp.169-186, 2006.

Bazilevskaya, G.A. (2000) Observations of variability in cosmic rays. Space Sci. Revs., v.94, pp.25-38.

Beer, J., Tobias, S. and Weiss, N. (1998) An active Sun throughout the Maunder minimum. Solar Phys., v.181, pp.237.

Bering, E.A. (1995) The global circuit: global thermometer, weather by produced or climate modulators, U.S. Nalt. Rep. Int. Union Geodesy Geophysics 1991-1994. Rev. Geophys., v.33, pp.845-862.

Bond, G., Kromer, B., Beer, J., Muscheler, R., Evans, M.N., Showers, W., Hoffmann, S., Lotti-Bond, R., Hajdas, I. and Bonani, G. (2001) Persistent solar influence on North Atlantic climate during the Holocene. Science, v.294, pp.2130-2136.

Callis, L. B., Boughner, R. E., Baker, D. N., Mewaldt, R. A., Blake, J. B., Selesnick, R. S., Cummings,J. R., Natarajan, M., Mason, G. M .and Mazur, J. E. (1996) Precipitating electrons: evidence for effects on mesospheric odd nitrogen. Geophys. Res. Lett., v. 23, pp.1901–1904.

Carslaw, K.S., Harrison, R.G. and Kirkby, J. (2002) Cosmic rays, clouds and climate. Science, v.298, pp.1732-1737.

Dorman, L. I., Iucci, N. and Villoresi, G. (1993) Possible monitoring of space processes by cosmic rays. Proc. 23th Int. Cosmic Ray Conf., Calgary 4, pp.695–698.

Eddy, J.A. (1976) The Maunder minimum. Science, v.192, pp.1189-1202.

Farrar, P.D. (2000) Do cosmic rays influence Oceanic cloud coverage : or is it only E1 Niño? Climate Change, v. 47, pp. 7-15.

Gray, L.J., Haigh, J.D. and Harrison, R.G. (2005) A review of the influence of solar changes on the Earth's climate. Hadley Centre Technical report no. 62.

Gringel, W., Rosen, J.M. and Hofmann, D.J. (1986) Electrical structure from 0-30 kilometers. In: *Krider, E.P. and Roble, R.G.(eds.),* Study in Geophysics, National Academic Press, Washington, D.C., pp166-182.

Hansen, J., Ruedy, R., Glascoe, J. and. Sato, M. (1999) GISS analysis of surface temperature change. J. Geophys. Res., v.104, pp.30997-31022,

Harrison, R.G. (2000) Cloud formation and the possible significance of charge for atmospheric condensation and ice nuclei. Space Sci. Rev. v.94, pp.381-396.

Harrison R.G. and Aplin, K.L. (2001) Atmospheric condensation nuclei formation and high-energy radiation. J. Atmos. Solar Terr. Phys., v.63, pp.1811-1819.

Harrison, R.G. and Ingram, W.J. (2005) Air-earth current measurements at Kew, London Atmos. Res., v.76, pp. 49-64.

Harrision, R.G. and Stephenson, D.B. (2006) Empirical evidence for a nonlinear effect of galactic cosmic rays and cloud. Proc. Roy. Soc., v.462, pp. 1221-1233.

Hilgers, A. and Daly, E.J. (1998) Space Weather Effects on Space Systems. *ESA WPP*, v.148, pp. 21–28.

Hoppel, W.A., Anderson, R.V. and Willett, J.C. (1986) Atmospheric Electricity in the planetary boundary layer. In: *Krider, E.P. and Roble, R.G.(eds.)*, Study in Geophysics, National Academic Press, Washington, D.C., pp 149-165.

Hoyt, D. V. and Schatten, K. (1998) Climate Impact of Solar Variability. In: *K. Schatten and A. Arking (eds.)*, Solar Phys., pp. 179-189

Jackman, C.H., Cerniglia, M.C., Nielsen, J.E., Allen, D J., Zawodny, J.M., McPeters, R.D., Douglass, A.R., Rosenfield, J.E. and Rood, R.B. (1995) Two dimensional and three-dimensional model simulations, measurements and interpretation of the influence of the October 1989 solar proton events on the middle atmosphere. J. Geophys. Res., v.100, pp.11 641–11 660.

Johansson, K. and Dyryklev, P. (1998) Space weather effects on aircraft electronics. *ESA WPP*, v.148, pp.29–34.

Jorgensen T.S. and A. W. Hansen (2000) Comments on "Variation of cosmic ray flux and global cloud coverage - a missing link in solar-climate relationships" by Henrik Svensmark and Eigil Friis-Christensen [Journal of Atmospheric and Solar-Terrestrial Physics 59 (1997) 1225- 1232], J. Atmos. & Solar Terr. Phys.,v. *62*, pp.73-77, 2000.

Kernthaler, S.C., Toumi, R. and Haigh, J.D. (1999) Some doubts concerning a link between cosmic rays fluxes and global cloudiness. Geophys. Res. Lett., v.27, pp.853-865.

Kniveton, D.R. and Todd, M.C. (2001) On the relationship of cosmic rays flux and precipitation. Geophys. Res. Lett., v.28, pp.1527-1530.

Kozin, I. D., Fedulina, I. N. and Chakenov, B. D. (1995) Prediction of natural ozone concentration variations on the basis of cosmic ray intensity data. Geomagn. Aeron., v.35, pp.423–425.

Kristjansson, J.E. and Kristiansen, J. (2000) Is there a cosmic rays signal in recent variations in global cloudiness and cloud radiative forcing? J. Geophys. Res., v.105, pp.11851-11863.

Kuang, Z.M., Jiang, Y.B and Yung, Y.L. (1998) Cloud optical thickness variations during 1983-1991: Solar cycle or ENSO? Geophys. Res. Lett., v.25, pp.1415-1417.

Kudela, K., Storini, M., Hoffer, M.Y. and Belov, A., (2000) Cosmic rays in relation to space weather. Space Sci. Revs., v.93, pp.153-174.

Laakso, L., Anttila, T., lehtinen, K.E.J., Aalto, P., Kulmala, M., Horrak, U., Paatero, J., Hanke, M. and Arnold, F. (2004) Kinetic nucleation and ions in boreal forest particle formation events. Atmos. Chem. Phys., v.4, pp.2353-2366.

Laštovicka, J. and Mlch, P. (1999) Is ozone affected by geomagnetic storms? Adv. Space Res., v.20 (5), pp. 631–640.

Lean, J., Beer, J. and Bradley, R. (1995) Reconstruction of solar irradiance since 1610: implication for climate change. Geophys. Res. Lett., v.22, pp.3195.

Lundstedt, H. (1998) Lund space weather model: status and future plans. *ESA WPP* 148, 107–112.

MacGorman, D.R. and Rust, W.D. (1998) The electrical nature of storms. Oxford University Press, New Yark.

Marcucci, M.F., Orsini, S., Candidi, M. and Storini, M. (1999) Auroral activity and antarctic stratospheric ozone. Phys. Chem. of the Earth, v. 24, pp. 141-146.

Marcz, F. and Harrison, R.G. (2003) Long-term changes in atmospheric electrical parameters observed at Nagycenk (Hungary) and the UK observatories at Eskdalemuir and kew. Ann. Geophys., v.21, pp.1-8.

Markson, R. (1981) Modulation of the Earth electric filed by cosmic radiation. Nature, v.291, pp.304-308.

Markson, R. and Muir M. (1980) Solar wind control of the Earth's electric field. Science, v. 206, pp.979-990.

Markson, R. and Price, C. (1999) Ionospheric potential as a proxy index for global temperatures. Atmos. Res., v.51, pp.309-314.

Marsh, N. and Svensmark, H. (2000) Low cloud properties influenced by cosmic rays. Phys. Rev. Lett., v.85, pp.5004-5007.

Neff, U., Burns, S.J., Mangini, A., Mudelsee, M., Fleitmann, D. and Matter, A. (2001) Strong coherence between solar variability and the monsoon in Oman between 9 and 6 kyr ago. Nature, v.411, pp.290-293.

Nehar, H.V. (1967) Cosmic rays particles that changed from 1954 to 1958 to 1965. J. Geophys. Res., v.72, pp.1527-1539.

Ney, E.P. (1959) Cosmic radiation and the weather. Nature, v.183, pp.451-452.

Pudovkin, M.I. and Veretenenko, S.V. (1995) Cloudiness decrease associated with Forbush decrease of galactic cosmic rays. J. Atmos. Terr. Phys., v.57, pp.1349-1355.

Ramanathan, Cess, R. R., Harrison, E.E., Minnis, P., Barkstrom, B.R., Ahmad, E. and Hartmann, D. (1989) Cloud-radiative forcing and climate: Results from the earth radiation budget experiments, Science, v.243, pp57-63.

Rogers, R.R. and M.K. Yau (1989) A short course in cloud Physics, 3rd ed, Pergamon, Rarntown, N.Y., 293 p.

Rossow, W.B. and Cairns, B. (1995) Monitoring changes in cloud's. J. Climate, v.31, pp.305-347.

Serrano C., Reis, A.H., Rosa, R. and Locio, P.S. (2006) Influence of cosmic radiation, artificial radioactivity and aerosol concentration upon the fir-weather atmospheric electric field in Lisbon (1995-1991). Atmos. Res., v.81, pp. 236-249.

Shea, M. A. and Smart, D. F. (1998) Space weather: the effects on operations in Space. Adv. Space Res., v.22, pp.29–38.

Shaviv, N.J. and Veizer, J. (2003) Celestial driver of phanerozoic Climate? GSA Today, v.13, pp. 4-10.

Shumilov, O. I., Kasatkina, E. A., Raspopov, O. M. and Henriksen, K. (1997) Impact of Forbush Decreases of Galactic CRs on Ozone Layer. Geomagn. Aeron., v.37, pp.24–30 (in Russian).

Singh, D.K., Singh, R.P. and Kamra, A.K. (2004) The electrical environment of the Earth's atmosphere: A review. Space Sci. Rev., v.113, pp.375-408.

Siingh, Devendraa, Singh, R.P., Kamra, A.K., Gupta, P.N., Singh, R., Gopalakrishnan, V. and Singh, A.K. (2005) Review of electromagnetic coupling between the Earth's atmosphere and the space environment. J. Atmos. Solar Terr. Phys., v.67, pp.637-658.

Siingh, Devendraa, Gopalakrishnan, V., Singh, R.P., Kamra, A.K., Singh, Shubah, Pant, Vimlesh, Singh, R. and Singh, A.K. (2007). The Atmospheric global electric circuit: An overview. Atmos. Res., v. 84, pp 91-110.

Stozhkov, Y.I. (2002) The role of Cosmic rays in the atmospheric process. J. Physics G: Nuclear and Particle Physics, v.29, pp.913-923.

Svensmark, H. and Friis-Christensen, E. (1997) Variation of cosmic rays fluxes and global cloud coverage-a missing link in solar-climate relationship. J. Atmos. Solar Terr. Phys., v.59, pp.1225-1232.

Tinsley, B.A. (1996) Correlations of atmospheric dynamics with solar wind induced changes of air-earth current density into cloud tops. J. Geophys. Res., v.101, pp.29701-29714.

Tinsley, B.A. (2000) Influence of solar wind on the global electric circuit and inferred effects on cloud microphysics, temperature and dynamics in the troposphere. Space Sci. Rev., v.94, pp.231-258.

Tinsley, B.A. and Deen, G.W. (1991. Apparent tropospheric response to Mev-Gev particles flux variations: A connection via electrofreezing of supercooled water in high level cloud? J. Geophys. Res., v. 96, pp.22283-22296.

Tinsley, B.A. and Heelis, R. A. (1993) Correlations of atmospheric dynamics with solar activity. Evidence for a connection via the solar wind, atmospheric electricity and cloud microphysics. J. Geophys. Res., v. 98, pp.10375-10384.

Tinsley, B.A., Rohrbaugh, R.P., Hai, M. and Beard, K.V. (2000) Effect of image charges on the scavenging of aerosol particles by cloud droplets, and on droplet charging and possible ice nucleation processes. J. Atmos. Sci., v.57, pp.2118-2134.

Todd, M.C. and Kniveton, D.R. (2001) Changes in cloud cover associated with Forbush decreases of galactic cosmic rays. J. Geophys. Res., v.106, pp.23031-23041.

Tripathi, S.N. and Harrison, R.G. (2001) Scavenging of electric field radioactive aerosols. Atmos. Environ., v.35, pp.5817-5821.

Tripathi, S.N. and Harrison, R.G. (2002) Enhancement of contact nucleation by scavenging of charged aerosol. Atmos. Res., v.62, pp.57-70.

Turco, R.P., J.X. Zhao and F.Yu. (1998) A new source of tropospheric aerosols: Ion-ion recombination. Geophys. Res. Lett., v.25, pp.635-638.

Yu, F. (2002) Altitude variations of cosmic rays induced production of aerosols: implications for global cloudiness and climate. J. Geophys. Res., v.107, pp1118.

Yu, F. and Turco, R.P. (2000). Ultrifine aerosol formation via ion-mediated nucleation. Geophys. Res. Lett., v.27, pp. 883-886.

Yu, F. and Turco, R.P. (2001) From molecular clusters to manoparticles: Role of ambient ionization in tropospheric aerosol formation. J. Geophys. Res., v.110, pp.4797-4814.

Yu, F., Turco, R.P., Karcher, B. and Schroder, F.P. (1998) On the mechanism contrilling the formation and properties of volatile particles in air craft wakes. Geophys. Res. Lett., v.25, pp.3839-3842.

Vikram Sarabhai (1942) The time distribution of cosmic rays. Proc. of Indian Academy of Sciences, Set A. v.15, pp. 89-104.

Viggiano, A.A. and Arnold, F. (1995) Chemistry and composition of the atmosphere. In: *H. Volland, (ed),* Atmospheric Electrodynamics, CRC Press, Boca Raton, USA, pp. 1-22.

Watanabe, T. (1999) Influence of solar activity on atmospheric environment over Syowa. Poster at the ISSI Workshop Cosmic Rays and Earth. March 21-26, 1999, Bern, Switzerland.

Wilcox, J.M., Schrrer, Ph., Svalgaad, L., Roberts, W.O., Olson, R.H. and Jenne, R.L. (1974) Influence of Solar magnetic sector structure on terrestrial atmospheric Vorticity. J. Atmos. Sci., v.31, pp.581-591.

Wilhelm, S. et al. (2004) Ion-induced aerosol formation: New insights from laboratory measurements of mixed cluster ions HSO_4-(H_2SO_4) a (H_2O) and H^+ (H_2SO_4) a (H_2O) w. Atmos. Environ., v. 38, pp.1735-1744.

Williams, E.R. (2005) Lightning and climate: A review. Atmos. Res., v.76, pp.272-287.

❑❑❑

17

Use of Interactions between NAO and MJO for the Prediction of Dry and Wet Spell in Monsoon Season

S.S.DUGAM
Indian Institute of Tropical Meteorology, Pune-411008, India
Email: dugam@tropmet.res.in

Abstract

The Indian summer monsoon exhibits prominent 30-40 day fluctuations with "active" periods of heavy rain interrupted by dry periods *i.e.* "Breaks". The circulation anomalies associated with active/break monsoon cover the entire Indian Ocean, remote tropics and North Pacific Ocean. A prolonged dry/wet period will result in severe drought/flooding, which have profound influences on the south Asia water cycle, agriculture and societal activity of more than one billion people. The atmospheric general circulation models have great difficulty in simulating the Intra-seasonal oscillation (ISO). Therefore, it is necessary to study the empirical relationship between various atmospheric processes, which are responsible for the ISO. In this paper, the analysis of North Atlantic Oscillation Index (NAOI) and Madden Julian Oscillation Index (MJOI) on daily scale is carried out in relation to daily Indian summer monsoon rainfall (June-September). The analysis is carried out for the period 1979-2001. Since the potential predictability limit for monsoon break is about 20 days, the 20 days running lag/lead correlation analysis between the NAOI and MJOI is found out for each year. It is observed that 20-day lag relationship between NAO and MJO is inverse and significant (0.1 level) and this relationship remains negative throughout the break monsoon period and in active phase it reverses. This twenty days lag relationship between NAO and MJO is potential predictor for break/ active monsoon conditions over the Indian region. The analysis is verified for major drought year 2002.

Introduction

In spite of governing industrialization, Indian economy still depends upon the rain-fed agricultural production. Moreover, the summer monsoon rainfall is also important for hydroelectric power generation and achieving drinking water requirements. Therefore, performance of the southwest monsoon over India plays a very crucial role in affecting the quality of life in Indian sub-continent. The southwest monsoon season (June-September) accounts for 80-90% of the annual rainfall of the country, out of these four months, more than 60% of the seasonal rainfall occurs during July and August, but if during this period the frequency of dry/wet spell prolonged it affects the agriculture production and create drought/floods situations. In 2002 July and 2004 monsoon season, the prolonged dry spells affected the entire monsoon seasonal rainfall. Therefore, rainfall during the months of July and August; particularly in month of July which is the rainiest month has a decisive role in determining the over all performance of the southwest monsoon and its subsequent impacts. Most of the operational and experimental forecasts based on statistical and dynamical models (Rajeavan, 2001) are unable to predict such a large-scale deficiency in July 2002. It has opened new challenges for forecasters/ researchers to examine the monsoon verifiability with new precursors and techniques for the prediction of dry/wet spells in monsoon. However, the state of art atmospheric general circulation models has great difficulty in simulating the monsoon cycle (Waliser *et al.,* 2003).

The active/break cycle in Indian summer monsoon rainfall is extensively studied by various scientists *viz.* Ramaswamy (1962); Krishnamurti and Bhalme (1976); Sikka and Gadgil (1980); Goswami and Shukla, (1984); Cadet and Daniel (1988); Gadgil and Asha (1992); Krishnan *et al.* (2000); Goswami and Mohan (2001); Annamalai and Slingo (2001); De and Mukhopadhyay (2002); Gadgil and Joseph (2003); Goswami and Xavier (2003); Srivastava *et al.* (2002) and many others. Recently, there has been growing interest in "sub-seasonal" forecast that have lead times on the order of weeks. The basis for developing and exploiting sub-seasonal predictions largely resides with phenomena such as NAO, MJO (Madden,1986; Madden and Julian, 1971;1994), Pacific North American (PNA) pattern, mid-latitude blocking and the memory associated with soil moisture, as well as modeling techniques that rely on both initial conditions and slowly varying conditions (*e.g.* tropical pacific SST).

The Indian summer monsoon exhibits prominent 30-40 day fluctuations with "active" periods of heavy rain interrupted by dry periods *i.e.* "Breaks" (Gadgil and Joseph, 2003; Krishnamurti and Bhalme,1976). Active and break episodes, characteristics of sub-seasonal variation of the Indian summer monsoon are associated with increased (decreased) rainfall over central and western India and decreased (Increased) rainfall over southern peninsula and eastern India (Singh *et al.*, 1992 ; Krishnamurthy and Shukla 2000). The circulation anomalies associated with active/break monsoon cover the entire Indian Ocean and influence remote tropics and North Pacific Ocean (Webster *et al.*, 1998). The intra-seasonal variation of rainfall (active-break cycle) is strongly coupled to ISO (Hartmann and Michelson, 1989; Webster *et al.*, 1998; Sperber *et al.*, 2000; Goswami and Mohan, 2001). A prolonged dry/wet period will result in severe drought/flooding, which have profound influence on the south Asia water cycle, agriculture and societal activity of over more than one billion people.

Previous studies have established that the active/ break conditions are triggered by organized northward propagation of heavy precipitating or cloud free zones from the equatorial region towards the continental landmass (*e.g.* Yasunari, 1980; Sikka and Gadgil,1980). However, a question where and how the convective anomalies that bring about active and break conditions are generated? is remained unanswered. There is some support for the idea that the upper-level divergent waves associated with MJO that circumnavigate the globe, could re-initiate convective anomalies over the Indian Ocean Lorenc, 1984; Lau and Chan, 1986). However, Salby and Hendon (1994) showed that the decorrelation time of MJO is less than 1 cycle. Hence, one event tend not to follow another. During boreal summer, the equatorial eastward propagating MJO weakens substantially (Hendon and Salby, 1994) and some northward propagating episodes are independent of MJO (Wang and Rui, 1990). Whether we could get signal from NAO about the weaking of the eastward propagating MJO, over the Indian Ocean remains to be examined to predict active/break cycle in monsoon season. The North Atlantic Oscillation is most important mode of variability in northern Hemisphere. It simply measures the strength of the westerly winds blowing across the North Atlantic Ocean between 40°N and 60°N. Although originally defined in terms of a regional see saw of sea-level pressure between the Icelandic low and Azores high, it is not a regional phenomenon but has hemisphere extent (Wallace, 2000; Hurrell and Van Loon, 1997). NAO is characterized by strong month to month and year to year variability but also has some decadal trends or inter-decadal variability (Ostermeier and Wallace 2003; Hurrell,1995). Many scientists (Dugam *et al.*, 1997; Kakade and Dugam, 2000; Chang *et al.*, 2001; Rajeevan, 2002; Goswami *et al.*, 2006) studied the links between NAO and Indian summer monsoon on inter-annual and decadal time scale.

Matthews and Kiladis (1999) have proposed tropical-extra-tropical interaction as a possible mechanism for the initiation of the convective cycle of MJO in the Indian Ocean. Propagation of extra tropical waves into the tropical Indian Ocean region has been shown to initiate the MJO in case studies (Hsu *et al.*, 1990) and simple modeling studies (Blade and Hartmann, 1993). It is known that the tropical-extra-tropical interaction through NAO and Pacific North American (PNA) pattern and coupling between the NAO and MJO through the upper level divergent wave that circumnavigate the globe, which could re-initiate convective anomalies over Indian Ocean (Wang *et al.*,2005). Now the question is "whether this divergent wave is perturbed by large-scale middle-latitude oscillation like NAO?" and can this relationship is useful for monitoring the dry/wet episode during the monsoon.

Data Details

The daily Madden Julian Oscillation index time series for the period 1979 to 2001 is taken from the Eric Maloney of Oregon State University. They have constructed these indices from the first 2 principal components of the band pass filtered 30-90 days 850 hPa zonal wind over the region 5°N-5°S, from the National Centers for Atmospheric Research Reanalysis.

Daily North Atlantic Oscillation index data have been taken from the website http://www.cpc.ncep.noaa.gov/ for the same period.

NCEP/NCAR reanalyzed monthly mean Outgoing Long Radiation (OLR) for 1958-2003 and divergences at 200hPa level data for 1958-1996 have been used.

We have used the daily rainfall data at 1°X1° resolution from the Indian Meteorological Department (IMD), based on 1803 stations (Rajeevan *et al.*, 2006) for the period 1951-2000.

Discussion

We have computed the 20-day lag-lead correlation between daily NAO and MJO index data for the monsoon period (June-September) since the potential predictability limit for monsoon break is about 20 days. The analysis is carried out for 23 years (1979- 2001). This lag-lead correlation analysis is compared with the daily Indian summer monsoon rainfall. Fig.1 (a, b, c) shows the 20-day running lag-lead relationship between NAO and MJO and its association with the 20-day mean rainfall anomalies for 1987 (drought year), 1997 (normal year) and 1988 (flood year). From the figure, it is observed that prior to the break (active) spell of monsoon rainfall the relationship between NAO and MJO is negative (positive) respectively. It is observed that 20-day lag relationship between NAO and MJO is inverse and significant (0.1% level) and this relationship remains negative throughout the break monsoon period and vice versa. This 20-day lag relationship between NAO and MJO may be used as potential predictor for dry/wet spells monsoon condition over Indian region.

The composite analysis for all drought years (1979, 1982, 1985, 1986 and 1987) and flood years (1983, 1988 and 1994) is also shown in fig. 2 (a, b). It is seen that 20- day lag-lead relationship of NAO and MJO with 20-day mean rainfall anomalies is going almost hand in hand.

It is known that there is tropical - extra-tropical interaction through NAO and Pacific North American (PNA) pattern. It is known that there is a coupling between the NAO and MJO, through the upper level divergent wave that circumnavigates the globe (Wang *et al.*, 2000; Ferranti *et al.*, 1990). This could re-initiate convective anomalies over Indian Ocean. The lag correlation analysis between the NAO and OLR and divergence at 200 hPa for the period 1958-2003 by NCEP/NCAR reanalysis data derived from the observations taken by polar-orbiting National Oceanic and Atmospheric Administration (NOAA) satellites explains the tropical - extra-tropical interaction (between NAO and MJO) and its association with rainfall activity during the monsoon period. Fig.3 (a, b) depicts the same. From this, it is seen that during the positive phase of NAO upper level divergence at 200hPa over Indian land/ocean region is negative. This means that when the NAO is stronger than normal upper level divergence field is weak, which affects the convention over Indian region.

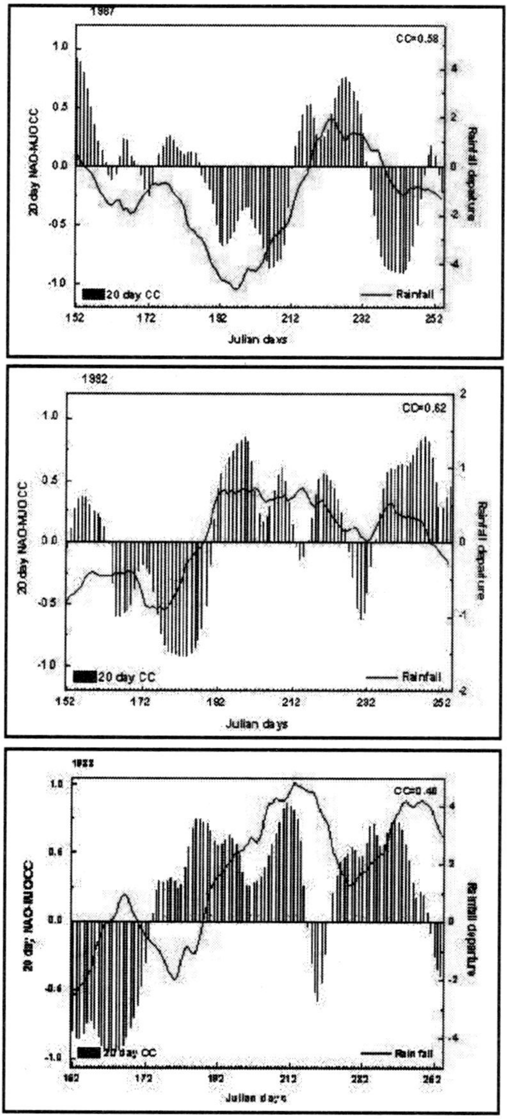

Fig. 1 : Association of 20-day lag-lead relationship between NAO and MJO with 20-day mean rainfall anomalies in monsoon season (June-September) (a) for the drought monsoon year1987 (b) for Normal monsoon year 1997 (c) for excess monsoon year 1988. The smooth running line represents the 20 day running mean rainfall anomalies and vertical bar represents lag-lead correlationship between daily NAO and daily MJO during the monsoon period.

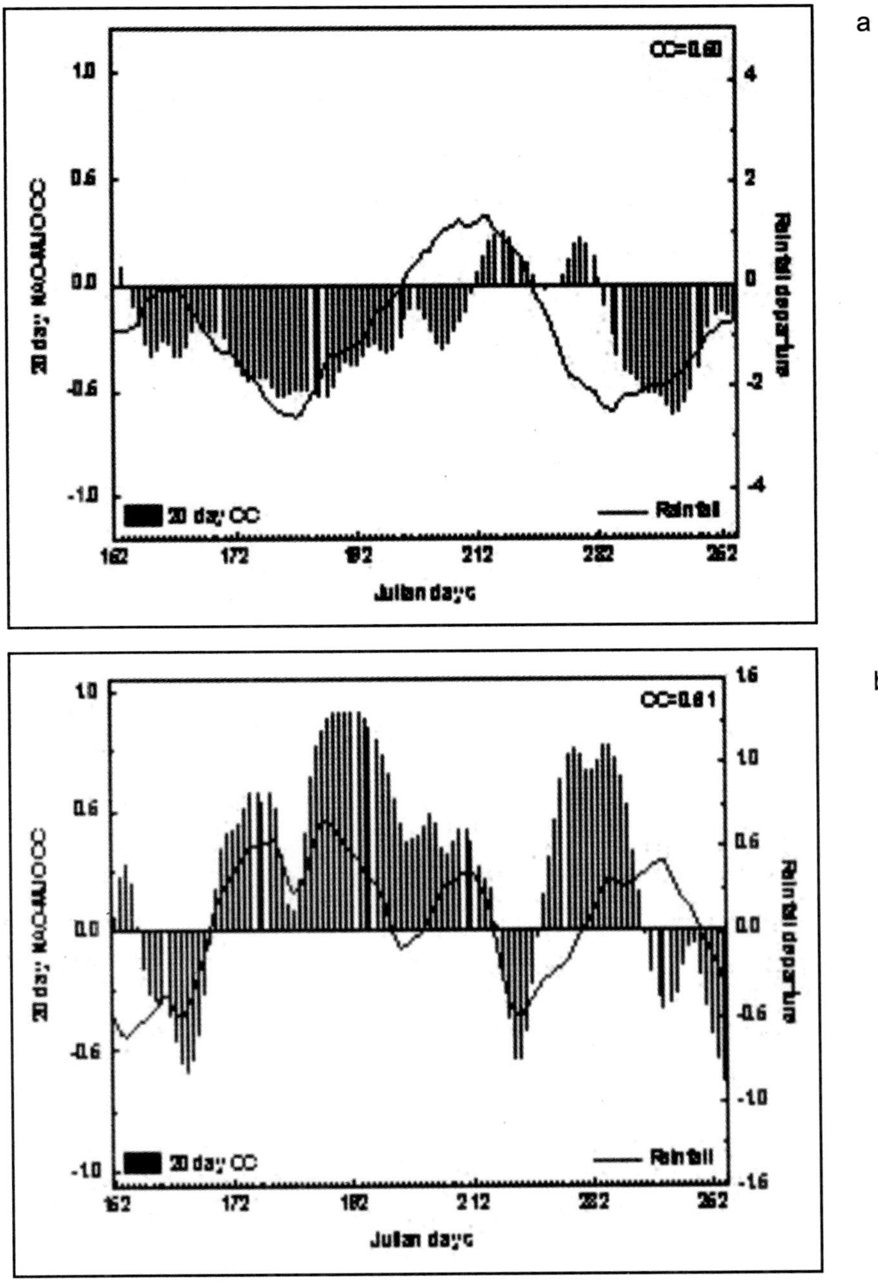

Fig. 2 : (a) Association of 20-day lag-lead relationship between NAO and MJO with 20-day mean rainfall anomalies in monsoon season (composite analysis of drought years) and (b) flood monsoon years. The smooth running line represents the 20 day running mean rainfall anomalies and vertical bar represents lag-lead correlationship between daily NAO and daily MJO during the monsoon period.

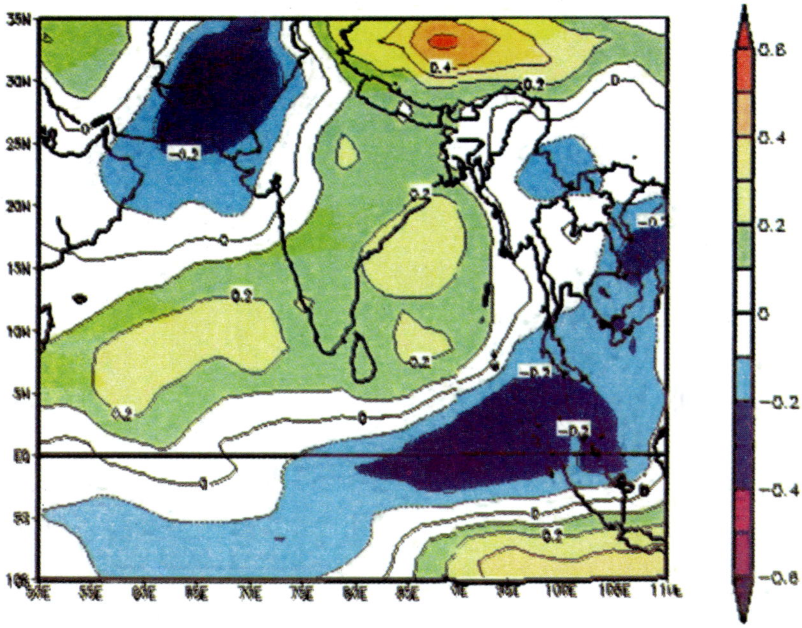

Fig. 3 : (a) The lag-lead relationship between NAO and Outgoing Long Radiation for the period 1958-1996 during the monsoon season (June- September).

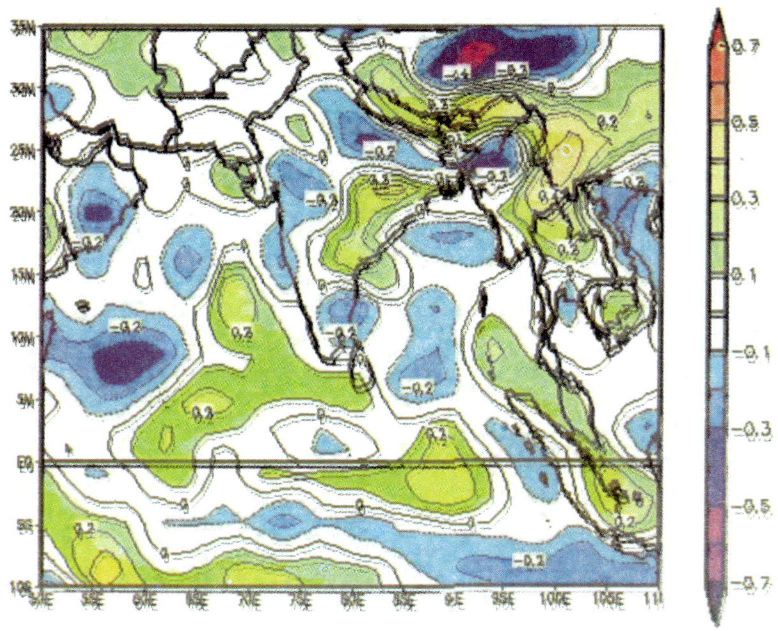

Fig. 3 : (b) The lag- leads relationship between NAO and divergence at 200 h Pa during monsoon season for period 1948-2003. Contours are correlation coefficients.

Probable physical linkage for using this relationship for predicting the dry/wet spell in monsoon period could be like this: the active/ break monsoon conditions are triggered by organized northward propagation of heavy precipitating or cloud free zones from the equatorial regions towards the continental land mass (Sikka and Gadgil, 1980). This northward propagating mode (NPM) over the Indian monsoon region was weaker (stronger) during drought (flood) (Krishnan *et al.*, 2000). It is also known that the MJO interacts with the NPM in the Indian monsoon region (Matthews and Kiladis, 1999). It is plausible that the different phases of NAO modulate the temperature gradients cold/warm in upper troposphere (Goswami *et al.*, 2006). Hence, it might be responsible for the increase in the period of the NPM. A longer period of NPM could possibly lead to longer monsoon break periods causing the major drought conditions and vice versa.

Conclusions

From this study following conclusions can be drawn:

1. 20-day lag-lead relationship between NAO and MJO is inverse and almost remains stable at the beginning and during the dry spell. It is direct and almost stable at the beginning and during the wet spell.

2. Lag-lead relationship analysis between NAO and OLR and 200 hPa divergence field over Indian region suggested that when NAO is stronger (weaker) then the OLR anomaly over the Indian region is positive (negative), which suppress (enhance) the convection.

3. This empirical association between them may be useful for prediction of dry/wet spell during the monsoon period.

Acknowledgment

The author is grateful to Prof. B.N. Goswami, Director, I.I.T.M. for providing necessary facilities for completing this study and to Dr. P. N. Mahajan, Head, Forecasting Research Division, for his encouragement and valuable suggestions. Thanks are due to the NOAA/ESRL Physical Sciences Division, Boulder Colorado for their Web site at http://www.cdc.noaa.gov/ and India Meteorology Department for providing the rainfall data.

References

Annamalai, H. and Slingo, J. M. (2001) Active/break cycles: diagnosis of the intraseasonal variability of the Asian summer monsoon. Climate Dynamics, v.18, pp.85-102.

Blade, I. and Hartmann, D.L. (1993) Tropical intraseasonal oscillations in a simple nonlinear Model. J. Atmos. Sci., v. 50, pp.2922-2939.

Cadet, K. L. and Daniel, P. (1988) Long-range forecast of the break and active summer Monsoon. Tellus, v. 40A, pp.133-150.

Chang, C. P., Harr, P. and Ju, P. (2001) Possible roles of Atlantic circulations on the weakening Indian monsoon rainfall ENSO relationship. J. Clim., v.14, pp. 2376–2380.

De, U. S. and Mukhopadhyay, R. K. (2002) Breaks in monsoon and related precursors. Mausam, v.53, pp.309-318.

Dugam, S.S., Kakade, S.B. and Verma, R.K. (1997) Interannual and long-term variability in North Atlantic Oscillation and Indian summer monsoon rainfall. Theoretical and Applied Climatology, v. 58, pp. 21-29.

Ferranti, L., Palmer, T. N., Molteni, F. and Klinker, E. (1990) Tropical- extra tropical interaction associated with 30-60 days oscillation and its impact on medium and extended range prediction. J. Atmos. Sci., v.47, pp.2177-2199.

Gadgil, S. and Asha, G. (1992) Intraseasonal variation of the Indian summer monsoon: observational aspects. Jap. Met. Society, v.70, pp.517-527.

Gadgil, S. and Joseph, P. V. (2003) On breaks of the Indian monsoon. Indian Acad. Sci. (Earth Planet. Sci.), v.112, pp.529-558.

Goswami, B. N. and Shukla, J. (1984) Quasi-periodic oscillation in a symmetric general circulation Model. J. Atmos. Sci., v. 42, pp.20-37.

Goswami, B. N. and Mohan, R. S. A. (2001) Intraseasonal oscillations and interannual variability of the Indian summer monsoon. J. Climate, v.14, pp.1180-1198.

Goswami, B. N. and Xavier, Prince K. (2003) Potential predictability and extended range prediction of Indian summer monsoon breaks. GRL, v. 30, No. 18, doi:10.1029/2003SGL017810, 2003.

Goswami, B. N., Madhusoodanan, M.S., Neema, C.P. and Sengupta, D. (2006) A physical mechanism for North Atlantic SST influence on the Indian summer monsoon. GRL, v.33, L02706, doi:10,1029/2005GL024803.

Hartmann, D.L. and Michelson, M.L. (1989) Intraseasonal periodicities in Indian rainfall J. Atmos. Sci., v.46, pp. 2838-2862.

Hendon, H.N. and Salby, M.L. (1994) The life cycle of the Madden-Julian Oscillations. J. Atmos. Sci., v.106, pp.447-462.

Hsu, H.H., Hoskins, B.J. and Jin, F.F. (1990) The 1985/86 intra-seasonal oscillation and the role of the extratropics. J. Atmos. Sci., v.47, pp.823-839.

Hurrell, J.W. (1995) Decadal trend in North Atlantic Oscillations and regional temperature and precipitation. Science, v.269, pp.676- 679.

Hurrell, J. W. and Van Loon, H., (1997) Decadal variations in climate associated with the North Atlantic Oscillations. Climate change, v. 36, pp. 301-326

Kakade, S.B. and Dugam, S.S. (2000) Simultaneous effect of NAO and SO on the monsoon activity over India. Geophysical Research Letters, v.27, pp.3501-3504.

Knutson, T.R., Weickmann, K.M. and Kutzback, J.E. (1986) Global scale intra-seasonal oscillations of outgoing long wave radiation and 250mb zonal wind during the Northern Hemisphere summer. Mon. Wea. Rev., v.114, pp.605-623.

Krishnamurti, T. N. and Bhalme, H.N. (1976) Oscillation of a monsoon systems, Part I observation aspects. J. Atmos. Sci., v.33, pp.1937-1954.

Krishnamurthy, V. and Shukla, J. (2000) Intraseasonal and interannual variability of rainfall over India. J. Clim., v.13, pp.4366-4377.

Krishnan, R., Zhang, C. and Sugi, M. (2000) Dynamics of breaks in the Indian summer Monsoon. J. Atmos. Sci., v. 57, pp.1354-1372.

Lau, K. M. and Chan, P. H. (1986), Aspect of the 40-50 day oscillation during northern summer as inferred from outgoing long wave radiation. Mon. Wea. Rev., v. 114, pp.1354-1367.

Lau, K. M. and Peng, L. (1987), Origin of low-frequency (intra-seasonal) oscillations in the tropical atmosphere. Part I: basic theory. J. Atmos. Sci., v. 44, pp. 950–972.

Lorenc, A. C. (1984) The evolution of planetary scale 200mb divergence flow during the FGGE year. Quart. J. Roy. Meteor. Soc., v.110, pp.427-441.

Madden, R. A. (1986) Seasonal variations of the 40-50 day oscillation in the tropics. J. Atmos.Sci., v.43, pp. 3138–3158.

Madden, R. A. and Julian, P. R. (1971) Detection of a 40-50 day oscillation in the zonal windin the tropical Pacific. J. Atmos. Sci., v. 28, pp.702–708.

Madden R.A. and Julian, P. R. (1994), Observations of the 40-50 day tropical oscillation -A review. Mon. Wea. Rev., v. 122, pp. 814-837.

Matthews, A. J. and Kiladis, G. N. (1999) The tropical extra tropical interaction between high-frequency transients and the Madden Julian Oscillation. Mon. Wea. Rev., v. 127, pp.661-677.

Ostermeier, G. M., and Wallace, J. M. (2003) Trends in the North Atlantic Oscillation-Northern Hemisphere annular mode during the twentieth century. J. Clim., v. 16, pp. 336–341.

Ramaswamy, C. (1962) Breaks in the Indian summer monsoon as a phenomenon of introduction between the easterly and the sub-tropical westerly jet stream. Tellus, v.14, pp.337-349.

Rajeevan, M. (2001) Prediction of Indian summer monsoon: Status, problems and prospects. Current Science, v.81,pp.101-107.

Rajeevan, M. (2002) Winter surface pressure anomalies over Eurasia and Indian summer monsoon. Geophys. Res. Lett., v. 29(10), 1454, doi: 10.1029/ 2001GL014363.

Rajeevan, M., Kale, J. D. and Bhate, J. (2005) High-resolution gridded daily rainfall data for Indian monsoon studies. Tech. Rep. 2, National Weather Service, Pune, India.

Rajeevan, M., Bhate, J., Kale, J. D. and Lal, B. (2006) High-resolution daily gridded rainfall data for the Indian region: Analysis of break and active monsoon spells. Curr. Sci., v.91(3), pp.296 306.

Salby, M.L. and Hendon, H.N. (1994) Intraseasonal behavior of clouds, temperature and motion in the tropics. J. Atmos. Sci., v. 51, pp. 2207-2224.

Sikka, D. R. and Gadgil, S. (1980) On the maximum cloud zone and the ITCZ over Indian longitudes during the southwest monsoon. Mon. Wea. Rev., v.108, pp.1840-1852.

Singh S.V., Kripalani, R.H. and Sikka, D.R. (1992) Interannual variability of the Maddan-Julian Oscillatios in Indian summer monsoon rainfall. J. Climate, v.5, pp. 973-979.

Srivastava, A. K., Rajeevan, M. and Kulkarni, R. (2002) Teleconnection of OLR and SST anomalies over Atlantic Ocean with Indian summer monsoon. Geophys. Res. Lett., v. 29(8),1284, doi: 10.1029/ 2001GL013837.

Srivastava, A. K., Rajeevan, M. and Kshirsagar, S.R. (2004) Role of the ITCZ over the North Indian Ocean and pre-Mei-Yu front in modulating July rainfall over India. J. Climate, v.17, pp.673-681.

Sperber, K. R., Slingo, J. M. and Annamalai, H. (2000) Predictability and the relationship between sub- seasonal and interannual variability during the Asian summer monsoon. Quart. J. Roy. Meteor. Soc., v.126, pp.2545-2574.

Waliser, D.E., Lau, K.M., Stern, W. and Jones, C. (2003) Potential predictability of the Madden-Julian Oscillation. Bull. Am. Meteorol. Soc. 84, pp. 33-50.

Wang, B., Webster, P. J. and Haiyan, Teng (2005) Antecedents and self-induction of active-break south asian monsoon unrevealed by satellites. GRL, v. 32, 204704, doi:10.1029/2004GL020996.

Wang, B. and Rui, H. (1990) Synoptic climatology of transient tropical intraseasonal convection anomalies: 1975-1985. Meteorol. Atmos. Phys., 44. pp. 43-61.

Wallace, J. M. (2000) North Atlantic Oscillation/annular mode: Two paradigms-one phenomenon, Q. J. R. Meteorol. Soc., v.126, pp.791–805.

Webster P. J. (1983) Mechanism of monsoon low-frequency variability: surface hydrological effects. J. Atmos. Sci., v. 40, pp. 2110–2124.

Webster P. J., Magaria V.O., Palmer, T.N., Shukla, J., Tomas, R.A., Yanai, M. and Yasunari, T. (1998) Monsoons processes, predictability and the prospects for prediction. J. Geophys. Res., v.103(C7), pp.14451-14510.

Weickmann, K .M., Lussky, G. R. and Kutzback, J. F. (1985) Intera-seasopnal (30-40days) fluctuations of outgoing long wave radiation and 250 stream function during northern winter. Mon. Wea. Rev., v.113, pp. 941-961.

Yasunari, T. (1980) A quasi-stationary appearance of 30 to 40 day period in the cloudiness fluctuations during the summer monsoon over India. J. Meteor. Soc. Japan, v.58, pp. 225-229.

Yasunari, T. (1981) Structure of an Indian summer monsoon system with around 40-day Period. Meteor. Soc. Japan, v.59, pp. 336–354.

□□□

18

Geothermal Energy Resources and its Potential in India

P.N.RAZDAN, R.K. AGARWAL and RAJAN SINGH
Geological Survey of India, Northern Region, Lucknow-226024, India
Email: pnrazdan@gmail.com

Abstract

Geothermal Energy is the vast reservoir of heat energy in the earth's interior, whose surface manifestations are the volcanoes, fumaroles, geysers, steaming grounds and hot springs. About 300 thermal springs are known to occur in India, falling in orogenic (Himalayan) as well as non-orogenic (Peninsular) province. Thirty one areas have been examined in detail and finally, shallow drilling has been done in sixteen areas. The temperatures of these springs range from 35°C to the boiling point of water. On the basis of enthalpy characteristics the geothermal systems in India, are classified into medium (100°C-200°C) and low enthalpy (<100°C) geothermal energy resources/systems. Medium enthalpy geothermal energy systems are associated with younger intrusive granites (as in Himalayas), major tectonic features/ lineaments and rifts and grabens. The Low enthalpy geothermal energy systems are associated with Tertiary tectonism and neo-tectonic activity. The primary and most important aspect of geothermal system is to understand the heat source and its capacity for exploitation. Studies carried out, so far, have clearly pointed that adequate geothermal potential in respect of direct heat utilization and power generation does exist at several geothermal sites. Reservoir simulation studies have suggested the possibility of generating over 3 MW electric power in Puga (J&K) field, if deeper levels are probed at least up to the depth of 500 m. The paper provides an overview on various geothermal fields of India and the status of exploration for future studies.

Introduction

The 'energy crisis' in the seventies and onward changed the approach and priorities in the field of discovery of new non-conventional energy resources. It enhanced the efforts by various countries to look for alternate indigenous and renewable energy resources. In the process wind energy, solar energy, geo-thermal energy and sea wave energies were tried and developed as non-conventional energy resources in various countries depending upon their requirements.

Thermal springs have been known to occur in India since centuries. At places, like Badrinath and Gangotri in Uttar Pradesh; Sohna in Haryana; Rajgir in Bihar; Bakreshwar in West Bengal and Ganeshpuri in West Coast, Maharashtra, temples have been built on thermal springs and thermal water was utilized by locals. Thermal springs have drawn attention, being the surface manifestation of the vast resources of geothermal energy at depth in the form of geothermal reservoirs. Italy, being the pioneer country produced electrical power from geothermal fluids, way back in 1904. At present the total estimated electric power production from geothermal energy is in the range of 8771 MWe all over the globe (Lund, 2004) , which includes U.S.A., Philippines, Mexico, Italy, Japan, Indonesia, New Zealand, El Salvador, China and Costa Rica. In the non-electrical sector, the thermal water is used in greenhouse cultivation, space heating, paper and pulp industry, agriculture industry, fish farming, tourism *viz.* bathing and swimming etc.

In India, almost all the three hundred thermally anomalous areas have been examined and assessed for geothermal potential. Major and representative Indian geothermal provinces are shown in the Fig.1. The temperatures of these springs range from 35° C to the boiling point of water at that height. Thirty one areas have been examined in detail and finally, shallow drilling has been done in sixteen areas. Systematic efforts to explore the geothermal energy resources commenced in 1973 with the launch of Puga geothermal project and gradually the exploration work was extended to cover Chhumathang, Ladakh, J&K;

Parbati valley, HP; Sohna, Haryana; West Coast, Maharashtra and Tattapani, Sarguja, Chhatisgarh. Six areas, four in Himalayas and two in Peninsula hold promise for utilization for either electrical power generation or direct heat applications on industrial scale. The deepest exploratory bore holes have been drilled up to a depth of 385m at Puga, Ladakh, J&K; 220 m at Chhumathang, Ladakh, J&K; 700m at Manikaran, Himachal Pradesh; 728m at Tapoban, Uttar Pradesh and 620m at Tattapani, Sarguja, Chhattisgarh. As a result, it was deduced that the total thermal discharge from springs and drill holes is 30 liters/sec and 250 tonnes/hr at Puga, Ladakh, J&K; 5 liters/sec and 50 tonnes/hr at Chhumathang, J&K, Ladakh; 15 liters/sec and 100 tonnes/hr at Manikaran, HP; 15 liters/sec and 150 tonnes/hr at Tapoban, Uttarkhand and 1liter/sec and 120 tonnes/hr at Tattapani, Chhattisgarh respectively (Geothermal Energy Resources of India, 2002 and references there in). The exploration endeavor also pointed to the favorable geological-hydrogeological setup in these geothermal fields, which have the possibility of encountering sizeable geothermal reservoirs which could sustain electrical power production on MW scale and non-electrical utilization projects on a commercial scale.

Geothermal System

On the basis of enthalpy characteristics, the geothermal systems in India, can be classified into medium enthalpy (100°C-200°C) and low enthalpy (<100°C) geothermal systems. These are described as follows:

Medium Enthalpy Geothermal Energy Systems

The medium enthalpy geothermal energy resources are associated with:

1. Younger intrusive granites as in Himalayas, *viz* Puga-Chumathang, Parbati, Beas and Satluj Valley geothermal fields.
2. Major tectonic features/lineaments such as the West Coast areas of Maharashtra; along the Son-Narmada-Tapi lineament zone at Salbardi, Tapi; Satpura areas in Maharashtra; Tattapani in Chhattisgarh and Rajgir-Monghyar in Bihar, Tatta and Jarom in Jharkhand and Eastern Ghat tracts of Orissa.
3. Rift and grabens of Gondwana basins of Damodar, Godavari and Mahanadi Valleys.
4. Quaternary and tertiary sediments occurring in a graban in the Cambay basin of West Coast.

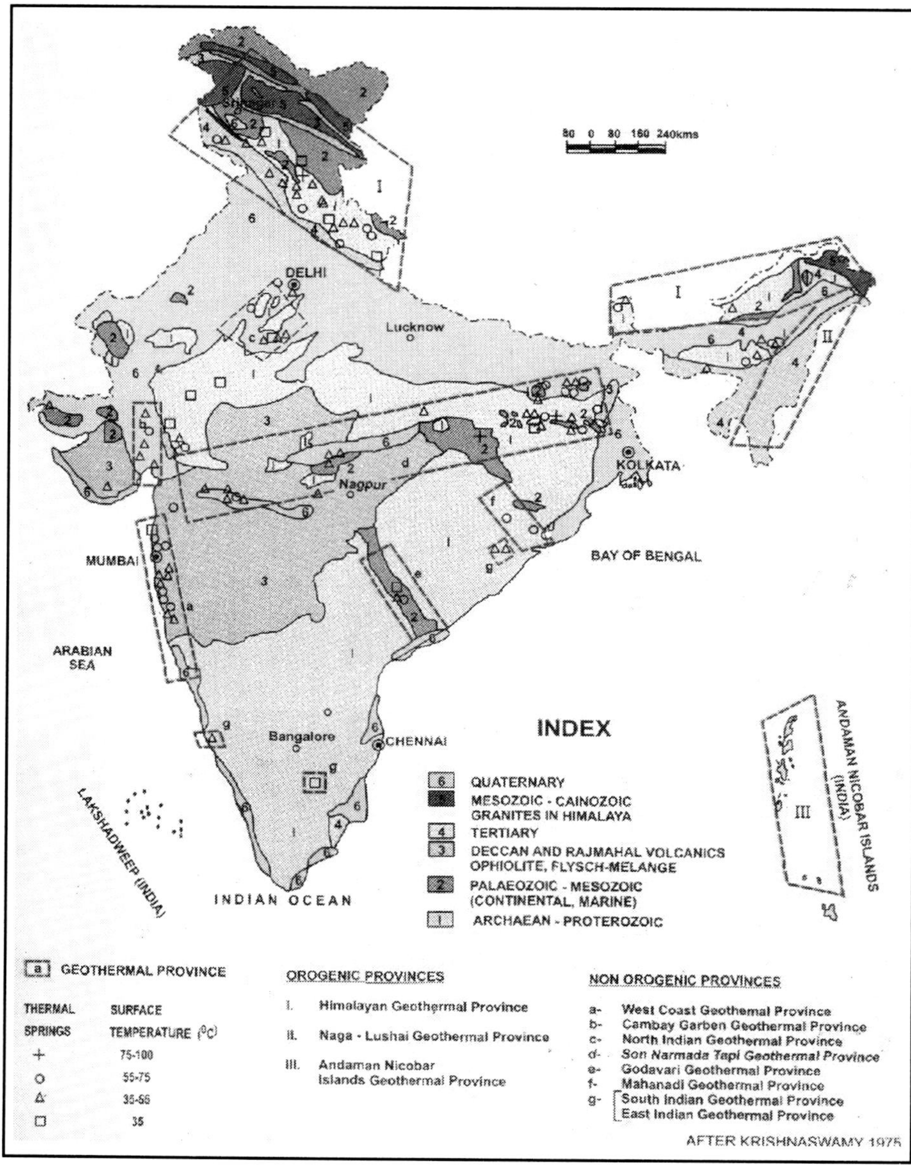

Fig.1: Major and representative Indian geothermal provinces

Low Enthalpy Geothermal Energy Systems

The low enthalpy geothermal energy systems are associated with:

1. Tertiary tectonism and neotectonic activity.
2. Shield areas with localized abnormal heat flow, which is normally very low.

Indian Geothermal Resources

Physiographically, India is divisible into three distinct geographical units *viz.* the Extra- Penninsular Region, Indo-Gangetic Plain and the Peninsular Region. A variety of rocks ranging in age from Archaean to Holocene are exposed in these geographical units. The thermal springs have been classified on the basis of their occurrence in specific geotectonic setups and have been grouped under different Geothermal Provinces. The geological and structural criteria that have been used in the identification of the prospective Geothermal Provinces are:

- Occurrence in an orogenic belt, which has undergone Cenozoic folding and upliftment
- Occurrence in structural depressions/grabens, associated with late Tertiary and Quaternary upliftment in non- orogenic belts
- Related to deep fault zones associated with recent seismicity
- Occurrence in the areas of Tertiary or Quaternary volcanic activity

The orogenic and non-orogenic geothermal energy resources of India are grouped as follows:

Orogenic Regions

The orogenic regions are further divided into different provinces:

1. Himalayan Geothermal province (Northwest and Northeast sub-provinces)
2. Andman-Nicobar Island geothermal province

Northwest Himalayan Geothermal Sub-Province

This sub-province has been divided into three belts on the basis of occurrences of thermal springs. These are the trans-Himalayan belt, the central Himalayan belt and the outer Himalayan belt. This sub-province encompasses the states of Jammu and Kashmir, Himachal Pradesh and Uttarakhand. The thermal springs occur at altitudes of 4000m to 4400m in Trans-Himalayan belt, at Chhumathang and Puga, while in central Himalayan belt, the altitude ranges from 1300m to 3000m in Parbati, Beas and Satluj Valleys in Himachal Pradesh and 1000m in outer Himalayan belts. Some of the important thermal fields are described below:

Puga Geothermal Field : The Puga area is located at a distance of about 180km from Leh in Ladakh Region of J&K across the Great Himalayan Range at an altitude of 4400m amsl. Hot spring region is located along the collided junction of two crustal plates, which were involved in the Himalayan Orogeny. Intense basic to ultrabasic, plutonic and sub-marine volcanism of middle to upper Cretaceous age (ophiolites suite of rocks) and several phases of widespread acid igneous activity from upper Cretaceous to upper Tertiary times have been identified.

Thermal manifestations in the form of hot springs (Fig.2), hot pools, sulphur condensates, borax evaporates have an aerial extent of 4 km. Hot springs with temperatures varying from 30°C to 84°C (boiling point at Puga area) and with discharge ranging up to 300 litres/min are present. The hottest thermal spring shows a temperature of 84°C and the maximum discharge from single spring is 5 liters/sec. In addition, hot patches and minor hot water seepages are also noticed in the thermal area. Typical cones of extinct hot spring deposit exist in the eastern part of the area. Borax evaporates occur as loose deposits on the surface while the sulphur condensate fills the joints and cracks (Fig.3).

Fig. 2: Geyser at Puga

Fig. 3: Sulphur deposit at Puga

Geophysical surveys have reported a major low resistivity zone (2-10 ohmm) in the southwestern part of the Puga geothermal area, which are indicative of the presence of thermal water in the sub-surface.

Thermal waters of Puga field are near neutral to weakly alkaline-pH 6 to 8.3. Major cations and anions, in the order of decreasing concentrations, are Na, K, Cs, Li, Mg, Rb, Fe, and HCO_3, Cl, SiO_2, B, SO_4, As, NH_3 respectively. The water is essentially $NaHCO_3Cl$ type. High concentration of Li, Rb and Cs also suggests indirectly that thermal waters have stayed for longer period underground permitting intensive water-rock interaction. Geochemical thermometers, based on the concentration of Na, K, Ca and Mg, have pointed to the possibility of having thermal fluids of 220°C-260°C temperatures at deeper levels whereas the oxygen isotope thermometry places this estimate at 180°C. Marked Oxygen-shifts depicted by Puga thermal water is a strong indication of presence of high temperature zones at depth. Stable isotope studies on oxygen and hydrogen carried out on thermal water have clearly pointed out that these waters are meteoric in origin.

The thermal logging of the geothermal wells indicates an abnormally high geothermal gradient generally from 0.35 to 2.5°C/m, although higher values up to 6.8°C/m have also been encountered locally. On the basis of the available thermal gradient data and a few thermal conductivity determinations, the minimum heat flow in the Puga valley has been estimated to be over 13 HFU; that is, about eight times the normal heat flow from the earth's crust.

A total of 34 boreholes ranging in depths from 28.5 m to 384.7 m have been drilled in Puga valley. The bore holes that struck steam-water mixture from shallow geothermal reservoir have yielded 10 to 15% of steam at a temperature up to 140°C and a pressure of 2 to 3 kg/cm^2. Of the 17 flowing bore holes, wellhead measurements was carried out on 8 bore holes indicating a total discharge of 190 tonnes/hr of water- steam mixture. The maximum discharge from a single drill hole is 30 tones/hr.

Experimental space heating and refinement of borax and sulphur have been successfully tried utilizing the geothermal fluids at Puga from the shallow wells drilled so far. The extraction plant could handle 1 tone of borax ore per day and the refining plant could handle 500 kg of borax per day. Space heating experiment carried out at Puga has proved quite successful. A prefabricated hut 5m x 5m x 2.5m was heated with thermal fluids and an excess temperature of 20°C over the ambient temperature could be maintained within the hut. Experimental space heating, green house cultivation and poultry farming have also been successfully conducted by the GSI and RRL (Jammu). The thermal waters have been used for hatching poultry and for growth of mushroom in a hut in 500 sq m areas by the RRL, Jammu. Puga fluid has one of the highest Cesium concentration (10 mg/l) in the world. Efforts were applied on to separate this valuable metal from geothermal fluid using the resin developed by BARC (Mumbai).

Puga geothermal field is the only field capable of producing either primary cycle electrical power or binary cycle power at present, though on a very small scale. However, reservoir simulation studies have suggested the possibility of generating over 3 MW electric power, if deeper levels are probed at least up to the depth of 500 m.

Chhumathang Geothermal Field : Chhumathang field is about 40 km north of Puga. It is located at a distance of 150 km from Leh, Ladakh; on Leh-Chushul road at an altitude of 4300 m. Geologically, area exposes a thick sequence of shallow marine to fluvial sediments deposited over granitic basement. These rocks range in age from Cenomanian to Miocene. The Chhumathang geothermal field is located in this belt.

S.P., magnetic, seismic refraction, and resistivity surveys have been carried out. AMT surveys indicated a low resistivity zone of 13-30 ohmm up to a depth of 300m. Geothermal area is characterized by spectacular development of spring deposits. These are in the form of carbonate plateaus up to 100 m diameter and attaining a height of 6-10 m. (Fig.4). Their presence gives clue to much vigorous thermal activity in the past in an area at least twice that of today's convective flow. Thermally anomalous area stretches for about 700 m along the right bank of Indus River in the form of hot springs, travertine deposits and stained ground. Hot springs with temperatures ranging between 30°C to 87°C (boiling point at Chhumathang) and many H_2S emitting hot spring water pools are the surface manifestations. The maximum temperature of the thermal spring recorded is 87°C with a discharge of 1.5 liters/sec. The cumulative discharge of all the springs is 200 lit/min.

Fig. 4: carbonate plateaus at Chhumathang field

Geochemical studies have indicated the existence of thermal fluid in the reservoir of about 150°C temperatures. The possibility, however, cannot be ruled out that thermal discharges are steam-heated re-equilibrated fluids and estimated reservoir temperatures pertain to shallower levels of latest chemical equilibrium. Six boreholes ranging in depth from 20-221 m have been drilled in the Chhumathang area. A temperature of about 109°C was recorded in flowing wells. Cumulative discharge from 4 flowing wells is about 50 tones/hr. The shallow geothermal bore holes drilled at Chhumathang have indicated very high temperature gradient ranging from 0.7°-2.5°C /m.

Chhumathang thermal waters resemble with Puga thermal waters but differ from them in having relatively higher pH and sulphate. Stable isotope studies on oxygen and hydrogen carried out on thermal waters have clearly pointed out their meteoric origin. Chemical thermometry has pointed out deep reservoir temperature of 150°C.

Geothermal energy was utilized at Chhumathang successfully for green house cultivation and space heating. A temperature of 20°C to 25°C could be maintained inside the green house with outside atmospheric temperature as low as -35°C to -40°C in winter. A variety of vegetables were grown.

Manikaran Geothermal Field : Manikaran geothermal field is located about 50 km east of Kulu, Himachal Pradesh at an altitude of about 1700 m. The field is located to the southwest of the Central Crystalline Axis, which runs along the great Himalayan range and comprises predominantly foliated gneiss intruded by younger granite, where profuse igneous activity has taken place during Tertiary period. The hot springs at Manikaran are located in the Manikaran quartzite (Devonian) beneath the Main Central Thrust. The granites intruding the Manikaran quartzite are possibly the source of heat of the hot-spring waters. The Manikaran quartzite is highly jointed, and these joints hold great significance as far as the movement of hot waters is concerned.

Geothermal activity at Manikaran occurs in the form of hot springs, over a distance of about 1.25 km on the right bank of Parbati river whereas on the left bank, this activity manifests itself over a distance of about 450 m. Thermal springs located on the right bank are in the temperature range of 34°C-96°C, whereas on the other side temperature ranges 28°C-37°C.

The geophysical surveys have indicated a zone of positive S.P. associated with low resistivity. Geothermally anamolous zone is characterized by apparent resistivity in the range of 30-100 ohmm. Thermal water is feebly alkaline with pH value ranging from 7.5 to 8.1. On the basis of dominant cations and anions, the Manikaran thermal water is $NaHCO_3Cl$ type and $NaCaHCO_3Cl$ type. A comparison of relative chloride, bicarbonate and sulfate contents shows that Manikaran water to be very similar to the thermal water discharging at Puga, a definitely higher temperature system, but without any obvious relationship. Geochemistry indicated a thermal water of about 120°±10°C in the sub-surface.

Nine boreholes have been drilled at Manikaran and the deepest being 707 m. Maximum discharge from a single spring is about 7 lit/sec. The cumulative thermal discharge from 8 bore holes is 100 tones/hr with temperature ranges from 45°C to 96°C. Distribution of major high temperature (96°C) thermal springs and bore holes are on the right bank as compared to lower temperatures (45°C) hot spring and drill hole on the left bank point to the inflow of thermal water from northern side.

As far as the utilization aspects are concerned, a 7.5 ton capacity cold storage plant based NH_4-absoption system was set up. A 5 KW pilot power plant based on binary cycle (ISO-Butane) principle was also made test run. In addition, the tourist complex could be developed in the region utilizing the geothermal fluids for hot water baths, spas etc. The thermal waters are seen to be free in respect of large scale corrosion and scaling.

Tapoban Geothermal Field : Tapoban Geothermal area is located in Dhauli river valley, a major tributary to Alaknanda River. It is approachable by an all weather 15 km long road from Joshimath, in Chamoli district, Uttrakhand at an altitude of about 1800 m. The surface manifestation is in the form of five hot springs spread over a distance of one km along the hill slopes on the left bank of Dhauli river. The highest temperature recorded is 65°C. The discharge from these springs varies between 0.83-9.22 lit/sec. No secondary deposition is observed. Only one spring has gaseous emanations. The thermal springs emerge through Tapoban quartzite belonging to the Helang Formation (Fig.5).

Fig. 5: Tapoban geothermal Field, Chamoli district, Uttrakhand

The rocks exposed in the area comprise quartzite with occasional schistose bands, a thick sequence of gneiss, schist, migmatites, phyllite, limestone and quartzite of the Tethyan sequence. Geophysical Surveys have indicated low resistivity (100 ohmm) zones extending down to 165 m depth.

Geochemical studies indicate that the thermal water is neutral with pH ranging from 6.9-7.2 .The cations in decreasing order of concentration are Ca, Mg, Na, K and HCO_3, SiO_2, SO_4, Cl, B and F. The water is of Ca-Mg-HCO_3 type. Only one spring has gaseous emanations. The gas sample collected comprises 78.6% CO_2, 18.96% N_2 and 2.4% O_2.

Secondary deposits were precipitated around the collar of the bore hole at Tapoban. The major mineral is aragonite, while quartz, alkali feldspar and mica are present in trace amount. The temperature of water was around 80°C.

The geothermal system provides an example of 'Normal Gradient Heating Model'. Chemically the thermal fluid at 92°C is indistinguishable from local cold springs, except for higher SiO_2 concentration in the former. The fluid descending from higher altitudes (at least 2 km above the discharge level) has attained heat under normal thermal gradient conditions. Lower temperatures and chemical inertness of quartzite have resulted in lack of water-rock interaction reflected in no significant change in fluid chemistry.

Four boreholes of the depth ranging from 291 to 728 m have been drilled in the area. All the boreholes struck artesian conditions. Cumulative discharge from four boreholes is about150 tons/hour. Highest measured temperature in the borehole discharge is 92°C.

Base temperature calculated by Na-K-Ca geochemical thermometry is 65°C. As indicated earlier, with the drop in temperature by 34°C, within a distance of 1.5 m from the bore hole, the chemistry has undergone a very significant change. Silica has precipitated so fast that the base temperature of around 150°C calculated on silica thermometry for the water samples has come down to around 90°C. Similarly ratios of Ca, Na and K have also changed drastically with the drop in temperature (causing quick precipitation).

Northeast Himalayan Geothermal Sub-Province

Thermal springs located in Sikkim are an integral part of "Himalayan Geothermal Province". Five thermal springs located at

Yumthang, Yumesamdong, Borong, Polot and Rishi are distributed in the North, South and West districts of Sikkim. The springs range in surface temperature from 38° to 59°C with moderate discharge and can only be applicable for direct utilization to promote non-conventional energy resources. The spring discharges have medicinal values due the presence of sulphur and boron contents. The discharges presently are used locally for drinking and bathing purposes.

Non-orogenic Regions

The different non-orogenic regions are given below:

1. Cambay graben geothermal province
2. Son-Narmada-Tapi graben geothermal province
3. West Coast geothermal province
4. Damodar valley geothermal province
5. Mahanadi valley geothermal province
6. Godavari valley geothermal province
7. North Indian Peninsular geothermal province
8. East Indian geothermal province
9. South Indian geothermal province.

Some of the important non-orogenic geothermal fields are described below:

Tattapani Geothermal Field: The Tattapani geothermal field is located 100 km northeast of Ambikapur, Sarguja district, Chhattisgarh. Tattapani is most promising geothermal field in the Peninsular India. Geothermal activity is seen in the east at Tattapani and in the west at Jhor, which are located 50 km apart. Thermal manifestation at Tattapani is very intense in an area of 0.05 sq. km with several hot spouts, hot water pools and a marshy land (Fig. 6). The surface manifestations show occurrence of white to dirty white deposit (identified as silica) and moderate to low gas activity. The temperature of the hot water varies from 50°C to 98°C. Surface flow of the individual springs is very low while the cumulative discharge is around 60 litres/ minutes. Indications of palaeo-geothermal activity are seen in the form of altered clay and hydrothermal deposits at a place 500m west of the thermal activity.

Fig. 6 : Geyser at Tattapani geothermal field, Sarguja district, Chhattisgarh

Tattapani geothermal field has two distinct lithological sequences belonging to Proterozoic and Gondwana Supergroup. Proterozoic rocks are widespread while Gondwana rocks are exposed at Tattapani and further west. Proterozoic rocks comprise grey and pink gneiss, augen gneiss, granulite, kyanite schist, sillimanite schist and hornblende gneiss. Phyllite, carbonaceous phyllite also forms part of the sequence. Gondwana Supergroup comprises green splintery shale; sandstone and shale alternations with coal streaks and plant fossils, reddish brown sandstone and grit bands.

Gravity, electrical, magnetic, resistivity, self potential and AMT surveys were conducted in area. The AMT surveys indicated a conductive zone in the sub-surface close to the hot spring area, which may correspond to the hot water bearing formation.

Two distinct types of waters have been identified, on the basis of proportion of cations and anions in Tattapani area *i.e.* Ca-Mg-HCO_3 and Na-HCO_3-Cl-SO_4 types. The thermal water from springs and bore holes show high fluoride content and low TDS. Higher concentration of fluoride is attributed to leaching of fluoride from hornblende of the hornblende granite. Sub-surface temperature using different thermometries is at variance with each other being 120° C to 150°C.

Hydrothermal alterations due to geothermal activity occur as encrustations around the hot springs, in hot water marsh, along joints, cracks and vug fillings in the country rock and breccia. Quartz, chalcedony, cryptocrystalline silica, stilbite, montmorillonite, traces of calcite and pyrite are present as the main alteration minerals. The hydrothermal mineral assemblage is in equilibrium within a temperature range of 100°± 10°C. Occurrence of low temperature alteration minerals in the high temperature zones suggests low water-rock ratio.

Twenty six boreholes have been drilled to depths ranging from 100 m to 620 m. Blow out conditions occurred in five wells. Bore hole GW-6 is the deepest well drilled to a depth of 500m. The shut-in temperature of the thermal fluids is around 112° C. The cumulative discharge from wells is 1600 liters/min. About 4 kg/cm^2 pressure has been measured by ONGC in all the five bore holes.

Based on the cumulative discharge of 1600 lit/min of thermal water (100° C) from five drill holes at Tattapani a joint project of ONGC-GSI-MPUVN (Madhya Pradesh Urja Vikas Nigam) to install a 300 kWe binary cycle power plant is under consideration. Keeping in view the remoteness of the area it is essential to utilize the thermal fluids for non- electrical purposes. Thermal waters can be used for Coccon boiling for extraction of silk thread, mushroom culture, cold storage, food processing, spa and tourist development, drying and canning of fish, rice bran oil and space heating during winter season.

West Coast Hot Spring Belt : Sixty thermal water springs occur at eighteen localities in the West Coast hot spring belt. This belt extends along the West Coast for a distance of about 350km from Koknere, north of Mumbai, to Rajapur in the south, with an average width of 20 km, (maximum width 50 km in the northern part and minimum of 10-15 km in the southern part). This belt falls in Thane, Raigad and Ratnagiri districts of Maharashtra. The NNW-SSE trending belt is hilly and covers basinal areas of seven westerly flowing rivers. The eastern boundary is marked by mighty scraps of Sahyadri Mountain commonly known as Western Ghats while its western margin is marked by coastline of the Indian Ocean. The thermal spring localities have been divided into three distinct clusters in the West Coast *viz.*

- Northern sector consists of six hot springs located at Koknere, Paduspada, Haloli, Sativli, Ganeshpuri and Akloli.
- Central sector consists of four hot springs located at Sov, Vadavli, Pali and Unahvre (Tamhane).

- Southern sector consisting of eight hot springs located at Khed, Unhavre (Khed), Aravli, Tural, Rajawadi, Sangameshwar, Math and Rajapur.

Geochemistry of all the hot spring waters except Rajapur, are Alkali chloride type. Rajapur hot spring water is bicarbonate type. Thermal waters have relatively high chloride content. The distinctly different range of chloride content of hot water suggests that occurrence and movement of thermal water is not connected with hydrologic cycle of surrounding cold water. Boron content except Unhavre (Khed), Khed (both 1 ppm), Rajapur, Math (both 0.5 ppm) hot springs is <0.5 ppm. Similarity of Cl/B ratio indicates that Math and Rajapur hot springs probably draw thermal water from reservoirs of similar hydrological condition. Cl/B value of Khed and Unhavre (Khed) hot springs suggest hydraulic continuity and their association with same geothermal system. Similar chloride and boron values form Tural-Rajwadi indicate their source in same geothermal system. Other hot springs having relatively high Cl/B ratio represent independent geothermal systems. Flourine values in thermal water of west coast are 1 to 3 ppm. It is similar to well known geothermal fields at Ice Land, USSR, and New Zealand. Lithium in thermal water at Unhavre (Tamhane), Vadavli, Pali, Ganeshpuri is <0.5 ppm. Lithium is more similar to thermal water in Basaltic province of Iceland. In other geothermal fields of the world, the lithium content is higher. Majority of geothermal systems in West Coast have minimum reservoir temperature of the order of 120 ± 10°C.

Twelve exploratory bore holes were drilled with depth ranging from 50 m to 500 m in West Coast areas. These included four bore holes at Ganeshpuri-Akloli, six at Unhavre (Khed), and two at Tural Rajwadi. Studies carried out, so far, have shown that adequate supplies of hot water can become available by drilling shallow production wells at Unhavre khed, Tural-Rajwari and Ganeshpuri-Akloli and Sativli thermal spring areas.

Ganeshpuri hot springs are already attracting large numbers of tourists for hot water bath. Development of "Sauna baths", animal husbandry, poultry farming, fresh water fish farming and greenhouse on a large scale for increase in the production of vegetables mushrooms under controlled conditions through multiple cropping can be thought of, which not only save the sizeable amount of fossil fuels but raise the economy of the Nation.

Conclusion

Almost all the 300 thermally anomalous areas have been examined. 31 areas have been studied in details, out of which, shallow drilling has been done only in 16 areas. Surface studies on Geology, Hydrogeology and Geochemistry (including isotope geochemistry) have been accomplished in all the areas. Sub-surface studies based on geophysical exploration and shallow to intermediate drilling have been accomplished in all the areas. The deepest exploratory boreholes drilled are in Puga (385m), Chhumathang (220m), Manikaran (700m), Tapoban (728m), Tattapani (620m) and West Coast (500m). Thermal discharges are at temperatures of 90°C to 140°C in the promising areas. Thermal fluids issuing from the boreholes have limited output. Electrical power production of KWe level only is possible from the thermal discharges from the exploratory boreholes drilled so far. However, reservoir simulation studies have suggested the possibility of generating over 3 MWe electric power in Puga geothermal field if deeper levels are probed at least up to the depth of 500m. Direct heat application on a moderate scale only, is possible from the thermal discharges available from the boreholes. Monitoring of some of the selected geothermal manifestations in northern part of India has been taken up systematically for geochemical studies since 2005. This will add to in synthesizing the enormous geothermal data generated during the last three decades and will also help in deciding the future course of action for the development of geothermal prospects in the country.

GSI has also synthesized most of the geothermal data, so far, carried out in the last three decades in the form of publications *i.e.* "Geothermal Atlas of India" in 1989 and subsequently "Geothermal energy resources of India" in 2002.

References

Geothermal Atlas of India (1989) Geol. Surv. India, Spl. Pub. No. 19, 143p.

Lund, John W. (2004) 100years of Geothermal Power Production. GHC Bull., September, 2004

Thussu, J.L. (2002) Geothermal energy resources of India. Geol. Surv. India, Spl. Pub. No. 69, 210p.

□□□

19

SFARF : Software for Classification of Fourier Shape of the Fold Around Tamadhaun District Almora, Uttarakhand

H. THOMAS[1], S. SUJATA[1] and A. CHANDEL[2]

[1] Department of Applied Geology, [2] Department of Computer Science and Applications, University of Sagar, Sagar-470003 (M.P.), India

Email: harelthomas@yahoo.com

Abstract

Fourier (harmonic) analysis is a very simplified graphical method of plotting fold shapes in two dimensions. The authors have developed software (SFARF) to directly assign the class of fold shapes on the basis of the value of the ratio b3 / b1 (Fourier Coefficient). The various fold forms obtained by the software around Tamadhaun area reveal that fold developed in quartzite occurring in the study area vary in the range between sine-waves and parabolas to almost sine-wave, whereas the folds which are developed in phyllite and schist rocks are statistically between sine-waves to almost chevron.

Introduction

The Tamadhaun area covering an area of about 120 sq. km (Fig.1) is situated in Almora district and can be approached by road from Kathgodam which is the nearest railway terminal. The area around Tamadhaun forms the southern limb of Dudatoli Syncline where the Dudatoli - Almora Crystalline (Heim and Gansser, 1939) and phyllite (Kumar *et al.,* 1974) are exposed (Fig 1). The superposed mesoscopic structures observed by the authors reveal that the area under investigation has suffered polyphase deformations (Shrivastava and Thomas, 1990; Thomas, 1991, 1993; Shrivastava *et al.*, 2003).

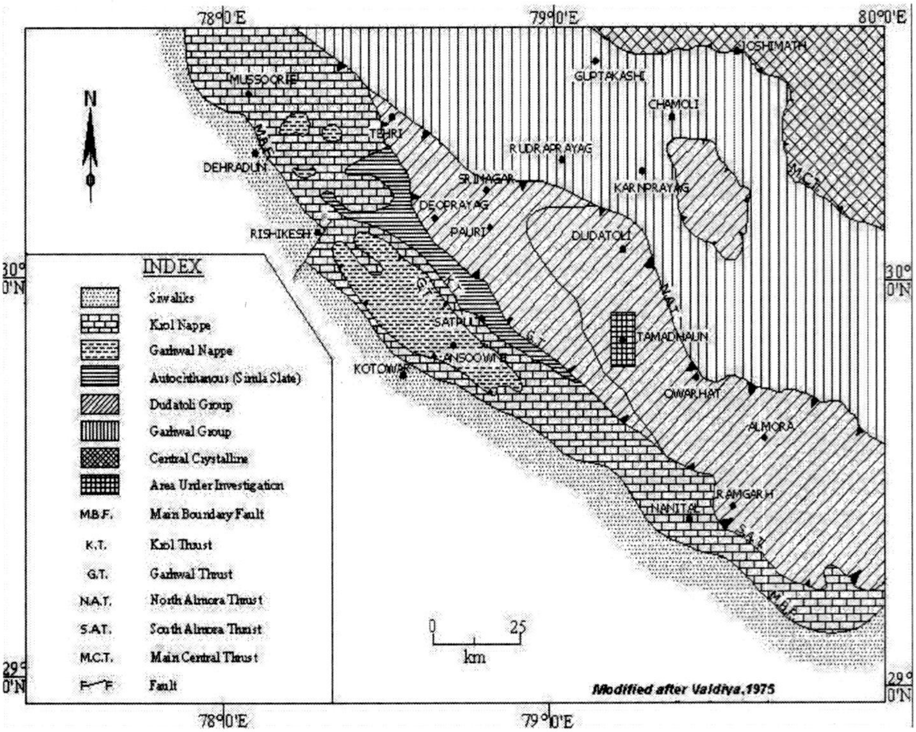

Fig. 1: Simplified geological map of Kumaon Himalaya showing area of study

The different folds developed in the area were studied with the help of Fourier (harmonic) analysis. One basic method used in mathematics for analyzing the curvature variation is to try a series of simple x – y functions and the technique is known as harmonic or Fourier analysis. Fourier (harmonic) analysis is a very simplified graphical method of plotting fold shapes on a two dimensional graphic (Hudleston, 1973). To express the slope of any curve, this analysis uses series of trigonometric functions, which is as follows:

$$f(x, y) = a_o + a_1 \mathrm{Cos}X + a_2 \mathrm{Cos}2X + a_3 \mathrm{Cos}3X + \ldots\ldots + b_1 \mathrm{Sin}X + b_2 \mathrm{Sin}2X + b_3 \mathrm{Sin}3X + \ldots \quad (1)$$

The curve is specified by various 'a' and 'b' parameters known as **Fourier Coefficients,** which refer to the amplitude of the various cosine and sine components.

To make the analysis we first select a sector of complete layer between an inflexion (i) line and a hinge line (h). The coordinate axis 'y' is chosen so that it passes through the inflexion point (i) and is parallel to the axial surface of the fold. The x–axis of the coordinate system passes through inflexion point (i) perpendicular to the y–axis so that this originated from point "I" (Fig. 2A) with this coordinate frame, all the a – type Fourier coefficient became zero, only the b – type Fourier coefficient can be represented in the function and it has given to the changes in shapes of successive quarter wave sectors they must also be zero. Therefore, the reduced Fourier series can represent fold sector.

$$f(x, y) = b_1 \operatorname{Sin} X + b_3 \operatorname{Sin} 3X + b5 \operatorname{Sin} 5X + \quad ..(2)$$

It is clear that the successive coefficients in this series become smaller and in mathematical terminology the series converges very rapidly. Only the first three coefficients provide significant input into the description of the wave shape, and the third of these takes on a very small value.

The length of the line defining quarter wave base (W) is first measured and then divided into three equal parts. A normal on each trisecting points of the quarter wave base line is drawn so that it cuts the curved sector of the fold. The length of each of the three normal D_1, D_2, & D_3 (Fig. 2A) are measured. The quarter wave base (W) is then converted into "p/2" units, by dividing "p/2 with W". The parameter Y_1, Y_2, & Y_3 are then calculated by multiplying (x = p/2W) with D_1, D_2, & D_3 respectively as follows:

$$\begin{aligned} Y_1 &= D_1 * X \\ Y_2 &= D_2 * X \\ Y_3 &= D_3 * X \\ \text{Where } X &= (p/2W) \end{aligned} \quad ..(3)$$

Stabler (1968) has proposed that Fourier coefficient b_1, b_2 and b_3 can be calculated by using equation (3) as follows:

$$\begin{aligned} b_1 &= (Y_1 + \text{Ö}3Y_2 + Y_3) / 3 \\ b_2 &= (2Y_1 - Y_3) / 3 \\ b_3 &= (Y_1 - \text{Ö}3Y_2 + Y_3) / 3 \end{aligned} \quad ...(4)$$

From the above equation b_2 and b_3 are providing the principal features of the fold shape and the positive values of b_3 reduce the amplitude effect of b_1, while negative values increases it (Huddleston,

1973). The other significant effect of the b_3 coefficient is to modify the shape of the sinusoidal b_1 components (Singh and Gairola, 1992; Thomas, 1993). Positive values of b_3 broaden the hinge region and tend to produce box - fold shapes (Thomas, 1993). Whereas negative values tend to steep and straighten the fold limbs and produces more chevrons like style (Thomas, 1993) (fig.2b).

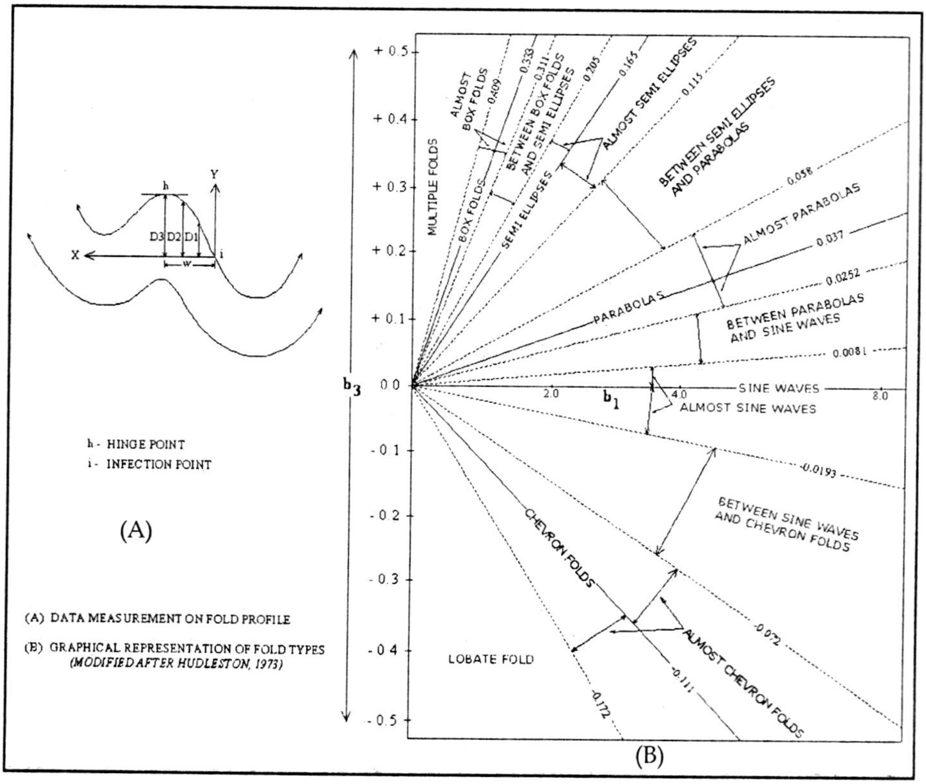

Fig. 2 : (A) Data measurement on fold profile, (B) Graphical representation of fold types (modified after Hudleston, 1973)

The fifth (b_5) Fourier component /harmonic is very small and exerts very little geometric effect on the style. Huddleston (1973) has given a graphical method for plotting fold shapes by representing graphically b_1 as abscissa and b_3 as ordinate (Fig.3). Each fold shape has a characteristic b_3/b_1 ratio viz. Chevron, Sinusoidal, Parabolic, Semi-ellipse and box folds have b_3/b_1 values as 0.11, 0.000, 0.037, 0.165 and 0.333 respectively (Huddleston, 1973). Singh and Gairola (1992) modified the Huddleston's (1973) classification of fold shapes from eight types of fold shapes to sixteen types.

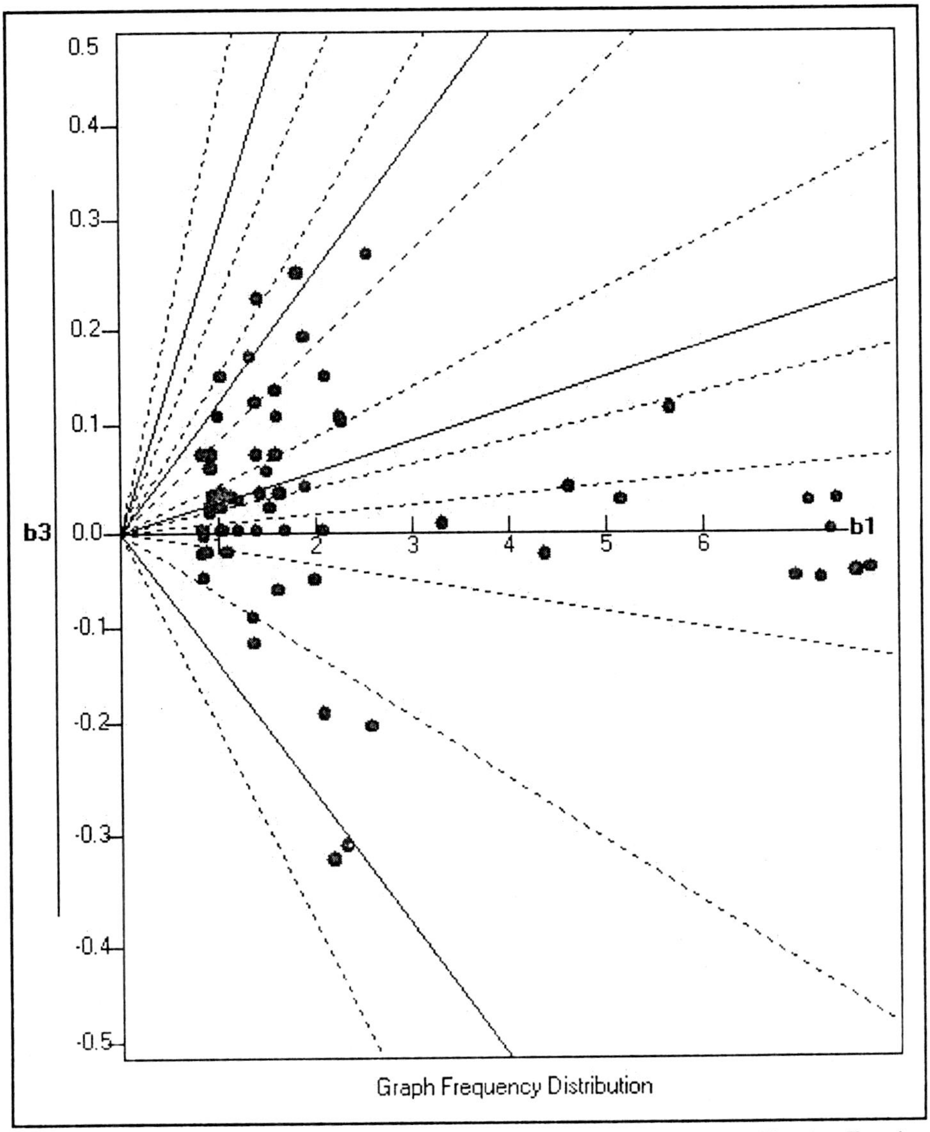

Fig. 3 : The graphical representation of various fold forms obtained by the Fourier analysis of the folds around Tamadhaun.

Thomas (1993) and Thomas and Thomas (2003) further modified the classification with replacement of cuspate fold by Lobate and developed a computer program in visual basic, which calculate the parameters b_1, b_3 and b_5 and ratios of b_3 and b_1. It directly assigns the class of fold shapes on the basis of the value of the ratio b_3 / b_1 (Table-1 and 2).

Table 1: The class of fold shapes on the basis of the value of the ratio b_3/b_1 (Thomas, 1993)

Class of Fold Shapes	Value of the ratio of b_3/b_1
Multiple fold	>0.409
Almost Box fold	<0.409 to >0.333 and <0.333 to >0.311
Box fold	0.333
Between Box fold and Semi ellipse	<0.333 to >0.205
Almost Semi-ellipse	<0.205 to 0.165 and <0.165 to >0.115
Semi ellipse	0.165
Between Semi ellipse and parabola	<0.115 to >0.058
Almost Parabola	<0.058 to > 0.037 and <0.037 to >0.025
Parabola	0.037
Between parabola and sine wave	<0.025 to >0.0081
Almost sine wave	<0.0081 to >0.00 and <0.00 to > -0.0193
Sine wave	0.000
Between sine wave and chevron fold	< - 0.0193 to > - 0.072
Almost chevron fold	<-0.072 to >- 0.111 and <-0.111 to >-0.172
Chevron fold	-0.111
Lobate fold	<-0.172

Table 2: Given values of D_1, D_2, D_3, W and results obtained by program b_3/b_1 and fold types

S.No.	D_1	D_2	D_3	W	b_3/b_1	Fold Type
1	0.90	1.20	1.30	0.70	0.117	Almost Semi-Ellipse
2	0.30	0.50	0.60	0.70	0.000	Sine Wave
3	0.65	1.15	1.20	0.85	0.026	Almost Parabola
4	0.65	0.90	0.95	0.70	0.111	Between Semi-Ellipse and Parabola
5	0.30	0.85	1.00	0.50	-0.144	Almost Chevron Fold
6	1.05	1.10	1.15	0.30	0.231	Between Box Fold and Semi-Ellipse
7	0.25	0.55	0.65	0.30	-0.081	Almost Chevron Fold
8	0.25	0.50	0.55	0.45	-0.030	Between Sine-Wave and Chevron Fold

Contd...

9	0.30	0.50	0.55	0.60	0.029	Almost Parabola
10	0.20	0.35	0.40	0.60	0.000	Sine Wave
11	0.15	0.50	0.55	0.30	-0.160	Almost Chevron Fold
12	0.45	0.70	0.80	1.00	0.041	Almost Parabola
13	0.40	0.70	0.85	0.95	-0.020	Between Sine-Wave and Chevron Fold
14	0.30	0.50	0.55	0.50	0.029	Almost Parabola
15	0.30	0.55	0.65	0.40	-0.026	Between Sine-Wave and Chevron Fold
16	0.30	0.40	0.42	0.60	0.127	Almost Semi-Ellipse
17	0.50	0.70	0.80	0.50	0.080	Between Semi-Ellipse and Parabola
18	0.30	0.50	0.60	1.00	0.000	Sine Wave
19	0.30	0.50	0.70	1.00	-0.054	Between Sine-wave and Chevron Fold
20	0.30	0.30	0.45	0.50	0.107	Between Semi-Ellipse and Parabola
21	0.08	0.11	0.12	0.35	0.102	Between Semi-Ellipse and Parabola
22	0.18	0.47	0.50	0.30	-0.094	Almost Chevron Fold
23	0.16	0.18	0.20	0.20	0.179	Almost Semi-Ellipse
24	0.30	0.35	0.40	0.30	0.153	Almost Semi-Ellipse
25	0.35	0.55	0.60	0.35	0.053	Almost Parabola
26	0.20	0.37	0.40	0.25	0.000	Sine Wave
27	0.35	0.55	0.70	0.60	0.000	Sine Wave
28	0.46	0.65	0.70	0.60	0.096	Between Semi-Ellipse and Parabola
29	0.20	0.40	0.48	0.75	-0.058	Between Sine-Wave and Chevron Fold
30	0.35	0.55	0.65	0.50	0.026	Almost Parabola
31	0.40	0.70	0.75	1.10	0.021	Between Parabola and Sine-Wave

Contd...

32	0.45	0.65	0.72	0.60	0.078	Between Semi-Ellipse and Parabola
33	0.25	0.45	0.50	0.78	0.000	Sine Wave
34	0.25	0.38	0.40	0.60	0.078	Between Semi-Ellipse and Parabola
35	0.45	0.70	0.80	1.10	0.041	Almost Parabola
36	0.20	0.30	0.35	0.60	0.047	Almost Parabola
37	0.25	0.50	0.60	0.85	-0.058	Between Sine-Wave and Chevron Fold
38	0.30	0.60	0.65	0.95	-0.025	Between Sine-Wave and Chevron Fold
39	0.40	0.65	0.70	1.30	0.045	Almost Parabola
40	0.45	0.68	0.80	1.00	0.041	Almost Parabola
41	0.40	0.55	0.60	1.00	0.102	Between Semi-Ellipse and Parabola
42	0.30	0.68	0.80	0.70	-0.088	Almost Chevron Fold
43	0.15	0.20	0.25	0.60	0.067	Between Semi-Ellipse and Parabola
44	0.08	0.12	0.15	0.55	0.023	Between Parabola and Sine Wave
45	0.10	0.20	0.24	0.65	-0.058	Between Sine-Wave and Chevron Fold
46	0.20	0.35	0.40	0.50	0.000	Sine Wave
47	0.22	0.35	0.40	0.45	0.033	Almost Parabola
48	0.60	0.85	0.90	0.90	0.101	Between Semi-Ellipse and Parabola
49	0.70	0.88	0.95	1.00	0.142	Almost Semi-Ellipse
50	0.50	0.85	0.95	0.80	0.017	Between Parabola and Sine-Wave
51	0.50	0.78	0.85	0.50	0.056	Almost Parabola
52	0.45	0.60	0.70	1.00	0.091	Between Semi-Ellipse and Parabola
53	0.30	0.45	0.55	0.75	0.031	Almost Parabola
54	0.30	0.60	0.75	1.20	-0.072	Between Sine-Wave and Chevron Fold

Contd...

55	0.25	0.49	0.51	0.80	-0.006	Almost Sine-Wave
56	0.20	0.40	0.50	0.80	-0.072	Between Sine-Wave and Chevron Fold
57	0.50	0.95	1.00	1.10	0.000	Sine Wave
58	0.35	0.55	0.65	0.80	0.026	Almost Parabola
59	0.60	1.10	1.35	1.00	-0.039	Between Sine-Wave and Chevron Fold
60	0.65	1.20	1.30	1.00	0.000	Sine Wave
61	0.70	1.10	1.20	1.10	0.053	Almost Parabola
62	0.60	1.00	1.05	0.95	0.044	Almost Parabola
63	1.10	1.65	1.85	1.75	0.060	Between Semi-Ellipse and Parabola
64	0.32	0.70	0.80	0.72	-0.069	Between Sine-Wave and Chevron Fold

2. SFARF (Software for Fourier Analyses of Rock Folds) HELP

Execution of the software is very simple and user friendly.

How to start

a. Click Start -> All programs -> SFARF- > SFARF

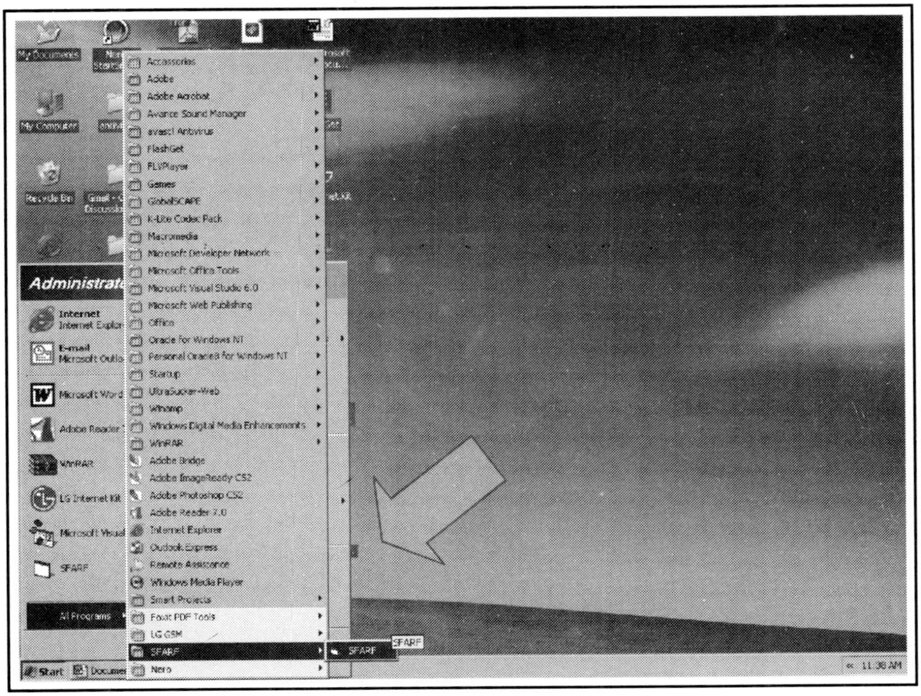

b. If you want to use any previous files then name the file with path and select **Load File**

c. If you want a new dataset file then click **New File**

d. Enter the data on the left top section

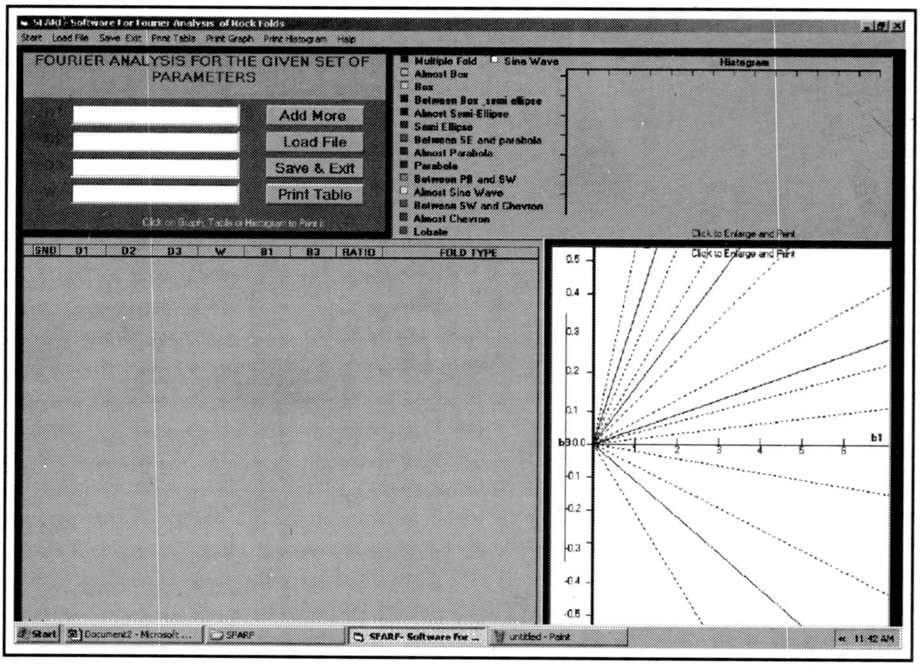

e. As you will enter the data the entries in the table below will occur automatically and by same time you can see the histogram showing up along with the graph plotting.

f. You can click on graph, histogram and table to enlarge it and print it

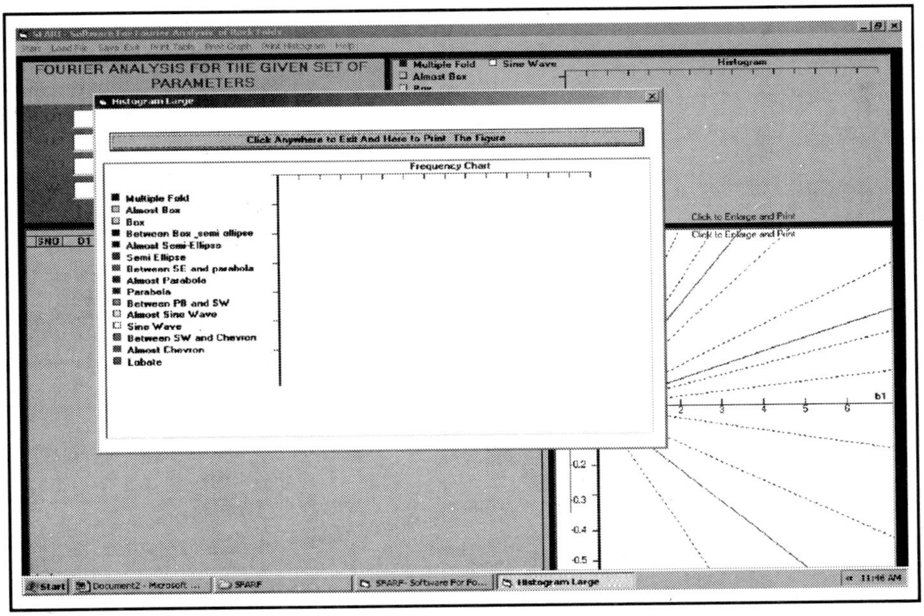

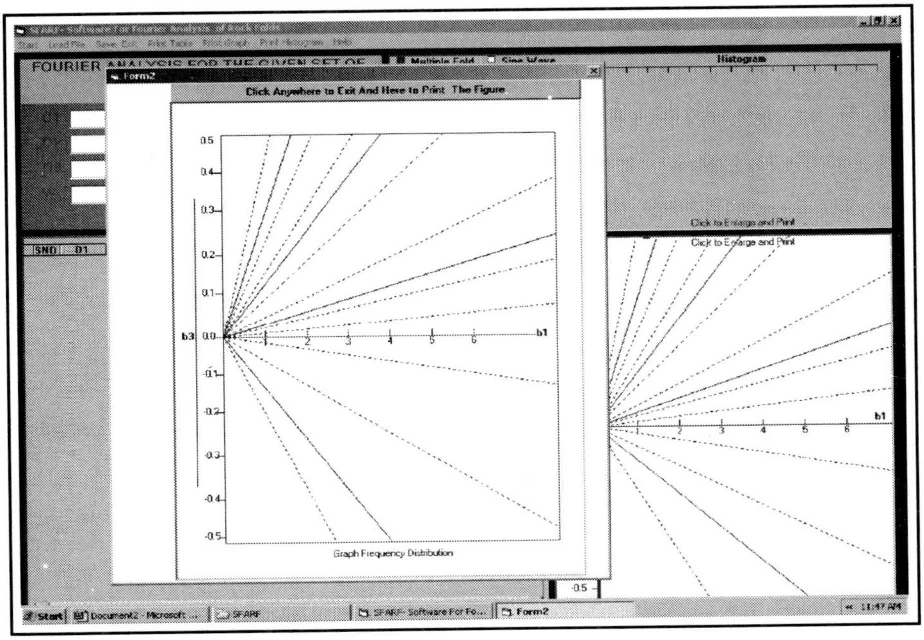

Click anywhere except the button on graph and histogram to return to the main screen.

When you click the print button for graph and histogram you are asked for the label you want to put on you can put any label you wish to.

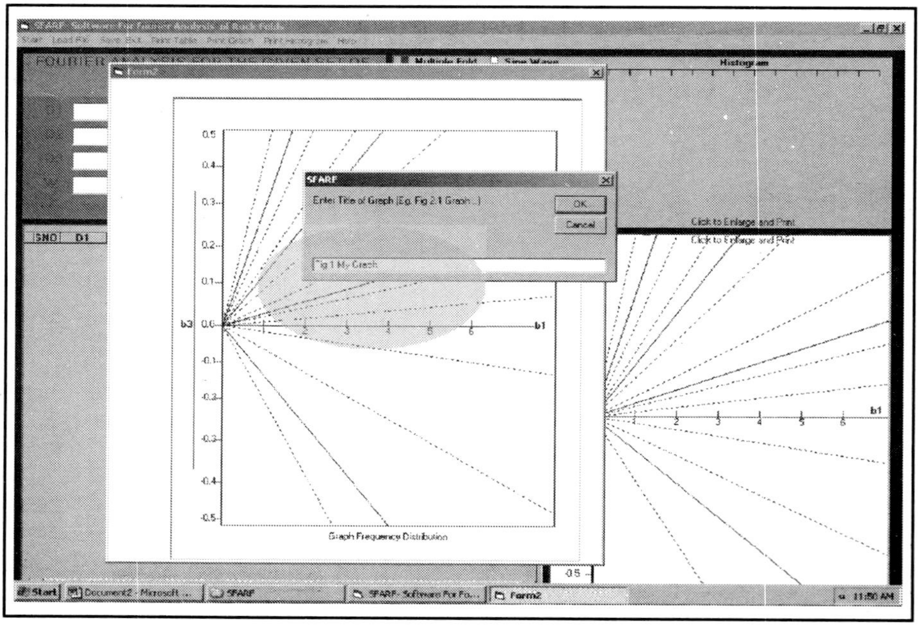

Printing the Table

There is a limitation to this software that you can print only one table in a session. So be sure that your entries are complete when you choose to print.

In other case you might need to save the file and restart the application, load the same file and print the table with new entries.

The files will be saved by default in the installation directory (by default C:\Program Files\SFARF)

Saving the File

Click on Save and Exit

On the dialogue box put the name of the file without extension file will be saved and you will return to the main screen.

You can also exit from the application from here.

Discussion

For validation of the software (SFARF), several data had been manually calculated (Thomas, 1993 and Thomas and Thomas, 2003) and reprocessed through this Software. It was observed that all the results are same as manually calculated (Table-2 and Fig. 4). The various fold forms obtained by the Software around Tamadhaun area reveal, that quartzite are varying in the range between sine wave and parabolas to almost sine wave, whereas the folds which are developed in phyllite and schist rock are statistically almost chevron.

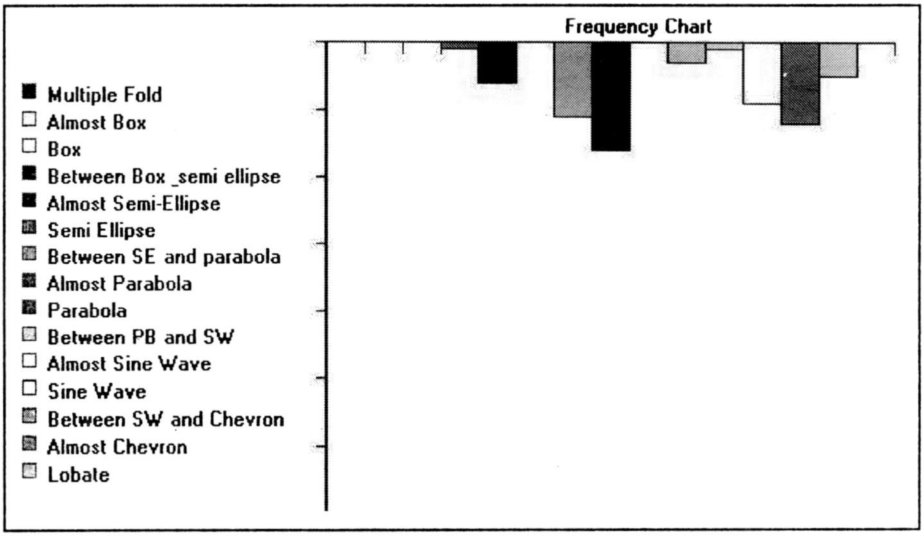

Fig. 4: Results obtained by Fourier analysis

Conclusion

The S/W developed for classification of fold shape is based on curvature variation *i.e,* Fourier analysis (harmonic) on the basis of (Hudleston, 1973 & Stabler, 1968) papers. This software is new and better than other software already available, because through this software, one can classify folds into sixteen types and simultaneously draw the two graphs (*i.e.* Fig. 3 & 4) and a table. The main aim of this paper is to publish the software for the geologists working in the field of structural geology.

Acknowledgement

The authors especially (HT) are thankful to DST for the FIST grant.

References

Heim, A. and Gansser, A. (1939) Central Himalayans geological observations of the Swiss expedition, 1936, Mem. Soc. Helv. Sci. Nat. v. 73(1), pp. 1-245.

Hudleston, P. J. (1973) Fold morphology and some geometrical implications of theories of fold developed. Tectonophysics, v. 16, pp. 1-46.

Kumar, R., Prakash, G. and Singh, K. N. (1974) Geology of the Deoprayag - Dwarhat area, Garhwal, Chamoli and Almora district, Kumaun Himalayas, U.P. Him. Geol., v. 4, pp. 321 - 346.

Singh, R. A. and Gairola, V. K. (1992) Fold shape analysis in the vicinity of North Almora Thrust in District Chamoli - Garhwal Himalaya. Him. Geol., v. 3 (2), pp. 111-116.

Shrivastava, H. B. and Thomas, T. (1990) Deformation and strain studies in a part of crystalline zone around Shalde, Kumaun Himalaya. Procee. Seminar on Evolution of Precambrian Crust in India, Calcutta, pp. 38.

Shrivastava, H. B., Thomas, T. and Tripathi, R. N. (2003) Petrographical approach to strain estimation in deformed rock, an example from granitic gneiss of a part of Almora synform Garhwal. Lesser Himalaya. Mem. Geol. Soc. India, v. 52, pp. 427- 446.

Stabler, C. L. (1968) Simplified Fourier analysis of fold shapes. Tectonophysics. v. 6(4), pp. 343-350.

Thomas, H. and Thomas, T. (2003) Fourier Shape of the fold developed around Tamadhaun, District Almora using a computer programm. Gond. Geol. Magz., Advances in Precambrian of Central India, Special Volume No.7, pp. 169-176.

Thomas, T. (1991) Structural studies in and around Shalde, Kumaun Himalaya. Abs. Indian Science Congress, Indore, pp. 24.

Thomas, T. (1993) A study of structure and tectonics of the area around Tamadhaun, district Almora, U.P. unpublished Ph.D. Thesis.

❑❑❑